Lecture Notes in Physics

365

M. Schumacher G. Tamas (Eds.)

Perspectives on Photon Interactions with Hadrons and Nuclei

Proceedings of a Workshop
Held at Göttingen, FRG
on 20 and 21 February 1990

Springer-Verlag
Berlin Heidelberg GmbH

Editors

Martin Schumacher
Zweites Physikalisches Institut, Universität Göttingen
Bunsenstraße 7–9, D-3400 Göttingen, FRG

Gabriel Tamas
DPhN/HE, CEN Saclay
F-91191 Gif-sur-Yvette Cédex, France

ISBN 978-3-662-13801-4 ISBN 978-3-540-46295-8 (eBook)
DOI 10.1007/978-3-540-46295-8

Originally published by Springer-Verlag Berlin Heidelberg New York in 1990
Softcover reprint of the hardcover 1st edition 1990

2158/3140-543210 – Printed on acid-free paper

Preface

The workshop on Perspectives on Photon Interactions with Hadrons and Nuclei was organized as part of the activities of the "Photon Club", which was created as a forum for physicists interested in the study of the nucleon and of nuclei with real photons. Goals of the Photon Club are to promote stronger connections between the members of the European Community and to help create collaborations on physics programmes already feasible with the existing facilities in Europe. The Photon Club is intended to be a forum for discussing the future of physics with real photons in Europe and for proposing the equipment needed for it. In this connection the Photon Club stresses the necessity of building a multi-GeV electron accelerator in Europe, taking into account the CEBAF project.

In accordance with the goals of the Photon Club the workshop was organized as follows. Part I was devoted to vector dominance and hard photon interactions taking place at large P_T. The talks on these themes span the energy range of real photons from a few GeV to about 100 GeV, supplementing the discussion by also taking into account the related aspects of virtual photons. This part of the workshop clearly explores the scientific prospects for the machines to come. For virtual photons and real photons in the small P_T domain the role of vector dominance appears to be well understood. Contributions were therefore of special interest where the largely unexplored direct interaction of the "bare" real photon with the quark content of the nucleon was discussed (E.P., J.M.L. and G.W.). With a different emphasis, but also directed to future developments, was the contribution of Part II in which the relativistic heavy ion as a source of equivalent photons and the physics goals to be tackled by this complementary technique were discussed.

The other parts of the workshop were devoted to fields of physics currently under study at intermediate-energy real-photon facilities. In Part III some emphasis was put upon the Compton effect, which is now beginning to be much better understood because of improved experimental data and because of recent progress in quark model calculations. Part IV contains contributions concerned with new developments in microscopic many-body theories of photonuclear reactions at intermediate energies.

The editors thank all the speakers and the participants for making this workshop a success.

Göttingen — M. Schumacher

Saclay — G. Tamas

31 May 1990

Contents

List of Participants

G. Anton	Physikalisches Institut der Universität Bonn Nußallee 12, D-5300 Bonn 1, F.R.G.
J. Ahrens	Max-Planck-Institut für Chemie, Abteilung Kernphysik, Saarstraße 23, D-6500 Mainz, F.R.G
A. Braghieri	Dipartimento di Fisica Nucleare e Teorica, Universita di Pavia, Via Bassi 6, I-27100 Pavia, Italy
N. Baron	Institut für Kernphysik der KFA Jülich , Postfach 1913, D-5170 Jülich, F.R.G.
G. Baur	Institut für Kernphysik der KFA Jülich , Postfach 1913, D-5170 Jülich, F.R.G.
C. Bennhold	Institut für Kernphysik der Universität Mainz, Postfach 3980, D-6500 Mainz, F.R.G.
S. Boffi	Universita degli Studi di Pavia, Dipartimento di Fisica Nucleare e Teorica, Via Bassi 6, I-27100 Pavia, Italy
D. Branford	Physics Department, JCMR, King's Buildings, Edinburgh University, Mayfield Road, Edinburgh EH 9 3JZ, U.K.
R. Carrasco	Departamento de Fisica Teorica, Facultad de Fisica, Universidad de Valencia, Burjassot (Valencia), Spain
P. Christillin	Istituto di Fisica, Piazza Torricelli, I-56100 Pisa, Italy
E. De Sanctis	Istituto Nazionale di Fisica Nucleare, Laboratori Nazionali di Frascati, Via E. Fermi 40, I-00044 Frascati, Italy
N. d'Hose	DPhN-SEPN, CEN Saclay, 91191 Gif-sur-Yvette, Cedex, France
A. De Graeve	Nuclear Physics Laboratory, Proeftuinstraat 86 B-9000 Gent, Belgium
J.W. Durso	Institut für Kernphysik der KFA Jülich , Postfach 1913, D-5170 Jülich, F.R.G.
M.G. Huber	Institut für Theoretische Kernphysik der Universität Bonn, Nußallee 12–14, D-5300 Bonn, F.R.G.
E. Kabuss	Max-Planck-Institut für Kernphysik, Saupfercheckweg 1, Postfach 10 39 89, D-6900 Heidelberg, F.R.G.
B. König	Institut für Kernphysik der Universität Mainz, Postfach 3980, D-6500 Mainz, F.R.G.
J.M. Laget	C.E.N. Saclay, DPhN/HE, L'Orme des Merisiers, F-91191 Gif-sur-Yvette, Cedex, France
M. Ludwig	II. Physikalisches Institut der Universität Göttingen, Bunsenstraße 7–9, D-3400 Göttingen, F.R.G.

A.I. L'vov	P.N. Lebedev Physics Institute, Academy of Sciences of the USSR, 117312 Moscow, USSR
B. Metsch	Institut für Theoretische Kernphysik der Universität Bonn, Nußallee 14–16, D-5300 Bonn, F.R.G.
G. Mondry	II. Physikalisches Institut der Universität Göttingen, Bunsenstraße 7–9, D-3400 Göttingen, F.R.G.
G. Müller	II. Physikalisches Institut der Universität Göttingen, Bunsenstraße 7–9, D-3400 Göttingen, F.R.G.
E. Oset	Departamento de Fisica Teorica, Facultad de Fisica, Universidad de Valencia, Burjassot (Valencia), Spain
E. Paul	Physikalisches Institut der Universität Bonn, Nußallee 12, D-5300 Bonn 1, F.R.G.
P. Pedroni	Dipartimento di Fisica Nucleare e Teorica, Universita di Pavia, Via Bassi 6, I-27100 Pavia, Italy
H. Petry	Institut für Theoretische Kernphysik der Universität Bonn, Nußallee 14–16, D-5300 Bonn, F.R.G.
V.A. Petrun'kin	P.N. Lebedev Physics Institute, Academy of Sciences of the USSR, 117312 Moscow, USSR
T. Pinelli	Dipartimento di Fisica Nucleare e Teorica, Universita di Pavia, Via Bassi 6, I-27100 Pavia, Italy
P. Rullhusen	II. Physikalisches Institut der Universität Göttingen, Bunsenstraße 7–9, D-3400 Göttingen, F.R.G.
P. Sauer	Institut für Theoretische Physik der Universität Hannover, Appelstraße 2, D-3000 Hannover 51, F.R.G.
A.M. Saruis	E.N.E.A, Via G. Mazzini 2, I-40138 Bologna, Italy
B. Schröder	University of Lund, Department of Physics, Sölvegatan 14, S-22362 Lund, Sweden
M. Schumacher	II. Physikalisches Institut der Universität Göttingen, Bunsenstraße 7–9, D-3400 Göttingen, F.R.G.
W.J. Schwille	Physikalisches Institut der Universität Bonn, Nußallee 12, D-5300 Bonn 1, F.R.G.
F. Smend	II. Physikalisches Institut der Universität Göttingen, Bunsenstraße 7–9, D-3400 Göttingen, F.R.G.
F. Stacley	CEN Saclay, DPhN/DIR, F-91191 Gif-sur-Yvette, Cedex, France
F. Steiper	Strahlenzentrum der J.L.-Universität , Leihgesterner Weg 217, D-6300 Giessen, F.R.G.
H. Ströher	II. Physikalisches Institut der Universität Giessen, Heinrich-Buff-Ring 16, D-6300 Giessen, F.R.G.
G. Tamas	C.E.N. Saclay, DPhN/HE, L'Orme des Merisiers, F-91191 Gif-sur-Yvette, Cedex, France
W. ten Brink	Institut für Theoretische Kernphysik der Universität Bonn, Nußallee 14–16, D-5300 Bonn, F.R.G.

L. Tiator — Institut für Kernphysik der Universität Mainz, Johann-Joachim-Becher-Weg 33, D-6500 Mainz, F.R.G.

G. Vagradov — Institut für Theoretische Physik der Universität Regensburg, Universitätsstraße 31, D-8400 Regensburg, F.R.G.

R. Van de Vyver — Nuclear Physics Laboratory, Proeftuinstraat 86, B-9000 Gent, Belgium

Th. Walcher — Institut für Kernphysik der Universität Mainz, Johann-Joachim-Becher-Weg 45, D-6500 Mainz, F.R.G.

M. Weyrauch — Institut für Kernphysik der Universität Mainz, Postfach 3980, D-6500 Mainz, F.R.G.

W. Weise — Institut für Theoretische Physik der Universität Regensburg, Universitätsstraße 31, D-8400 Regensburg, F.R.G.

G. Wormser — Laboratoire de l'Accelerateur Lineaire, Université de Paris-Sud, Bâtiment 200, F-91405 Orsay, France

This book was processed by the author using the TEX macro package from Springer-Verlag.

I.

VECTOR MESON DOMINANCE AND POINTLIKE COUPLING OF THE PHOTON

VECTOR MESON DOMINANCE*
- selected topics -

G. Piller and W. Weise

Institute of Theoretical Physics
University of Regensburg
D - 8400 Regensburg, W. Germany

Abstract: We summarize and discuss phenomena related to the vector meson dominance (VMD) of electromagnetic hadron currents. This includes VMD implications for pion and nucleon electromagnetic form factors and shadowing effects seen in the interactions of real and virtual high energy photons with nuclei. In particular, we discuss deep inelastic lepton-nucleus scattering at small values of the Bjorken variable $x = Q^2/2M\nu$.

1. INTRODUCTION: ELECTROMAGNETIC CURRENTS OF HADRONS

Hadrons interact with photons through the electromagnetic currents of their quark constituents. The fundamental electromagnetic current is

$$J_\mu(x) = \bar{q}(x)\gamma_\mu Q q(x) \tag{1}$$

where $q(x)$ are the quark fields. For $N_f = 3$ flavours,

$$q = \begin{pmatrix} u \\ d \\ s \end{pmatrix}, \tag{2}$$

and Q is the quark electric charge,

$$Q = \frac{1}{2}(B + S + \tau_3) \tag{3}$$

which involves the baryon number $B = 1/3$, strangeness $S(= -1$ for the s-quark, 0 for u,d-quarks) and isospin $\tau_3(\pm 1$ for u,d-quarks, 0 for the s-quark). At high energies and

*Work supported in part by BMFT grant 06 OR 762

momentum transfers in the multi-GeV range, QCD is perturbative and the elementary current quarks are directly visible. At low energies, where hadrons (rather than quarks) are the relevant degrees of freedom, the currents (1) are realized in the form of vector mesons ($\rho, \omega, \phi, ...$). This is the basic idea behind the Vector Meson Dominance (VMD) model.

2. ELECTRON- POSITRON ANNIHILATION INTO HADRONS

Consider current matrix elements $< 0 \mid J_\mu \mid \; hadrons >$, which describe the formation of hadronic states out of the vacuum through electromagnetic interactions. These matrix elements are measured in the process $e^+e^- \to$ hadrons. The data are shown in Figs. 1, 2. At e^+e^- center-of-mass energies $\sqrt{s} < 1\ GeV$, the $e^+e^- \to$ hadron spectrum is visibly dominated by the $\rho(770)$, $\omega(783)$ and $\phi(1020)$. The region $\sqrt{s} > 1$ GeV extends into the domain of heavy quarks with a continuum plateau and heavy vector mesons of the J/Ψ and Υ families. The fact that the ρ, ω and ϕ completely dominate the low energy electromagnetic excitation spectrum suggests the current field identity [1]

$$J_\mu^{e.m.}(x) = \sum_{V=\rho^0,\omega,\phi} \frac{m_V^2}{f_V} V_\mu(x), \tag{4}$$

where $V_\mu = \rho^0_\mu$, ω_μ and ϕ_μ are the respective vector meson fields, m_V are their masses and f_V their decay constants (empirically, $f_\rho \simeq 5.3, f_\omega \simeq 15.2$ and $f_\phi \simeq 13.4$). The current-field identity (4) is the basis of a remarkably successful phenomenology. For example, the accuracy of the model in predicting various radiative decay widths of mesons is generally at the 10 % level (see ref. [2] for a recent survey). We continue here by summarizing briefly the role which VMD plays in hadron electromagnetic form factors at $\mid q^2 \mid \leq 1\ GeV^2$.

3. VMD PHENOMENOLOGY AT WORK: PION FORM FACTOR

A well-known demonstration of the degree to which VMD works is the pion form factor $F_\pi(q^2)$. It is measured to high accuracy in the spacelike region $q^2 < 0$ by pion-electron scattering experiments [3] and in the timelike region $q^2 > 0$ by $e^+e^- \to \pi^+\pi^-$. The data are shown in Fig. 3. The timelike region exhibits the strong dominance of the ρ^0 resonance.
The pion form factor can be written as a dispersion relation,

$$F_\pi(q^2) = 1 + \frac{q^2}{\pi} \int_{4m_\pi^2}^{\infty} dt \frac{Im\ F_\pi(t)}{t(t - q^2 + i\epsilon)}. \tag{5}$$

Here the imaginary part $Im\ F_\pi(t)$ represents the $\pi^+\pi^-$ mass spectrum with its strong

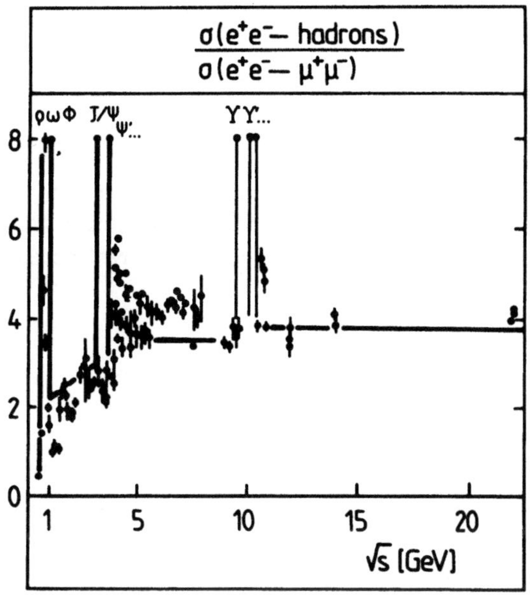

Figure 1: Ratio of the total cross section for the e^+e^- amplitude into hadrons to the muon pair production cross section $\sigma(e^+e^- \to \mu^+\mu^-) = \frac{4\pi\alpha^2}{3s}$.

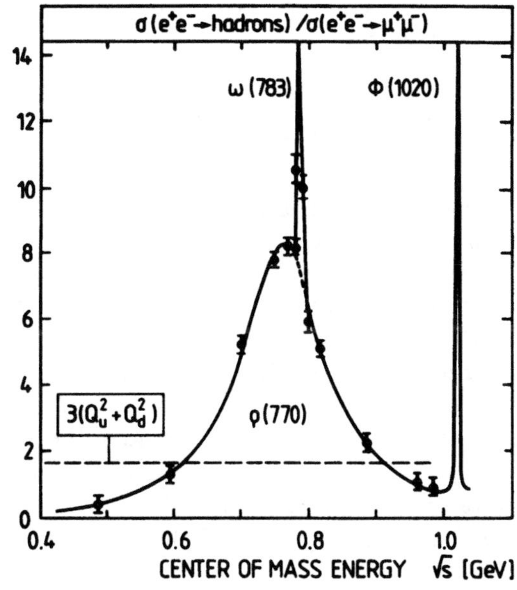

Figure 2: Low energy part of the ratio $\sigma(e^+e^- \to hadrons)/\sigma(e^+e^- \to \mu^+\mu^-)$, showing the dominance of vector mesons (ρ, ω, ϕ).

resonant peak at $t = m_\rho^2 \simeq 0.6\ GeV^2$. The electromagnetic pion radius is given by

$$< r_\pi^2 > = \left(\frac{6dF_\pi}{dq^2}\right)_{q^2=0} = \frac{6}{\pi}\int_{4m_\pi^2}^{\infty}\frac{dt}{t^2} Im\ F_\pi(t). \tag{6}$$

Hence the pion size as seen by a photon is determined by the low mass part of the $J^\pi = 1^-$, isospin I = 1 mesonic spectrum. The VMD assuption is that this spectrum is dominated by the ρ meson. The current-field identity (4) restricted to the ρ meson implies $Im\ F_\pi(t) = \pi m_\rho^2 \delta(t - m_\rho^2)$, and one finds

$$< r_\pi^2 >^{1/2} = \frac{\sqrt{6}}{m_\rho} \simeq 0.63 fm, \tag{7}$$

which is remarkably close to the measured radius $< r_\pi^2 >^{1/2} = (0.66 \pm 0.01)\ fm$ [4]. Corrections to the naive VMD model, mainly related to a realistic description of the 2π continuum have been discussed in the literature and must of course be taken into account in a more detailed quantitative analysis.

4. HADRONIC STRUCTURE OF THE PHOTON

Consider a photon of energy ν and momentum $\vec{q}$, i.e. with squared four-momentum $q^2 = \nu^2 - \vec{q}^{\,2} = -Q^2$. Its propagator, including vacuum polarization effects, can be written in terms of a spectral representation:

$$D_{\mu\nu}(q^2) = -g_{\mu\nu}\left[\frac{Z}{q^2 + i\epsilon} - \int_{\mu_0^2}^{\infty}\frac{d\mu^2}{\mu^2}\frac{\Pi(\mu^2)}{q^2 - \mu^2 + i\epsilon}\right] + (q_\mu q_\nu - \ terms). \tag{8}$$

The polarization function $\Pi(\mu^2)$ describes the mass spectrum of intermediate states that contribute to vacuum polarization. In the following, we consider only hadronic vacuum polarization, the process shown in Fig. 4 by which a photon creates virtual quark-antiquark pairs which evolve into hadrons. The lightest hadronic system that can be formed in this way is $\pi^+\pi^-$. Hence the threshold in the integral in eq. (8) is $\mu_0^2 = 4m_\pi^2$. The Z factor gives the reduced probability of finding the photon in a "bare" state, and the $q_\mu q_\nu$ terms guarantee gauge invariance.

The spectral function $\Pi(\mu^2)$ is a measured quantity. Consider the process $e^+e^- \to$ hadrons $\to e^+e^-$ at a c.m. energy $\sqrt{s}$. Its amplitude is proportional to the hadronic polarization part of the photon propagator $D_{\mu\nu}(s)$. The cross section $\sigma(e^+e^- \to$ hadrons) is then proportional to the imaginary part of this amplitude, and hence to Π (s).

One finds

$$\begin{aligned}\Pi(s) &= \frac{s}{\pi e^2}\sigma(e^+e^- \to hadrons) \\ &= \frac{1}{12\pi^2}\frac{\sigma(e^+e^- \to hadrons)}{\sigma(e^+e^- \to \mu^+\mu^-)},\end{aligned} \tag{9}$$

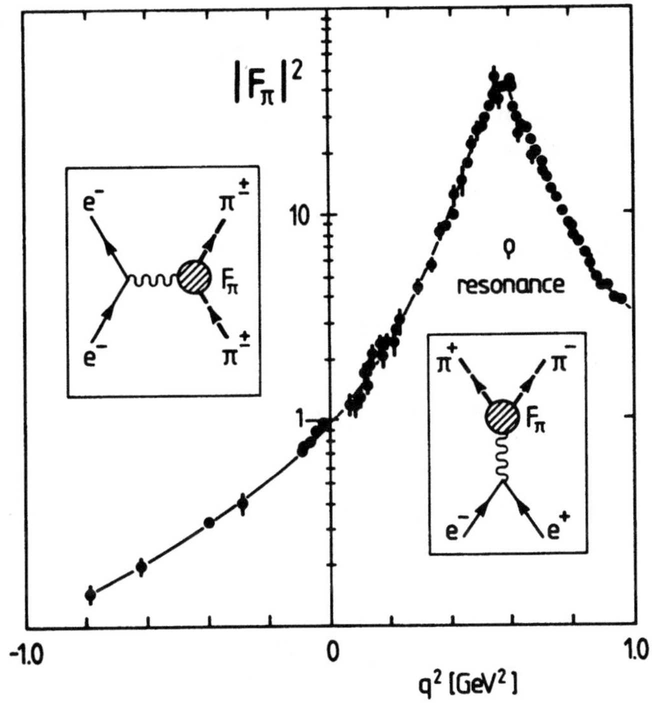

Figure 3: Pion Formfactor; Data from ref. [3].

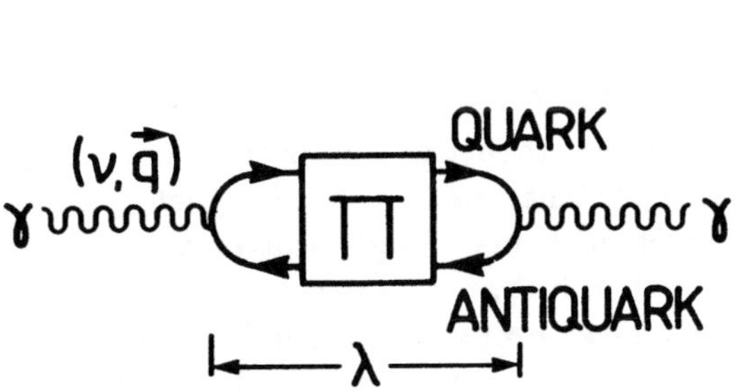

Figure 4: Hadronic vacuum polarisation process.

where the cross section ratio on the right hand side is just the one shown in Figs. 1,2. We now introduce a length scale which is of crucial importance for our further discussions: the propagation length (or coherence length) λ of a hadronic fluctuation when it appears in the vacuum polarization process, Fig. 4. Let μ be the mass of this hadronic intermediate state. Then the lifetime of the fluctuation is $\hbar/\Delta E$ where $\Delta E = \sqrt{\vec{q}^2 + \mu^2} - \nu$, the difference between the energy of the hadronic state and the photon energy. Using $Q^2 = \vec{q}^{\,2} - \nu^2$ and assuming that the fluctuation travels with a velocity close to the speed of light (i.e. $\mu << \nu$), we have

$$\lambda \simeq \frac{2\nu}{Q^2 + \mu^2}. \tag{10}$$

As an example, consider a real photon with $Q^2 = 0$ and $\nu = 10$ GeV. Then a ρ meson intermediate state with $\mu = m_\rho \sim 4\ fm^{-1}$ has a coherence length $\lambda > 6 fm$, comparable to the sizes of heavy nuclei.
For later purposes when discussing virtual photons in deep inelastic scattering, it is useful to rewrite λ in terms of the Bjorken scaling variable $x = Q^2/2M\nu$ (in the lab. frame), where M is the nucleon mass:

$$\lambda \simeq \frac{1}{Mx}\left(\frac{Q^2}{Q^2 + \mu^2}\right). \tag{11}$$

For $Q^2 >> \mu^2$ we see that $\lambda \simeq 1/Mx$ starts to reach nuclear length scales when $x < 0.1$. One therefore expects that processes such as deep inelastic muon scattering at small x will be strongly influenced by phenomena related to the sizeable coherence length of photon-like hadronic fluctuations.

5. SHADOWING

Consider the Compton scattering of a real or virtual photon from a nucleon or nucleus. We prefer a description in the laboratory frame in which the photon has energy ν and squared four-momentum Q^2. A hadronic component of the photon with mass μ then travels over a distance $\lambda = 2\nu/(Q^2 + \mu^2)$. If this coherence length becomes comparable to typical nuclear dimensions, then the Compton scattering process will undergo the lab. frame time ordering sketched in Fig. 5: the photon converts into hadrons which then scatter on the target nucleons. Single scattering leads to $\sigma_{\gamma A} = A\sigma_{\gamma N}$, i.e. the γ-nuclear cross section is then just an incoherent sum of the cross sections for individual nucleons. For sufficiently large propagation length λ, however, multiple scattering can occur, which leads to $\sigma_{\gamma A} = A^\alpha \sigma_{\gamma N}$ with $\alpha < 1$. This is the so-called shadowing effect.

5.1 Shadowing in Total Photon-Nucleus Cross Sections

For real photons, nuclear shadowing is a well established phenomenon at energies $\nu > 5$ GeV. Note that at these energies, the coherence length of vector mesons is

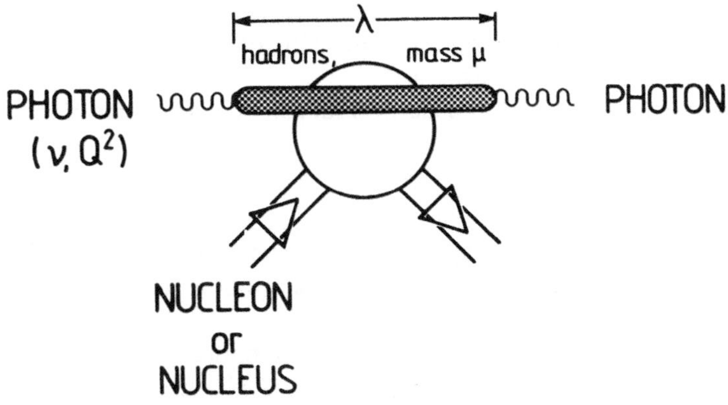

Figure 5: Illustration of high energy Compton scattering on nucleons or nuclei through scattering of hadronic intermediate states.

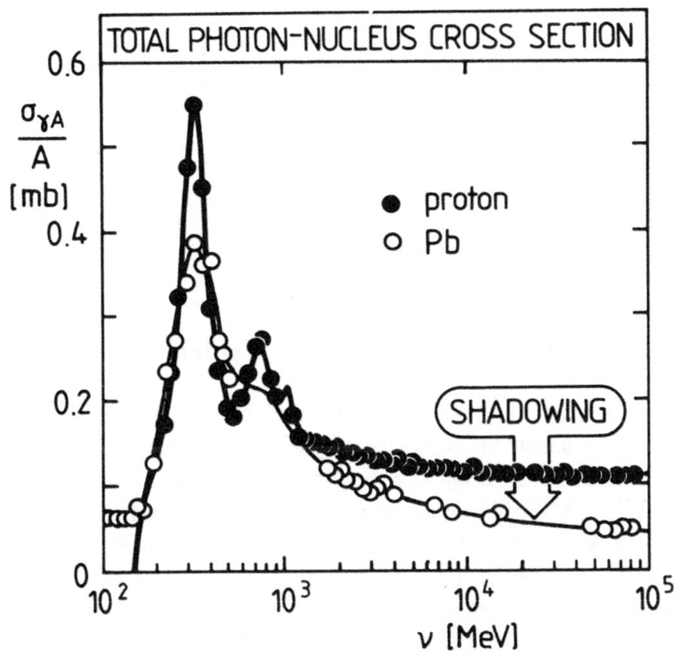

Figure 6: Comparison of the total photon–proton cross section and the photonuclear cross section for ^{208}Pb. Data from ref. [4].

$\lambda_{\rho,\omega} \simeq 2\nu/m^2_{\rho,\omega} > 3\ fm$, i.e. it already exceeds the average distance $d \simeq 2\ fm$ between nucleons in the center of nuclei. Hence double scattering of vector mesons from nucleons in the nucleus starts to become relevant.
The empirical data above $\nu > 5$ GeV can be parametrized in the form [5]

$$\sigma_{\gamma A} \simeq A^{0.9} \sigma_{\gamma N}. \tag{12}$$

An example is given for ^{208}Pb in comparison with the free proton in Fig. 6. We observe there that the shadowing starts to develop at photon energies around 2 GeV and stabilizes above 5 GeV.
The VMD interpretation of this effect has been subject of much work in the mid seventies [5, 6]. The basic picture is that of Fig. 5, but now with the hadronic intermediate states dominated by the vector mesons $V = \rho, \omega, \phi, ...$; the VMD principle then implies

$$\frac{\sigma_{\gamma A}}{\sigma_{\gamma N}} = \frac{\sum_V \sigma_{VA}/f_V^2}{\sum_V \sigma_{VN}/f_V^2} \tag{13}$$

where σ_{VN} and σ_{VA} are the vector meson-nucleon and -nucleus cross sections. One can connect these cross sections by multiple scattering theory and finds (for the example of a Gaussian nuclear density):

$$\sigma_{VA} = A\sigma_{\gamma N}\left[1 - \frac{3}{8\pi}\frac{A\sigma_{VN}}{R_A^2} + ...\right], \tag{14}$$

where R_A is the nuclear radius. At the same time, the analysis of vector meson photoproduction experiments on the proton [6] leads to $\sigma_{\rho N} \simeq \sigma_{\omega N} \simeq 25$ mb and $\sigma_{\phi N}/\sigma_{\rho N} \sim m_\rho^2/m_\phi^2$. This gives already a rough idea about the effects which generate the "shadow" in eq. (12).

5.2 Shadowing in Deep-Inelastic Scattering on Nuclei

Deep inelastic muon scattering on a variety of nuclear targets has also revealed a pronounced shadowing effect at small values of the Bjorken variable $x = Q^2/2M\nu$. Recent measurements in the kinematical range $0.3\ GeV^2 < Q^2 < 3.2\ GeV^2$ (EMC [7]) and $1\ GeV^2 < Q^2 < 20\ GeV^2$ (NMC [8]) show a systematic reduction of the nuclear structure functions $F_2^A(Q^2, x)$ with respect to A times the free nucleon structure function $F_2^N(Q^2, x)$ at $x < 0.1$. This observation has stimulated a great deal of theoretical activity, with attempts to describe this shadowing effect either in terms of quark-gluon dynamics on the light cone [9–11] or in complementary hadronic language [12–14].
Our aim here is to investigate the role of vector meson dominance phenomena in these processes. We point out [15] that VMD in its simplest form is by far not sufficient to understand the behaviour of the free nucleon structure function F_2^N at small x, whereas it may still be an important factor in the shadowing mechanism [15].
Let us first return to the propagation length (11) of hadronic fluctuations of the photon. We have already mentioned that $\lambda \doteqdot 1/Mx = 0.21 fm/x$ reaches nuclear dimensions

for $x < 0.1$. Consider now a description of deep-inealstic lepton scattering on nucleons or nuclei in the laboratory frame with the target at rest. In this frame the basic diagrams representing the interaction of the virtual photon with the nucleon involve time orderings as shown in Figs. 7a and b: the photon either hits a quark in the target which then picks up the large incident momentum and energy, or the photon converts into a quark-antiquark pair which subsequently interacts with the target. At small x, the pair production term Fig. 7b dominates. This can be seen by comparing the energy denominators of the processes illustrated in Fig. 7 in standard "old fashioned" perturbation theory. Let m_q be a typical (constituent) quark mass. For large energy transfer $\nu >> m_q$ and $\nu^2 >> Q^2$ the characteristic energy difference for process (a) is $\Delta E_a \equiv E_a(t_2) - E_a(t_1) \sim -m_q + \frac{m_q^2+Q^2}{2\nu}$. In process (b), let the quarks in the produced $q\bar{q}$-pair carry momenta $\vec{q}/2 + \Delta\vec{q}$ and $\vec{q}/2 - \Delta\vec{q}$, respectively. We assume that the relative momentum of the pair is small compared to its total momentum: $|\Delta\vec{q}| << |\vec{q}| = \sqrt{Q^2+\nu^2}$. In fact $|\vec{q}|$ is of order 10 GeV or larger in our kinematical range, whereas the typical hadronic scale which governs the relative motion of the $q\bar{q}$-pair implies $|\Delta\vec{q}| < 1$ GeV. Therefore we find for the energy difference in process (b): $\Delta E_b \equiv E_b(t_2) - E_b(t_1) \simeq \frac{4m_q^2+Q^2}{2\nu}$. Consequently the ratio of magnitudes of the quasifree amplitude A_a and the pair amplitude A_b behaves roughly like

$$\left|\frac{A_a}{A_b}\right| \sim \frac{\Delta E_b}{\Delta E_a} \sim \frac{Q^2}{2m_q\nu}\left[1 + \frac{4m_q^2}{Q^2}\right]. \tag{15}$$

This ratio is evidently small compared to unity for $x < 0.1$. Hence the $q\bar{q}$-pair mechanism, Fig. 7b, dominates at small x. The further evolution of the process will then lead to the mechanism in which a hadronic state, carrying photon quantum numbers, strongly interacts with the nucleon.

This picture implies that in deep-inelastic scattering at small x, the virtual photon primarily interacts with a nucleon or nucleus through the scattering of its hadronic components on the target. Nuclear shadowing is expected to occur when the coherence length λ of the hadronic intermediate state exceeds the average distance between two nucleons in the nucleus, i.e. when double scattering starts to occur. With $d \simeq 1.8$ fm and $\lambda > d$ this implies $x < 1/Md \simeq 0.1$ as a condition for shadowing.

5.2.1 Free Nucleon Structure Function

The free nucleon structure function $F_2^N(Q^2, x)$ (introduced here as the average of proton and neutron structure functions) can be written

$$F_2^N(Q^2, x) = \frac{Q^2}{\pi e^2}\sigma(\gamma^* N) \tag{16}$$

in terms of the virtual photon-nucleon cross section (the sum of longitudinal and transverse cross sections). Following the previous discussions we assume that, at small x, the virtual photon interacts with the nucleon by first converting into a $q\bar{q}$ - pair which then propagates and interacts strongly with the nucleon. This leads to the following

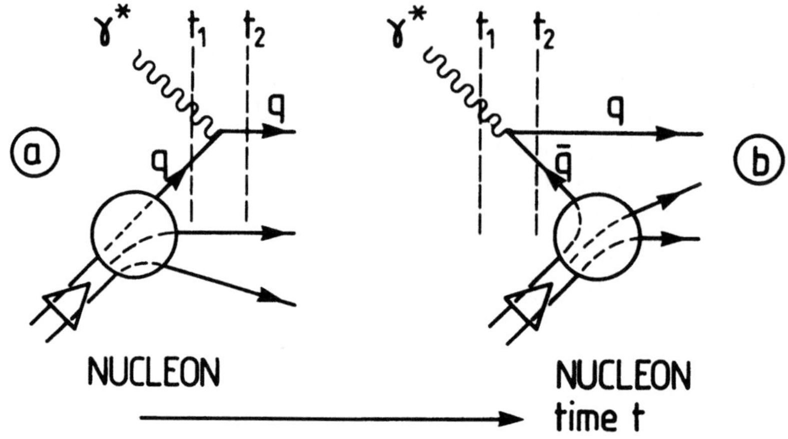

Figure 7: Time orderings in a lab. frame description of virtual photon–nucleon interactions. The active quark (or antiquark) is denoted by q (or $\bar{q}$).

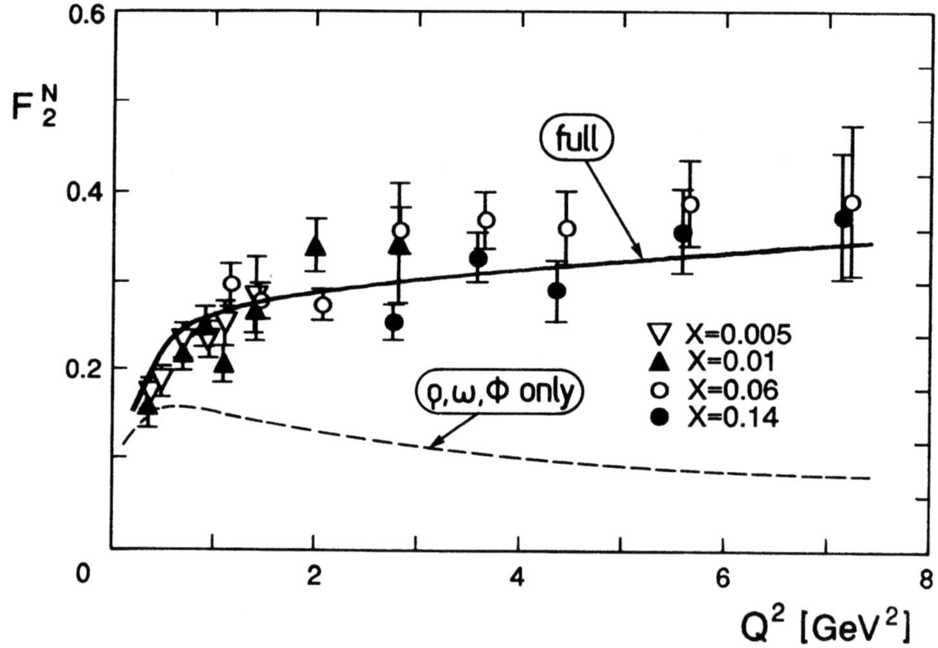

Figure 8: Free nucleon structure function for small values of x. Data from ref. [17]. The solid curve is obtained from eq. (17) with $\sigma_N = 16\,mb \cdot GeV^2/\mu^2$. The dashed line incorporates the low mass vector mesons (ρ, ω, ϕ) only.

approximate form of the structure function, valid at small x:

$$F_2^N(Q^2, x) = \frac{Q^2}{\pi} \int_{4m_\pi^2}^{\infty} d\mu^2 \frac{\mu^2 \Pi(\mu^2)}{(\mu^2 + Q^2)^2} \sigma_N(\mu^2; s \simeq Q^2/x). \tag{17}$$

Here $\Pi(\mu^2)$ is the mass spectrum of hadronic components of the photon, as given by eq. (9). The factor $(\mu^2 + Q^2)^{-2}$ is the squared propagator of these fluctuations, and $\sigma_N(\mu^2; s)$ is the effective cross section for the scattering of a hadronic state of mass μ from the nucleon. The energy variable in σ_N is $s = 2M\nu + M^2 - Q^2 \simeq Q^2/x$.
The effective cross section σ_N summarizes all interactions of hadronic components of the photon with the composite nucleon. For small x the hadron-nucleon center-of mass energy $\sqrt{s}$ is in the multi-GeV range where cross sections are flat. We therefore use an average σ_N, independent of s. The leading dependence of σ_N on the hadron mass μ should satisfy the following criteria:

i) For large Q^2, the structure function F_2^N should have the proper scaling behaviour, i.e. it should become independent of Q^2. This requirement is fulfilled with the ansatz $\sigma_N = const./\mu^2$. By inspection of the integral (17) it is indeed obvious that such a μ^{-2} dependence of σ_N leads to scaling.
ii) Empirical vector meson-nucleon cross sections follow roughly a $1/\mu^2$ rule.

We see this from table 1 where the ansatz

$$\sigma_N(\mu^2) = \frac{16mb \cdot GeV^2}{\mu^2} \tag{18}$$

is compared with the available vector meson cross sections deduced from photoproduction data.

	$\sigma_{VN}[mb]$ empirical	from eq. (18)
ρ, ω	27 ± 3	26
ϕ	12 ± 3	15
J/Ψ	2 ± 1	2

Table 1: Vector meson-nucleon cross sections: empirical values [6, 16] (left) compared with the ansatz, eq. (18)

The simple ansatz (18), together with the empirical $\Pi(\mu^2)$ from $\sigma(e^+e^- \to \text{hadrons})$ according to eq. (9), reproduces the nucleon structure function at small x remarkably well. This is demonstrated in Fig. 8. Note that the ansatz (18) makes F_2^N independent of x for small x. This is consistent with the data [17] within (admittedly sizeable) error bars. At the same time Fig. 8 clearly shows that the standard vector meson dominance model is unable to reproduce the Q^2 dependence of F_2^N. When the spectrum $\Pi(\mu^2)$ is cut off at $\mu^2 = m_\phi^2 \sim 1\ GeV^2$ so that only the low-mass vector mesons contribute, a major part of the strenght in F_2^N is obviously missing for $Q^2 > 1\ GeV^2$: the $q\bar{q}$-continuum with masses $\mu > 1$ GeV is very important to reproduce F_2^N at small x.
We note that in principle, F_2^N of eq. (17) could have been written in the more general form of a double dispersion relation [18] in order to account for off-diagnoal diffractive dissociation processes (such as $VN \to V'N$ where V and V' are two different vector meson states). However, the authors of ref. [13] have pointed out strong destructive interferences between such on- and off-diagonal terms. Remaining effects are then thought to be absorbed in the parametrization of the effective cross section $\sigma_N(\mu^2)$.

5.2.2 Nuclear Structure Function

Next we describe the nuclear structure function F_2^A at small x using exactly the same model as previously developed. We write

$$F_2^A(Q^2, x) = \frac{Q^2}{\pi} \int_{4m_\pi^2}^{\infty} d\mu^2 \frac{\mu^2 \Pi(\mu^2)}{(\mu^2 + Q^2)^2} \sigma_A(\mu^2; s), \tag{19}$$

where σ_A is now the effective hadron-nucleus cross section. Given the large photon energy ν it is justified to use the eikonal approximation and employ Glauber mutiple scattering theory to connect σ_A and σ_N.
A typical feature at high energy is the smallness of $Re\ f_N/Im\ f_N$, the ratio of real and imaginary parts of the forward hadron-nucleon amplitudes; we assume this ratio to be negligible. We also neglect diffractive dissociation terms $VN \to V'N$ in the multiple scattering series. This could lead to an error of about 5 % in σ_A due to inelastic screening effects [19].
An important point is that the propagation length λ of intermediate hadronic states has to be taken into account properly. This is done by an extension of Glauber theory as described by Gribov [20]. The finite coherence length $\lambda = 2\nu/(Q^2 + \mu^2)$ enters in a characteristic phase factor. It first appears in the double scattering term as $exp[i(z_1 - z_2)/\lambda]$ and has an obvious interpretation: if λ is large compared to the distance $d = z_1 - z_2$ between two nucleons in the target, the phase factor is close to unity and double scattering occurs with full strength. However, if $\lambda << d$, then this phase factor rapidly oscillates and double scattering is suppressed.
Altogether we end up with the following expression:

$$\sigma_A(\mu^2; s) = A\sigma_N(\mu^2; s) +$$

$$+\sum_{n=2}^{A}\left(-\frac{1}{2}\right)^{n-1}\binom{A}{n}\left[\frac{\sigma_N(\mu^2;s)}{A}\right]^n Re\left\{\int d^2b\, dz_1...dz_n\, \rho_n(\vec{b}, z_1, ..., z_n)\, e^{i(z_1-z_n)/\lambda}\right\}. \tag{20}$$

Here ρ_n is the n-particle nuclear density with coordinates taken along the path of the scattering hadron, i.e. parallel to the z-axis at an impact parameter $\vec{b}$. We expand ρ_n up to terms linear in the two-body correlation function $\Delta(\vec{r}, \vec{r}\,') = \rho_2(\vec{r}, \vec{r}\,') - \rho(\vec{r})\rho(\vec{r}\,') \equiv g(\vec{r}, \vec{r}\,')\rho(\vec{r})\rho(\vec{r}\,')$ where $\rho(\vec{r})$ is the one-body density:

$$\rho_n(\vec{b}, z_1, ..., z_n) = \rho(\vec{b}, z_1)...\rho(\vec{b}, z_n) + \sum_{all\ permutations} \Delta(\vec{b}, z_1, z_2)\, \rho(\vec{b}, z_3)...\rho(\vec{b}, z_n). \tag{21}$$

In practice we use for the correlation factor $g(|\vec{r} - \vec{r}\,'|)$ a simple parametrization which approximates the nuclear G-matrix [21]. We also use realistic nuclear density distributions $\rho(r)$ which fit elastic electron scattering data.
With these ingredients we have calculated [15] the ratio

$$R(Q^2, x) = \frac{F_2^A(Q^2, x)}{A\, F_2^N(Q^2, x)} \tag{22}$$

of the nuclear and free nucleon structure functions for $x < 0.1$ and compare with recent data for ^{12}C and ^{40}Ca as obtained by the NMC collaboration at CERN. These data at small x cover the range $1\ GeV^2 < Q^2 < 20\ GeV^2$ with a 200 GeV incident muon beam energy. Results are shown in Fig. 9. We find that the pronounced shadowing observed at $x < 0.1$ can in fact be reproduced quite well in our simple model. Note that in this calculation, no additional free parameters have been introduced beyond the effective cross section $\sigma_N = 16mb \cdot GeV^2/\mu^2$ which enters in F_2^N for the free nucleon. The shadowing effect turns out to be dominated (see Fig. 9) by the low-mass vector meson components of the spectrum $\Pi(\mu^2)$; their propagation length λ is large compared to the nuclear size. Components of large mass μ, even though they are very important in the free structure function F_2^N, are much less effective in the shadowing process since their propagation length is shorter; in addition the $1/\mu^2$ behaviour of the effective cross section σ_N suppresses multiple scattering for large μ.
Much discussion has recently been devoted to the Q^2 dependence of the shadowing effect [10, 14]. We show in Fig. 10 a comparison of our calculations with the earlier EMC data [7] which involve lower Q^2 values ($0.3\ GeV^2 < Q^2 < 3.2\ GeV^2$) than the new NMC data. We do this also in order to show the smooth Q^2 dependence of our model, which turns out to be close to a logarithmic one for $1\ GeV^2 < Q^2 < 10\ GeV^2$. For Q^2 much larger than $10\ GeV^2$, the shadowing effect disappears, as one would expect.
At $Q^2 = 0$, the present model reproduces quite well the nuclear shadowing observed with real photons at $\nu = 50 - 100$ GeV (see section 5.1 and ref. [22]): we find $\sigma_{\gamma A} = A^{0.91}\sigma_{\gamma N}$ in this energy range.

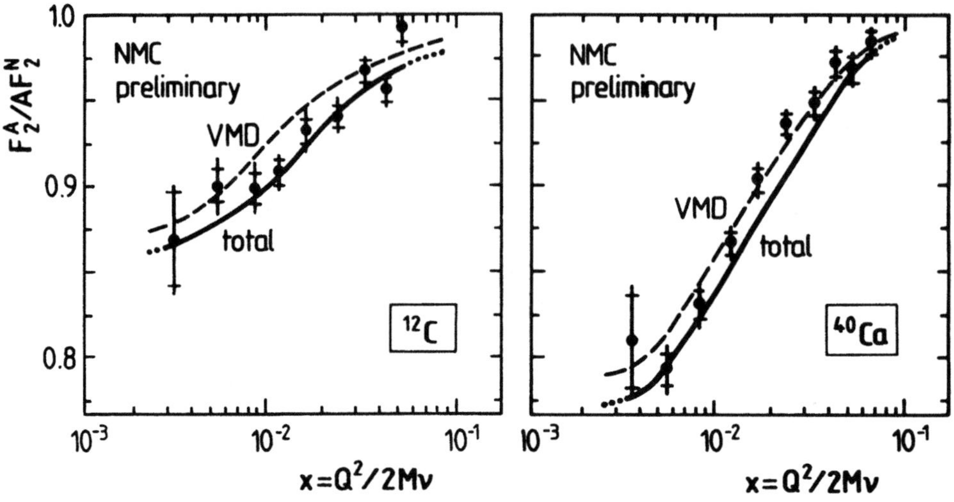

Figure 9: Ratio of nuclear and free nucleon strucure function at small x for ^{12}C and ^{40}Ca. Preliminary data of the NMC collaboration [8]. The solid curve is calculated using eqs. (19, 20). The dashed curves are obtained by including only the ρ, ω and ϕ mesons in the mass spectrum $\Pi(\mu^2)$.

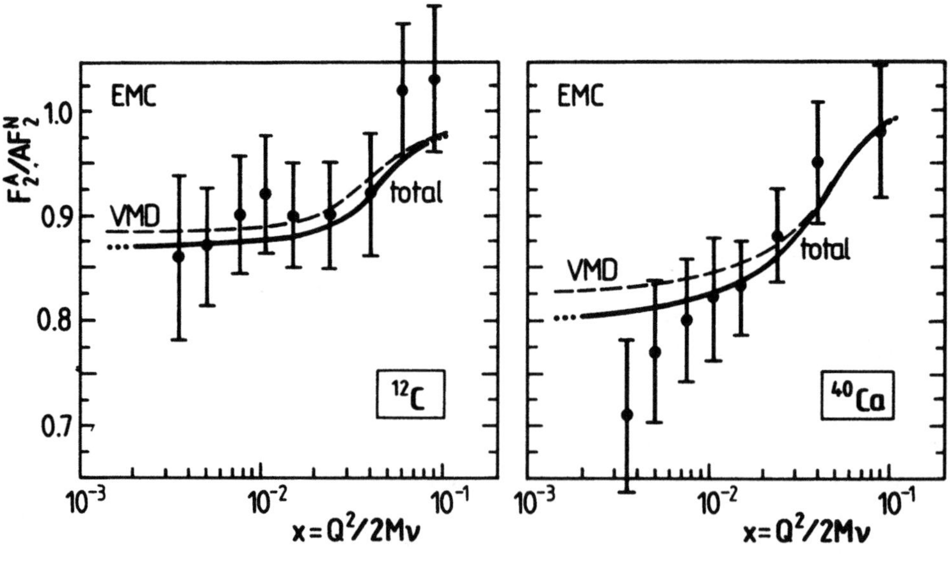

Figure 10: Same as Fig. 9, but for the EMC data [7].

6. SUMMARY

Vector mesons play an important role in the electromagnetic interactions of hadrons. They govern hadron form factors at $q^2 < 1\ GeV^2$; they are responsible for a large part of the pronounced hadronic shadowing effects seen with real photons in the multi-GeV range. One of our primary aims in this presentation was to explore to what extent vector meson dominance still plays a role in deep-inelastic scattering at small x where sea-quarks become important. This question has been subject of some debate [10, 14]. We find that the free nucleon structure function at small x can not at all be reproduced by the "naive" version of VMD, with only ρ, ω and ϕ included. It is of crucial importance to incorporate the higher mass quark-antiquark continuum in the description. On the other hand, the shadowing phenomenon in the nuclear structure function at $x < 0.1$ is still governed by the low-mass vector meson states, at least in our simple model. It is of interest to explore the possible connections between the phenomenological hadronic description presented here and the QCD-based approaches of refs. [9–11].

We would like to thank Gerry Brown, John Durso, Jerry Miller, Kazuo Takayanagi and Mark Strikman for illuminating discussions. One of us (W.W.) enjoyed the kind hospitality of the Theory Group at GSI (Darmstadt) during the preparation of these notes.

References

1. M. Gell-Mann and F. Zachariasen, Phys. Rev. **124** (1961) 953;
   J.J. Sakurai, Currents and Mesons, Univ. of Chicago Press (1969)
2. J. W. Durso, Phys. Lett. B **184** (1987) 348
3. S. R. Amendolia et al., Nucl. Phys. B **277** (1986) 168
4. J. Ahrens, Nucl. Phys. A **446** (1985) 229; and refs. therein
5. G. Grammar and J. D. Sullivan, in: Electromagnetic Interactions of Hadrons, Vol. 2, A. Donnachie and G. Shaw, eds., Plenum, NY (1978), p. 195;
   W. Weise, Phys. Reports **22** (1974) 53
6. T. H. Bauer, R. D. Spital, D. R. Yennie and F. M. Lipkin, Rev. Mod. Phys. **50** (1978) 261
7. M. Arneodo et al. (EMC Collaboration), Phys. Lett. B **211** (1988) 493
8. NMC preliminary data (E. M. Kabuss, these proceedings)
9. E. L. Berger and J. Qui, ANL-HEP-CP-88-42;
   J. Qiu, Nucl. Phys. B **291** (1987) 746;
   A. H. Mueller and J. Qiu, Nucl. Phys. B **268** (1986) 427
10. J. D. Bjorken, in: Particles and Detectors: Festschrift for Jack Steinberger, K. Kleinknecht and T. D. Lee, eds., Springer Tracts in Mod. Phys. **108** (1986) 17
11. F. E. Close, J. Qiu and R. G. Roberts, Phys. Rev. D **40** (1989) 2820;
    F. E. Close and R. G. Roberts, Phys. Lett. B **213** (1988) 91
12. G. Shaw, Phys. Lett. B **228** (1989) 125;
    P. Ditsas and G. Shaw, Nucl. Phys. B **113** (1976) 246;
    D. Schildknecht, Nucl. Phys. B **66** (1973) 398
13. C. L. Bilchak, D. Schildknecht and J. D. Stroughair, Phys. Lett. B **214** (1988) 441
14. L. Frankfurt and M. Strikman, Nucl. Phys. B **316** (1989) 340;
    S. J. Brodsky and H. J. Lu, Phys. Rev. Lett. **64** (1990) 1342;
    J. Kwiecinski and B. Badelek, Phys. Lett. B **208** (1988) 508
15. G. Piller and W. Weise, Univ. Regensburg preprint TPR–90–38 (1990)
16. J. J. Aubert et al., Nucl. Phys. B **213** (1983) 1
17. M. Arneodo et al. (EMC collaboration), Nucl. Phys. B **333** (1990) 1
18. H. Fraas, B. J. Read and D. Schildknecht, Nucl. Phys. B **86** (1975) 346
19. P. V. R. Murthy et al., Nucl. Phys B **92** (1975) 269
20. V. N. Gribov, Sov. Phys. JETP **30** (1970) 709
21. G. E. Brown, S. O. Bäckman, E. Oset and W. Weise, Nucl. Phys. A **286** (1977) 191
22. D. O. Caldwell et al., Phys. Rev. Lett. **42** (1979) 553

SIGNS OF VECTOR MESON DOMINANCE IN THE TOTAL PHOTOABSORPTION CROSS SECTION

J. Ahrens
Max-Planck-Institute for Chemistry
P.O.Box 3060, D-6500 Mainz, West-Germany

Abstract

Features of the total photoabsorption cross section are presented that are used as arguments for the vector meson dominance model. These features are the similarities in the absorption of hadrons and photons by the proton and the shadowing effect seen in the total photon absorption of complex nuclei.

1. Introduction

This article is a comment on W.Weise's paper contributed to this workshop [1]. The absorption of photons by the proton and by nuclei has been investigated over a wide range of mass numbers and photon energies. Although all data lack high precision some features have become clearly visible [2].

The absorption of hadrons and photons by the proton shows similarities. This is taken as an indication of the photon's hadronic nature. In the same sense the decrease of the absorption cross section per nucleon with increasing mass number, the shadowing effect, is interpreted.

In the vector meson dominance model [1] it is assumed that a high energy photon, when interacting with matter, fluctuates into a vector meson. In the interaction the photon then behaves like a meson. The cross section of the photon is then given by the coupling strength of the photon to the vector meson and the cross section of the vector meson. Features of this model can be illustrated by looking at different lengths and dimensions in this context.

The mean free path length of a particle in nuclear matter is given by

$$\lambda = \frac{m}{\rho \sigma_{tot}}.$$

ρ is the density of nuclear matter, σ_{tot} the total absorption cross section for the particle and m the mass of the absorbing particle. In a good approximation we have

$$\lambda / fm \approx \frac{100}{\sigma_{tot}/mb}.$$

When the photon fluctuates into a vector meson the length of this fluctuation is limited by the mismatch of energy or momentum. The space time difference over which the fluctuation can go is for real photons

$$\delta \approx \frac{2\omega}{\mu^2},$$

where ω is the photon energy and μ the rest mass of the vector meson. Again to a good approximation we have

$$\delta / fm \approx 400 MeV \frac{\omega}{\mu^2}.$$

In figure 1 these lengths are compared with the dimension of nuclei. The mean free path length of the photon was determined using the total absorption cross section of the photon on the proton, the mean free path length of the meson was estimated using the average absorption cross section for the π^+– and the π^-–meson on the proton. For references see [2].

It can be seen that in contrast to the photon the meson has a mean free path length that is significantly

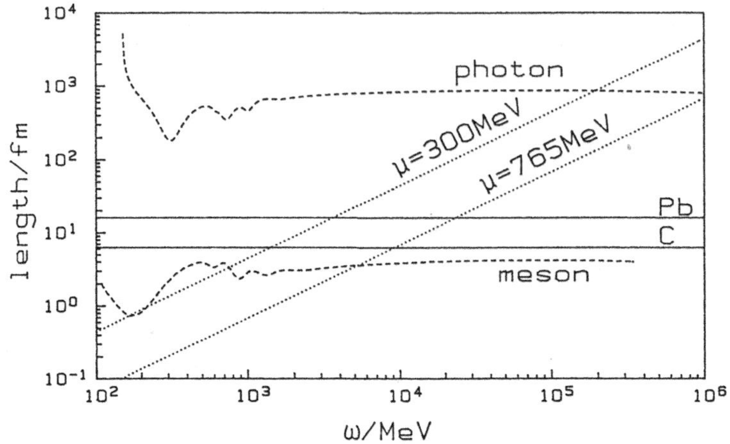

Figure 1: ***Lengths and dimensions in the context of the vector meson dominance model. Dashed lines are mean free path lengths for the photon and for a meson. The full lines indicate the dimensions of the C and the Pb nuclei. The dotted lines give the fluctuation lengths for μ = 300 MeV (upper line) and 765 MeV.***

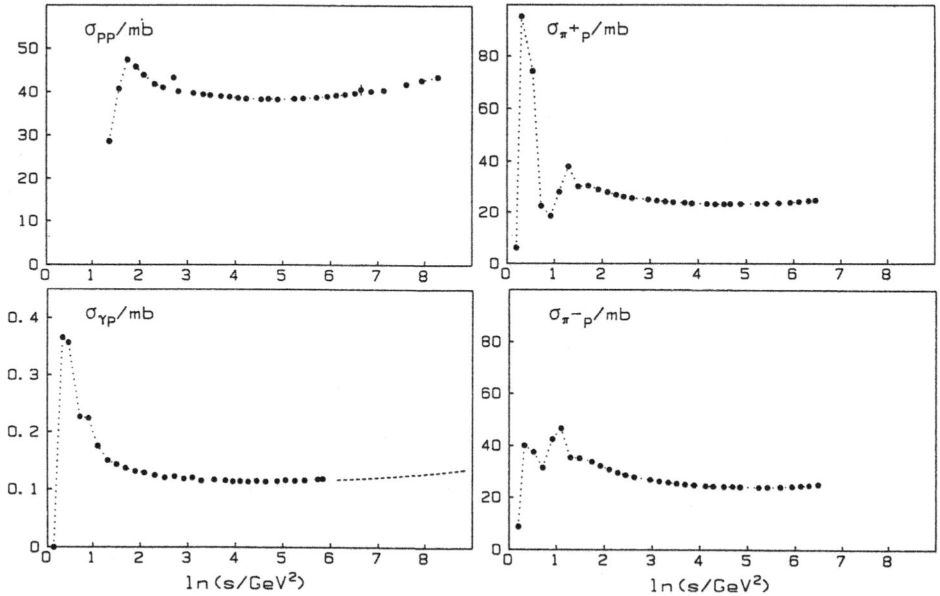

Figure 2: ***Measured cross sections for the absorption of protons, photons, and pions by the proton***

smaller than the dimension of a nucleus. For the meson a nucleus is a "thick target". The length of the fluctuation, δ, increases with energy. In the figure it is given for the approximate mass of two pions, 300 MeV, and for the mass of the ρ-meson μ_ρ = 765 MeV. For several GeV the fluctuation length becomes comparable with the mean free path length of the meson and with the dimension of nuclei.

2. The absorption cross sections of the proton

In figure 2 the measured cross sections for the absorption of photons, pions and protons by the proton are shown as a function of $ln(s/GeV^2)$. s is the c.m. energy squared. The points shown are averages over energy bins of the data found in literature. Since the energy bins are comparativily large the low energy structures appear obscured. There are similarities in the cross sections above $s = 7$ GeV². These similarities are attributed to the hadronic nature of the photon during the interaction.

It has been proposed [3] to describe these cross sections by a function of the form $\sigma(s) = A + Bln^2(s/s_0)$ which is the shape predicted as the Froissart bound [4]. Since the photon absorption cross section has not been measured to energies as high as in the case of the hadrons it can be estimated for higher energies by using pieces of information from the hadron-hadron cross section. A reasonable extrapolation of the photon absorption cross section is thus given by

$$\sigma_{\gamma,p}(s)/mb = 0.137 - 1.04 \cdot 10^{-2} ln(s/GeV^2) + 1.14 \cdot 10^{-3} ln^2(s/GeV^2)$$

This is illustrated by a dashed line in figure 2.

3. The shadowing effect

The total absorption cross section for nuclei has been investigated in the Δ-resonance region for several mass numbers ranging from the proton to ^{238}U. It turns out that the cross section per nucleon is approximately the same for nuclei from Be to U. The proton cross section is different, the deuterium cross section is already close to the Be to U "universal" curve. These features are illustrated in figure 3. The proton and the deuteron cross sections are approximated by smooth lines.

In the region of the higher baryon resonances, i.e. from 500 MeV to 2 GeV only a few data are available. There are data for the proton and deuterium and with low precision for Be and C.

Above 2 GeV the shadowing effect starts to become visible. Figure 4 shows cross sections for deuterium and Cu normalized to the proton cross section. In this energy domain the cross section per nucleon is no longer independent of the mass number but decreases with increasing mass number. It can be written as

$$\sigma_{tot}^{A} = A^{\alpha} \sigma_{tot}^{proton}$$

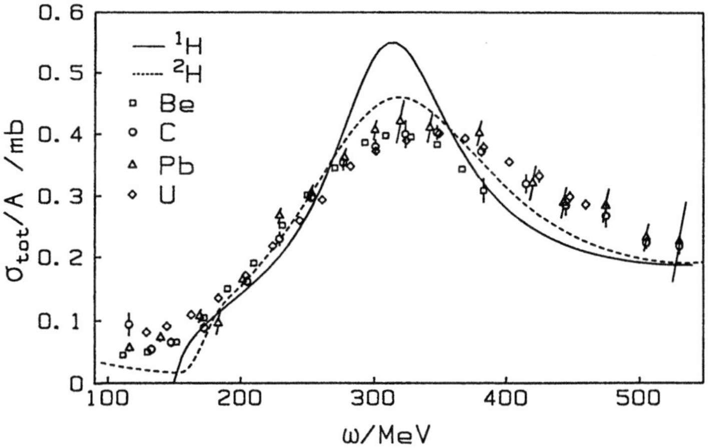

Figure 3: *Total photon absorption cross section per nucleon in the Δ-resonance region for the proton, deuterium and several complex nuclei.*

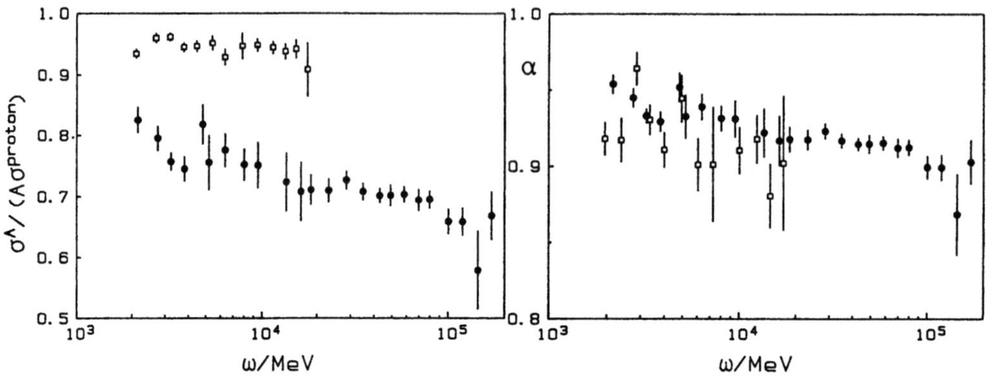

Figure 4: ***Shadowing effect shown for deuterium, open box, and Cu, full circle. Left – cross sections per nucleon normalized to the proton cross section, right – α from*** $\sigma^A_{tot} = A^\alpha \sigma^{proton}_{tot}$.

In figure 4 α as deduced from measurements is shown as a function of the energy. $\alpha \approx 0.9$ is found. The data indicate that α is gradually decreasing with energy and within the precision of the data not depending on A.

4. Conclusion

In this article features of the total photon absorption cross section are shown that can be understood in the frame of the vector meson dominance model. The pieces of information shown here are to supplement the contribution by W.Weise presented during the same workshop. The similarity of the cross section shapes for the absorption of hadrons and photons by the proton is understood by assuming that the photon converts into a vector meson before it interacts with the proton. In the shadowing effect the short mean free path length of the photon's hadronic fluctuation becomes visible. The photon no longer probes the volume of the nucleus as would be expected from the size of the absorption cross section.

References

[1] W.Weise, contribution to this workshop
[2] J.Ahrens, Nucl.Phys. **A466** (1985) 229c
[3] E. Leader and U. Maor, Phys. Lett. **43B** (1983) 505
[4] M. Froissart, Phys. Rev. **123** (1961) 1053

The Shadowing Effect in the Compton Scattering Cross Section

P. Rullhusen

II. Phys. Institut , Univ. Göttingen , D-3400 Göttingen , Fed. Rep. of Germany

1 Introduction

The presence of hadronic components of the photon is observed in total photo-absorption cross sections on heavy nuclei at energies of several GeV. In contrast to measurements in the energy region of the Δ resonance – where a scaling behaviour of the total photo-absorption cross section with the mass number A of the nucleus is observed – at photon energies of several GeV the measured cross sections are considerably smaller than predicted by assuming the same scaling behaviour with mass–number A at these higher energies. This "shadowing effect" can be understood in the framework of the vector dominance model (VDM) of photoprocesses.

The forward Compton scattering amplitude is related to the total photo–absorption cross section via the optical theorem and a dispersion relation. Therefore, at high energies Compton scattering close to the forward direction is also strongly affected by vector–meson contributions to the photon propagator. From a comparison of the cross section measured on a complex nucleus with that measured on free nucleons the shadowing effect can be studied as a function of momentum transfer. Extrapolation to $t = 0$ allows for a direct comparison with the shadowing effect observed in photo–absorption measurements. In addition , from the diffractive pattern of the scattering cross section information on the nuclear size is obtained , as seen by the photon at high energies .

In the following the experimental results on high–energy Compton scattering are reviewed. No attempt has been made to review also the extensive related work on vector–meson photoproduction. Only few experiments using polarized photons are discussed in the last section revealing the potentialities to study spin characteristics in such diffractive reactions. The review articles of Bauer et al. [1] and Leith [2] have been extensively used in the preparation of this manuscript and are strongly recommended for further reading.

2 Compton scattering amplitudes

Compton scattering, i.e. the elastic scattering of photons, is described in lowest–order approximation as a two–step process strongly related to photo–absorption. Higher–order QED corrections (i.e. Delbrück scattering) are not of relevance in the range of momentum transfers considered here and can safely be neglected, as will be shown later. At forward direction the scattering amplitude f and the total photo–absorption cross section $\sigma_{\gamma t}$ are related by the optical theorem and a once–subtracted dispersion relation :

$$\mathrm{Im}\, f(\omega) = \frac{\omega}{4\pi}\sigma_{\gamma t}(\omega) \tag{1}$$

$$\mathrm{Re}\, f(\omega) = f_T + \frac{\omega^2}{2\pi}\,\mathcal{P}\int_0^\infty \frac{\sigma_{\gamma t}(\omega')}{\omega'^2 - \omega^2}\, d\omega' \quad . \tag{2}$$

In (2) the classical Thomson scattering amplitude f_T has been used for the long–wavelength limit :

$$\lim_{\omega\to 0} f(\omega) = f_T \quad . \tag{3}$$

In photo–absorption measurements on complex nuclei the cross section $\sigma_{\gamma t}(A)$ is smaller than the sum of A independent nucleon cross sections $\sigma_{\gamma t}(N)$ ("shadowing"):

$$\sigma_{\gamma t}(A) = A_{eff}\, \sigma_{\gamma t}(N) \tag{4}$$

with $A_{eff} < A$. In a first approximation , A_{eff} may be assumed to be energy–independent. Neglecting also the small contribution from f_T to the real part of the scattering amplitude, one obtaines from (1),(2) and (4) the A–dependence of the forward Compton scattering cross section $\sigma_{\gamma\gamma} = |\, f\,|^2$:

$$\frac{d\sigma_{\gamma\gamma}(A)}{dt}(t \to 0) \approx A^2_{eff}\, \frac{d\sigma_{\gamma\gamma}(N)}{dt}(t \to 0) \quad . \tag{5}$$

Figure 1 shows a more realistic prediction for the shadowing effect expected to be observed in a heavy nucleus. In Fig.(1a) smooth fits [3] to photo–absorption data measured on H and Pb are given. These data have been used to evaluate numerically the forward scattering amplitude using (1) and (2) and the resulting scattering cross sections are shown in Fig. 1b. The sizeable effect of shadowing expected to be seen in Compton scattering off heavy nuclei is apparent.

For a comparison of values for A_{eff}/A obtained from Compton scattering with similar values obtained from photo–absorption measurements the phase $\alpha = \mathrm{Re}\, f\, /\, \mathrm{Im}\, f$ has to be known . Therefore, in (2) reasonable extrapolations of $\sigma_{\gamma t}(\omega)$ for $\omega \to \infty$ have to be made. Figure 1c shows the phase $\alpha(\omega)$ calculated from the empirical fits of Fig.1a . In the high–energy domain the scattering cross section is dominated by the imaginary part of the amplitude, even for heavy nuclei like Pb. Therefore, the phase α has to be known to only 10% accuracy in order to arrive at a 5% level of accuracy in the prediction of forward-scattering cross sections from photo–absorption data and uncertainties in the extrapolation of $\sigma_{\gamma t}(\omega)$ for $\omega \to \infty$ do not affect seriously the result.

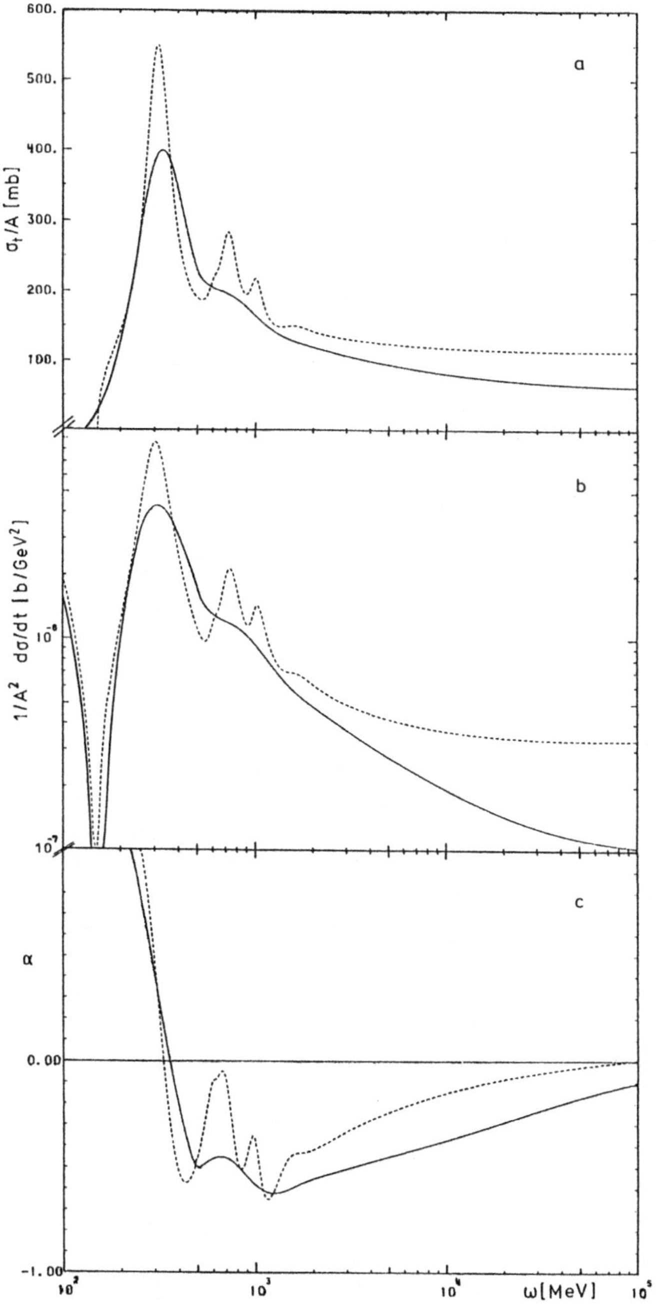

Fig. 1. a) Smooth fits [3] to photo-absorption cross sections measured on H (dotted) and Pb (solid) scaled by $1/A$. b) Forward Compton scattering cross sections scaled by $1/A^2$ calculated from data shown in (a) using (1) and (2) . In (a) and (b) the differences in the two curves for energies above 2 GeV are an indication of the shadowing effect. c) Phase $\alpha = \mathrm{Re}f\,/\,\mathrm{Im}f$ of forward Compton scattering amplitudes used in the calculations of (b)

3 Compton scattering from the nucleon

By far most of the experimental work on Compton scattering at high energies has been dedicated to the proton. Table 1 summarizes the experiments carried out at photon energies ω above pion threshold.

Table 1. Compton scattering from the proton above pion threshold

		$\omega\,[GeV]$	$\Theta^{c.m.}\,[^\circ]$	$-t\,[GeV^2]$
G. Bernardini et al.	[4]	0.20 – 0.30	90 – 140	0.05 – 0.20
J. W. de Wire et al.	[5]	0.30 – 0.80	60 – 120	0.08 – 0.20
P. S. Baranov et al.	[6]	0.25	55 – 150	0.03 – 0.15
R. F. Stiening	[7]	0.50 – 0.90	90	0.20 – 0.50
D. R. Rust et al.	[8]	0.40 – 1.40	90	0.20 – 1.00
P. S. Baranov et al.	[9]	0.25	55 – 150	0.03 – 0.15
P. S. Baranov et al.	[10]	0.21 – 0.25	70 – 150	0.02 – 0.15
E. Gray, A. O. Hanson	[11]	0.20 – 0.30	90 – 142	0.10 – 0.20
G. Barbiellini et al.	[12]	0.32	90	0.12
G. Buschhorn et al.	[13]	5.00 – 7.00	9 – 22	0.06 – 0.60
R. L. Anderson et al.	[14]	5.00 – 17.0	6.5 – 30	0.06 – 1.10
A. M. Boyarski et al.	[15]	8.0 , 16.0	3.5 – 12.5	0.01 – 0.17
G. Buschhorn et al.	[16]	2.20 – 7.00	8.0 – 38	0.06 – 0.85
G. Buschhorn et al.	[17]	3.5	15 – 40	0.10 – 0.70
H. Genzel et al.	[18]	0.20 – 0.40	50 – 130	0.05 – 0.32
S. Kabe et al.	[19]	0.70 – 0.80	90	0.40 – 0.47
J. S. Barton et al.	[20]	0.60 – 1.40	60 – 140	0.18 – 1.65
H. Alvensleben et al.	[21]	2.2	10	0.027
M. Deutsch et al.	[22]	0.50 – 4.50	15 – 115	0.12 – 1.00
H. Genzel et al.	[23]	0.20 – 0.40	50 – 130	0.03 – 0.33
L. Criegee et al.	[24]	5.0 , 6.0	1.5 – 10	0.002 – 0.08
K. Toshioka et al.	[25]	0.40 – 1.00	60, 90	0.20 – 0.60
M. A. Shupe et al.	[26,27]	2.00 – 6.00	45 – 128	0.70 – 4.30
T. Ishii et al.	[28]	0.30 – 1.20	70 – 130	0.10 – 1.30
M. Jung et al.	[29]	0.70 – 1.00	40 – 130	0.10 – 1.00
Y. Wada et al.	[30]	0.40 – 1.00	100, 130	0.30 – 0.80
J. Duda et al.	[31]	1.20 – 1.70	35 – 80	0.20 – 1.00
Y. Wada et al.	[32]	0.40 – 1.10	150, 160	0.30 – 1.40
T. Ishii et al.	[33]	0.90 – 1.20	50 – 130	0.20 – 1.30

In connection with shadowing we are primarily concerned with measurements at $\omega \gtrsim 2\,GeV$ and small momentum transfer. In this energy–momentum range the Compton scattering cross section is directly related to the vector meson photoproduction cross sections in the framework of the vector dominance model. Figure 2 shows the results of Compton scattering experiments carried out at DESY and SLAC in comparison with ρ_o meson photoproduction data [2]. The similarity in the t–dependence of both reactions is evident.

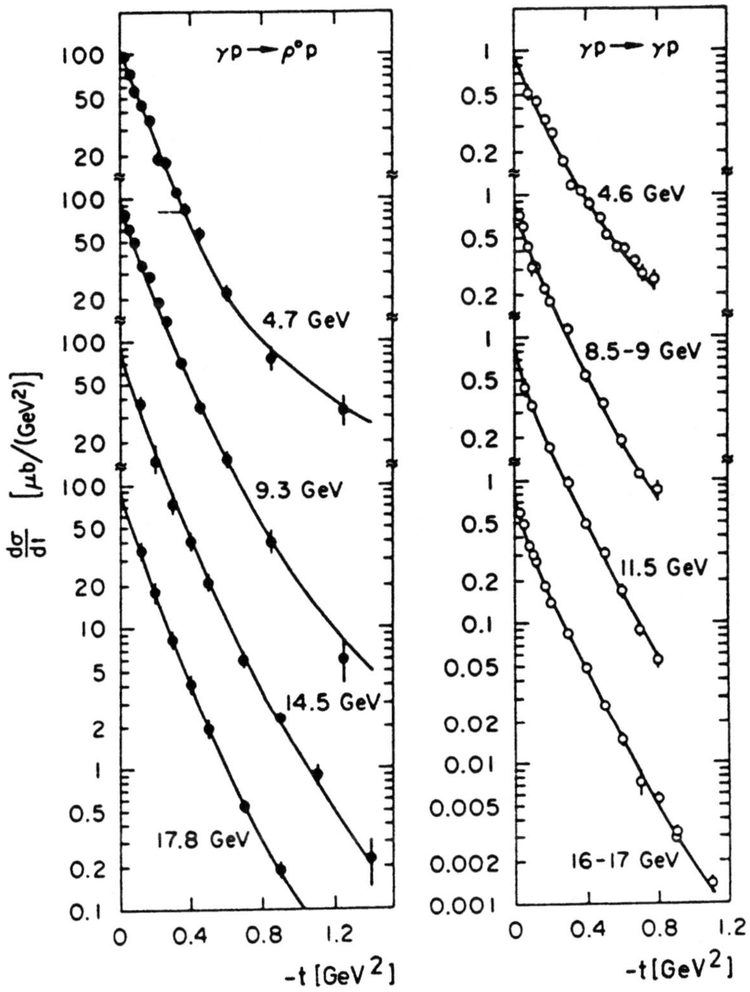

Fig. 2. ρ_o **photoproduction and Compton scattering from hydrogen at high energies [2]**

For a quantitative comparison wit VDM predictions, the Compton scattering cross sections have to be extrapolated to $t = 0$. This can be achieved by means of simple exponential fits. The extrapolated forward scattering cross sections have been found [24] in good agreement with predictions from the optical theorem (1) and the dispersion relation (2) but higher by about 40% than VDM predictions using a sum-rule approach and experimentally determined photoproduction cross sections for ρ , ω and ϕ mesons and coupling constants from storage ring experiments . The remaining discrepancies can be reduced by extending the VDM to the GVDM including vector-mesons of higher masses [24].

Compton scattering cross sections for the neutron may be deduced from scattering experiments on the deuteron applying suitable corrections for Fermi motion, screening of one nucleon by the other, and binding effects. These corrections, though small in absolute size, are nevertheless very important , since they are of the same magnitude as the difference in the proton and neutron cross sections , which are of the order of 5% at $10\,GeV$ [2]. A careful analysis [35] of the proton and deuteron Compton scattering cross sections in terms of isovector exchange amplitudes indicates a vanishing or at least very small $\Delta T = 1$ contribution, in agreement with the observed small difference in the proton and neutron photo-absorption cross sections.

4 Compton scattering from complex nuclei

Most of the data on Compton scattering from hydrogen and deuterium have been measured using a spectrometer to detect the recoiling nucleus in coincidence with the elastically scattered photon. By this technique the inelastic photon background due mainly to the decay of π^o and η mesons produced in the target can largely be suppressed. At very small momentum transfer and in scattering experiments on heavy nuclei, however, the energy of the recoiling nucleus is too small to be detected and the experiment has to rely on the detection of the scattered photon. This makes the experimental situation much more difficult than in the case of scattering from hydrogen and deuterium and only a few experiments have been carried out up to now at high energies. Table 2 summarizes Compton scattering experiments on nuclei with mass numbers $A \geq 2$ at photon energies above pion threshold.

Table 2. Compton scattering from nuclei with $A \geq 2$ above pion threshold

		A	$\omega\,[GeV]$	$\Theta^{c.m.}$ [°]
E. Eisenhandler et al.	[34]	6, 74	1.0 – 1.3	90
A. M. Boyarski et al.	[15]	2	8, 16	3.5 – 12.5
L. Criegee et al.	[24]	1, 2	5, 6	1.5 – 10
L. Criegee et al.	[35]	9 – 197	5, 6	1.5 – 10
V. S. Arakyan et al.	[36]	2, 4	2 – 6	90
E. Hayward, B. Ziegler	[37]	6, 82	0.15 – 0.4	150
E. J. Austin et al.	[38]	4	0.18	130
A. S. Aleksanyan et al.	[39]	4	1.8 – 3.8	90
E. J. Austin et al.	[40]	4	0.26 – 0.36	130

As in the case of the nucleon , we will be concerned with measurements at $\omega \gtrsim 2\,GeV$ only. Figure 3 shows the results obtained at DESY on heavy nuclei at photon energies of $3\,GeV$ and $5\,GeV$ [35]. The data were taken at very small scattering angles using a pair spectrometer to detect the scattered photons. Photon background from inelastic reactions was calculated by a Monte Carlo method and has been subtracted from the data. The solid lines in Fig.3 show fits to the data, obtained from an optical model calculation of coherent scattering cross sections including direct photon scattering and contributions from intermediate vector–mesons . At large momentum transfers t (open points in Fig.3) an incoherent contribution which vanishes at $t \to 0$ was added , with an overall scaling factor. The sum of both contributions was fitted to the low–t data (solid points in Fig.3). In the fitting procedure the nuclear radii were taken as free parameters and the values obtained from the fits were found to be in good agreement with those obtained from high–energy electron scattering [35].

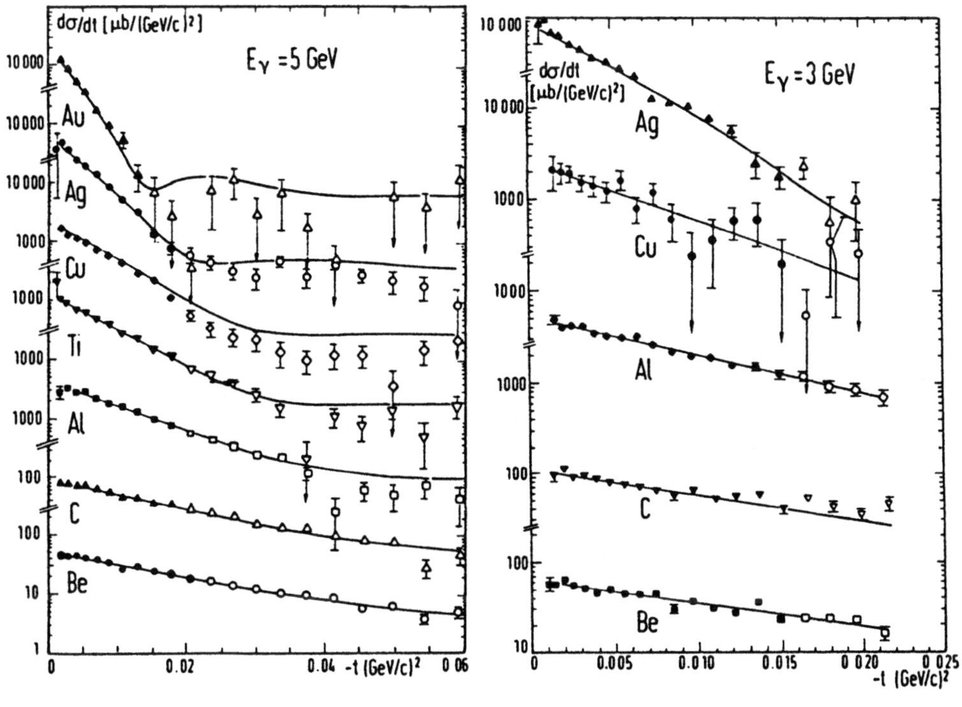

Fig. 3. Small–angle Compton scattering cross sections from complex nuclei [35]

From the data of Fig.3 the forward differential cross sections were extrapolated and information on the shadowing effect on the total photo–absorption cross section $\sigma_{\gamma t}(A)$ was obtained using (1) and estimating the phases α in an optical model. The results of this calculation are shown in Fig.4 together with measured total cross sections [41-43]. The dashed curves indicate no shadowing, i.e. $\sigma_{\gamma t}(A) = A\ \sigma_{\gamma t}(N)$.

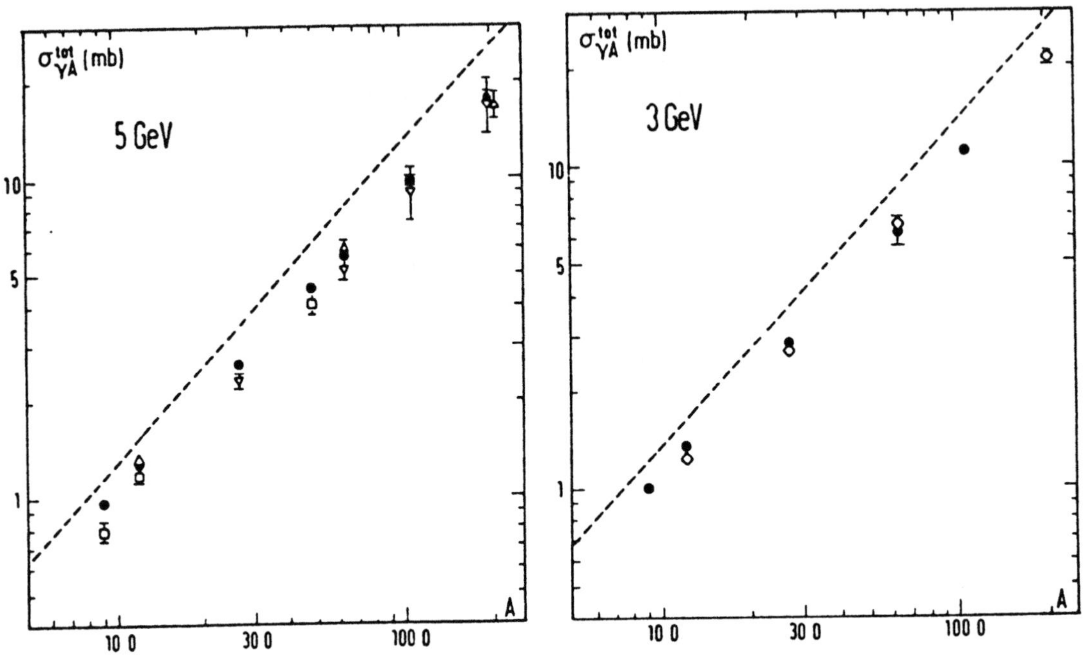

Fig. 4. Total photon cross sections calculated from forward Compton scattering (solid) [35] , compared to measured total cross sections (open) [41-43]. Dashed curves: predictions for the case of no shadowing , i.e. $\sigma_{\gamma t}(\mathrm{A}) = \mathrm{A}\ \sigma_{\gamma t}(\mathrm{N})$

The shadowing of 20–30 % observed in the Compton scattering experiments is in very close agreement with the effect observed in the total–absorption measurements. The A–dependence of the absorption cross sections at fixed photon energy as well as the energy–dependence of A_{eff}/A agree well in both methods. The observed shadowing effect was also found [35] in qualitative agreement with VDM predictions obtained using experimentally measured photoproduction cross sections of vector mesons .

The experiments from the DESY group [35] have clearly demonstrated the possibility to use Compton scattering as a tool to investigate shadowing effects, complementary to total–absorption measurements. In the extrapolation to $t=0$, however, care has to be taken not to include in the experimental data higher–order QED corrections (i.e. Delbrück scattering) which are dominant close to the forward direction and show a very strong t–dependence . Fortunately, in the energy domain and momentum range of interest here, these QED contributions are not important and can safely be neglected even in nuclei of large charge–number Z. Figure 5a shows the t–dependence observed in photon scattering experiments on Au at $\omega=1\text{–}7\ GeV$ and $|t|<10^{-4}\ GeV^2$ [44]. The solid line in Fig.5a was calculated in the impact–factor approximation [45] and is in very good agreement with the experimental data. In Fig.5b the same approximation has been used to predict Delbrück cross sections from Au at larger t–values. At $|t|\ =\ 10^{-2}\ GeV^2$ the

Delbrück scattering cross section is of the order of $3 \times 10^{-5}\ b/GeV^2$, to be compared with $10^{-2}\ b/GeV^2$, taken from Fig.3 .

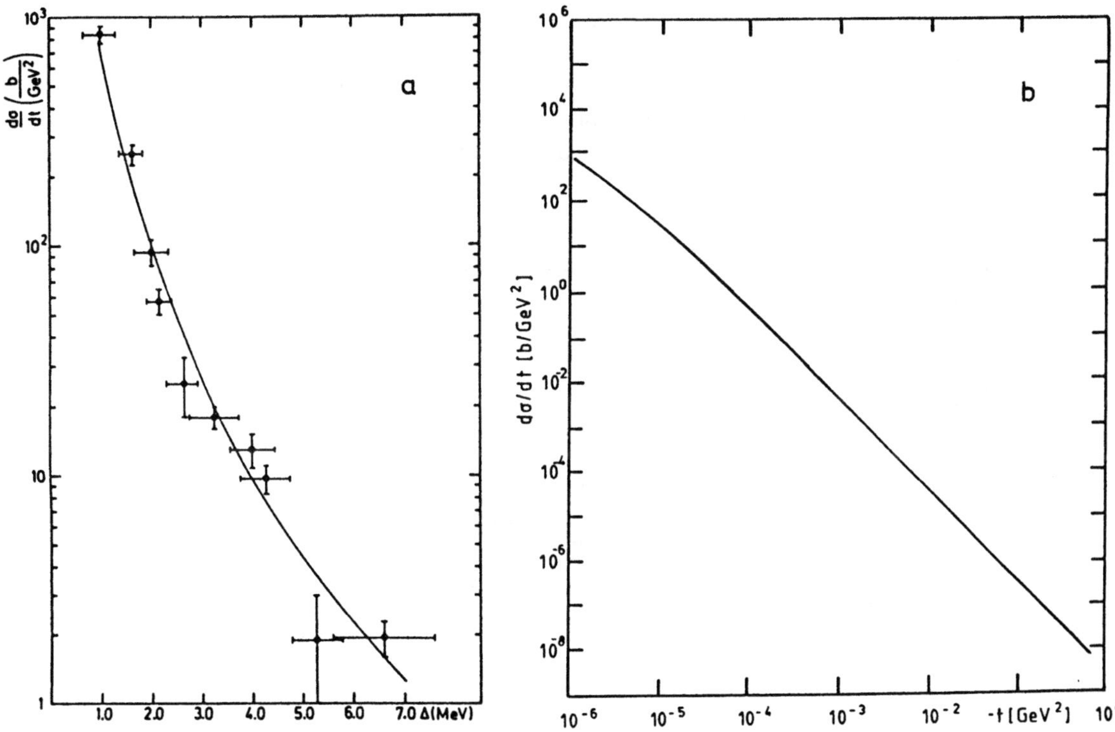

Fig. 5. a) Elastic photon scattering from Au at very small momentum transfer $-t = \Delta^2$ [44] Solid line: Delbrück scattering cross sections calculated in the impact–factor approximation [45] b) Predictions for Delbrück scattering cross sections at large momentum transfer calculated in the impact–factor approximation

5 Polarized photons

The investigation of asymmetry ratios of cross sections in photonuclear reactions using polarized photons gives valuable additional information on the nature of the processes under consideration. Large improvements have been made in the last years on new techniques to produce linearly polarized quasimonochromatic photon beams of high energies. Off–axis tagging , back–scattering of laser light from high–energy electrons , and coherent bremsstrahlung produced in single crystals have been applied successfully to study contributions of different multipolarity in photon scattering and in photoproduction measurements.

In Fig.6 the results of Compton scattering experiments on the proton at energies $\omega = 3.2 - 3.7 GeV$ are shown [46]. The asymmetry $\Sigma = \frac{\sigma_\perp - \sigma_\parallel}{\sigma_\perp + \sigma_\parallel}$ of the Compton scattering cross sections measured at $\Theta = 90°$ for linear photon polarization parallel and perpendicular to the scattering plane, respectively , is sensitive to the spin dependence of the scattering process. The data are compared to predictions of three different models which have been discussed in connection with vector meson photoproduction , i.e. the 0^+–exchange model, s–channel helicity concervation (SCHC), and the spin–independent model (SIM). Best agreement was obtained with predictions assuming a momentum–independent asymmetry as predicted by the SCHC model, similar to the observations made in photoproduction measurements of ρ_0 mesons with linearly polarized photons [47-49] , indicating that the ρ_0 meson produced in these reactions is linearly and transverse polarized as the photon.

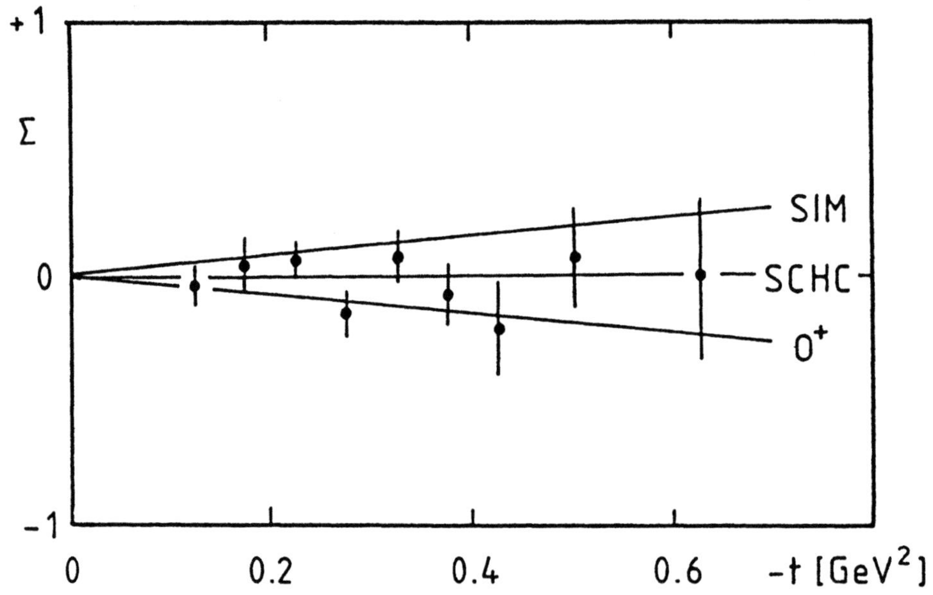

Fig. 6. Asymmetry $\Sigma = \frac{\sigma_\perp - \sigma_\parallel}{\sigma_\perp + \sigma_\parallel}$ observed in Compton scattering experiments [46] with plane–polarized photons in comparison with predictions from different models : 0^+–exchange , s–channel helicity concervation (SCHC) , and spin–independent model (SIM)

6 Summary

Compton scattering experiments carried out at photon energies of several GeV have shown the characteristic t–dependence predicted by the VDM and observed in vector–meson photoproduction. The shadowing effects observed in scattering experiments on complex nuclei are in agreement with those seen in total–absorption measurements. The nuclear sizes seen by the high–energy photons are the same as seen in electron scattering experiments.

Comparison of scattering experiments on hydrogen and deuterium have allowed to separate the amplitudes for isospin exchange $\Delta T = 0, 1$. The observed isospin ratios have shown a vanishing or at least very small $\Delta T = 1$ contribution in agreement with the small differences in the total absorption cross sections of the proton and the neutron, extracted from measurements on hydrogen and deuterium.

Experiments using linearly polarized photon beams have demonstrated the possibility to study the spin structure of the scattering process. The asymmetries observed in the Compton scattering experiments indicate helicity concervation in the s–channel , in agreement with the observations made in ρ meson photoproduction.

References

1. T. H. Bauer, R. D. Spital and D. R. Yennie: Rev. Mod. Phys. **50** 261 (1978)
2. D. W. G. S. Leith: "High–Energy Photoproduction: Diffractive Processes", in Electromagnetic Interactions of Hadrons, ed. by A. Donnachie and G. Shaw (Plenum Press, New York and London 1978) Vol. 1,pp. 345–440
3. J. Ahrens, priv. comm.
4. G. Bernardini et al.: Il Nuovo Cim. **18** 1203 (1960)
5. J. W. de Wire et al.: Phys. Rev. **124** 909 (1961)
6. P. S. Baranov et al.: Sov. Phys. JETP **14** 1219 (1962)
7. R. F. Stiening: Phys. Rev. Lett. **10** 536 (1963)
8. D. R. Rust et al.: Phys. Rev. Lett. **15** 938 1965
9. P. S. Baranov et al.: Sov. Phys. JETP **23** 243 (1966)
10. P. S. Baranov et al.: Sov. J. Nuc. Phys. **3** 790 (1966)
11. E. Gray, A. O. Hanson: Phys. Rev. **160** 1213 (1967)
12. G. Barbiellini et al.: Phys. Rev. **174** 1665 (1968)
13. G. Buschhorn et al.: Phys. Lett. **33B** 241 (1970)
14. R. L. Anderson et al.: Phys. Rev. Lett. **25** 1219 (1970)
15. A. M. Boyarski et al.: Phys. Rev. Lett. **26** 1600 (1971)
16. G. Buschhorn et al.: Phys. Lett. **37B** 207 (1971)
17. G. Buschhorn et al.: Phys. Lett. **37B** 211 (1971)
18. H. Genzel et al.: Lett. Nuovo Cim. **4** 695 (1972)
19. S. Kabe et al.: Nucl. Phys. **B50** 17 (1972)
20. J. S. Barton et al.: Phys. Lett. **42B** 297 (1972)
21. H. Alvensleben et al.: Phys. Rev. Lett. **30** 328 (1973)
22. M. Deutsch et al.: Phys. Rev. **D8** 3828 (1973)
23. H. Genzel et al.: Z. Phys. **A279** 399 (1976)
24. L. Criegee et al.: Nucl. Phys. **B121** 31 (1977)
25. K. Toshioka et al.: Nucl. Phys. **B141** 364 (1978)
26. M. A. Shupe et al.: Phys. Rev. Lett. **40** 271 (1978)
27. M. A. Shupe et al.: Phys. Rev. **D19** 1921 (1979)
28. T. Ishii et al.: Nucl. Phys. **B165** 189 (1980)
29. M. Jung et al.: Z. Phys. **C10** 197 (1981)
30. Y. Wada et al.: Il Nuovo Cim. **63A** 57 (1981)
31. J. Duda et al.: Z. Phys. **C17** 319 (1983)
32. Y. Wada et al.: Nucl. Phys. **B247** 313 (1984)
33. T. Ishii et al.: Nucl. Phys. **B254** 458 (1985)
34. E. Eisenhandler et al.: Phys. Lett. **24B** 347 (1967)
35. L. Criegee et al.: Nucl. Phys. **B121** 38 (1977)

36. V. S. Arakyan et al.: Sov. J. Nuc. Phys. **33** 808 (1981)
37. E. Hayward, B. Ziegler: Nucl. Phys. **A414** 33 (1984)
38. E. J. Austin et al.: Phys. Rev. Lett. **57** 972 (1986)
39. A. S. Aleksanyan et al.: Sov. J. Nuc. Phys. **45** 628 (1987)
40. E. J. Austin et al.: Phys. Rev. Lett. **61** 1922 (1988)
41. V. Heynen et al.: Phys. Lett. **34B** 651 (1971)
42. D. O. Caldwell et al.: Phys. Rev. **D7** 1362 (1973)
43. G. R. Brookes et al.: Phys. Rev. **D8** 2826 (1973)
44. G. Jarlskog et al.: Phys. Rev. **D8** 3813 (1973)
45. H. Cheng, T. T. Wu : Phys. Rev. **D5** 3077 (1972)
46. L. Criegee et al.: Phys. Lett. **37B** 211 (1971)
47. L. Criegee et al.: Phys. Lett. **28B** 282 (1968)
48. J. Ballam et al.: Phys. Rev. Lett. **24** 960 (1970)
49. G. Diambrini–Palazzi et al.: Phys. Rev. Lett. **25** 478 (1968)

The Shadowing Effect in Deep Inelastic Lepton Scattering

Eva-Maria Kabuß

Max-Planck-Institut für Kernphysik, D-6900 Heidelberg, Fed. Rep. of Germany

Abstract: The discovery of the EMC effect triggered new measurements of cross section ratios in deep inelastic scattering of charged leptons. These measurements include also the kinematic region where shadowing was predicted, but could not be established beyond doubt by previous experiments. A review is given on recent results, mainly from the EMC, and models developed to describe these data. In addition new preliminary results from the NMC for He/D, C/D and Ca/D are presented. The high precision of the NMC data in a wide x range allows to check various models in detail. At present, none of the models provides a satisfactory description in the whole measured kinematic range.

1. Introduction

In measurements with *real* photons the cross section per nucleon σ_A for bound nucleons was found to be smaller than the cross section σ_N for free nucleons. This effect is usually interpreted as shadowing due to the hadronic components of the photon [1].

In deep inelastic scattering the interaction of *virtual* photons with nuclei is studied. The basic graph for this interaction is the single photon exchange diagram shown in fig.1 [2]. It can be described in terms of the four momentum transfer squared $-Q^2$, the energy transfer $\nu = E - E'$, the Bjorken scaling variable $x = Q^2/2M\nu$ and the relative energy transfer $y = \nu/E$. E is the energy of the incoming, E' the energy of the scattered lepton and M is the proton mass.

The double differential cross section per nucleon in the one photon exchange approximation reads

$$\frac{d^2\sigma}{dxdQ^2} = \frac{4\pi\alpha^2}{Q^4}\left(\left[1 - y - \frac{xyM}{2E}\right]\frac{F_2(x,Q^2)}{x} + y^2 F_1(x,Q^2)\right). \tag{1}$$

The structure functions $F_1(x,Q^2)$ and $F_2(x,Q^2)$ describe the internal structure of the nucleon. $F_1(x,Q^2)$ and $F_2(x,Q^2)$ are related to $R(x,Q^2)$, the ratio of the absorbtion cross sections for longitudinally and transversely polarized photons

$$R = \frac{\sigma_L}{\sigma_T} = \frac{(1 + 4M^2x^2/Q^2)F_2 - 2xF_1}{2xF_1}. \tag{2}$$

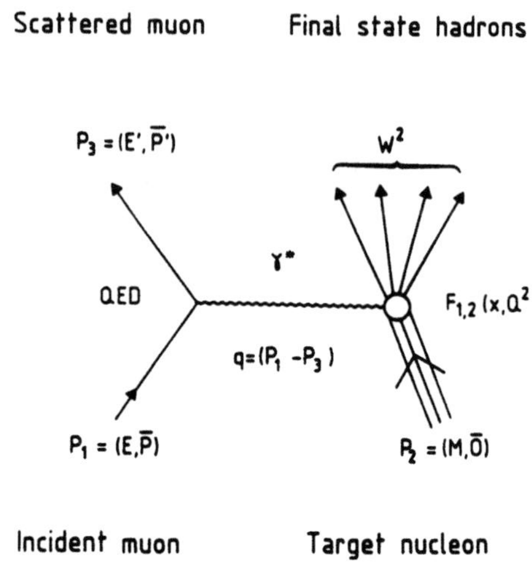

Fig. 1. Schematic picture for the deep inelastic muon nucleon scattering process

In terms of R the cross section (1) reads

$$\frac{d^2\sigma}{dxdQ^2} = \frac{4\pi\alpha^2}{Q^4}\frac{F_2(x,Q^2)}{x}\left(1 - y - \frac{xyM}{2E} + \frac{y^2}{2}\frac{1+4M^2x^2/Q^2}{1+R(x,Q^2)}\right). \tag{3}$$

In the parton model F_2 represents the sum over the quark and antiquark distributions $q(x,Q^2)$ and $\bar{q}(x,Q^2)$ weighted by the square of the charge

$$F_2(x,Q^2) = x\sum_i e_i^2(q_i(x,Q^2) + \bar{q}_i(x,Q^2)). \tag{4}$$

Recent measurements of R^A show that $R(x,Q^2)$ is independent of the atomic mass number A of the target nucleus, at least in the kinematic range $x > 0.17$ [3]. Thus the ratio of the cross sections for different nuclei A_1 and A_2 is equal to the ratio of structure functions $F_2^{A_1}/F_2^{A_2}$, where F_2 is the structure function per nucleon.

The fluctuation length for real photons into hadrons is given by $d \sim \frac{\nu}{M_V^2}$, where M_V is the mass of the hadronic component. In charged lepton scattering the exchanged virtual photon has a mass squared of $-Q^2$. Therefore one expects for the fluctuation length

$$d \sim \frac{\nu}{M_V^2 + Q^2} \sim \frac{1}{x} \quad (\text{for large } Q^2). \tag{5}$$

Shadowing should appear when the fluctuation length is large, therefore it should occur at large ν. For large Q^2 the fluctation length becomes proportional to $1/x$ so that shadowing is expected at small x.

2. Measurements of Cross Section Ratios

For a long time the expected shadowing effect could not be experimentally established beyond doubt due primarily to the large systematic errors caused by uncertainties in the radiative corrections [1,4]. The measured cross sections include, besides single photon exchange, contributions from higher order QED processes such as Bremsstrahlung. These have to be calculated and corrected for, in order to determine the cross section for single photon exchange. The corrections are particularly large at high y, and therefore introduce large uncertainties in the interpretation of the data especially at small x ($x < 0.1$) in low energy electron scattering. For muon scattering, where the beam energies are usually higher, these corrections are much smaller.

In 1983 it was shown by the EMC collaboration that the scattering from nucleons in iron and deuterium is significantly different in the whole measured x range [5] indicating that the quark distributions in bound nucleons are different from those in free nucleons. This observation has led to many theoretical and experimental activities (for a review see [6]).

Figure 2 shows a compilation of measurements of cross section ratios for $A \approx 57$ [7-10].

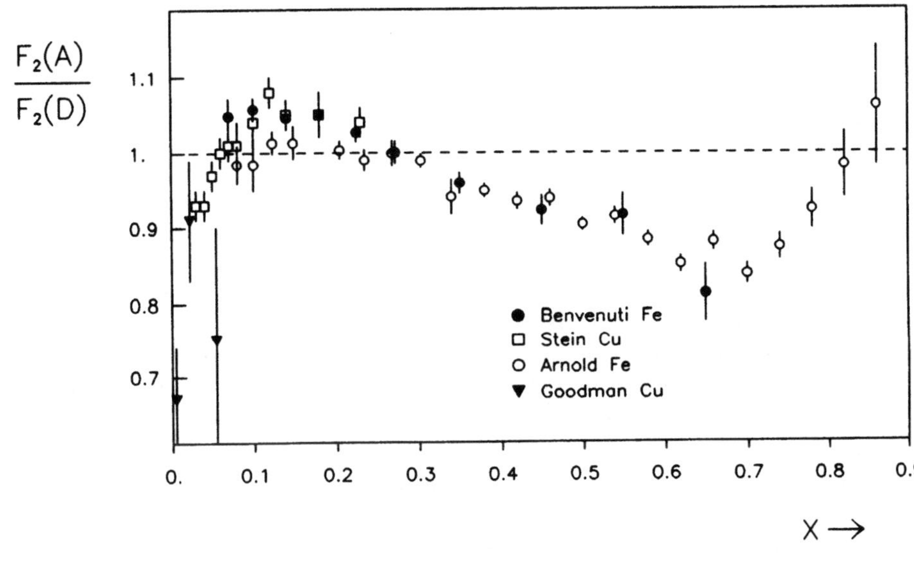

Fig. 2. The EMC effect for $A \approx 57$ (Benvenuti [8], Stein [9], Arnold [7] and Goodman [10])

At $x > 0.3$ a drop of the ratio below unity is observed. The A dependence of the drop has been measured at SLAC [7] and was found to be roughly proportional to $\log A$. No Q^2 dependence was found within errors. At $x > 0.7$ the influence of the Fermi motion becomes visible, causing an increase of the ratio. At $x \approx 0.1$ the situation is less clear. There seems to be some enhancement above 1, but the precision of the data is not sufficient to establish the size of the effect. The enhancement is much smaller than in the original EMC measurement. At very small x there seems to be a drop below 1 again, that might be interpreted as the onset of shadowing.

In 1988/89 new results on cross section ratios were published by the EMC from a measurement on D, C, Cu and Sn [11] in the x range from 0.03 to 0.6. The mean Q^2 increases from 4 GeV^2 at the lowest x to 32 GeV^2 at the highest x. The EMC also did a dedicated experiment at very small x and Q^2 to study exclusively the shadowing region in the range $0.2\,\mathrm{GeV}^2 < Q^2 < 8\,\mathrm{GeV}^2$ and $0.002 < x < 0.17$ for D, C and Ca [12]. The results of both measurements are shown in fig.3 on a logarithmic x scale.

The data establish for the first time clearly the onset of shadowing in deep inelastic muon scattering and show a strong increase of shadowing with the mass number A. In addition the data extrapolate nicely to the real photon results at similar values of ν [13]. The results are also compared with earlier results from other muon experiments [9,10,14] obtained for similar A. The data are consistent with a small enhancement above 1 for x between 0.1 and 0.3. The cross over point between shadowing and the enhancement shifts towards larger values of x when A increases.

The Q^2 dependence in two x bins for the EMC data of ref.12 is shown in fig.4 for C and Ca. Within errors no Q^2 dependence is observed even at low Q^2.

3. Models for Shadowing

The new results from the EMC triggered many theoretical activities to explain the shadowing effect [15-22]. Attempts to describe the data were made using both vector meson dominance (VMD) and parton mechanisms.

Vector meson dominance models. In VMD models one assumes that it is predominantly the hadronic component of the virtual photon which interacts with the target. Generally, the Q^2 and x dependence is predicted in these models.

Simple ρ dominance models predict that shadowing vanishes very quickly with increasing Q^2 in contradiction with the data. Generalised VMD models take into account not only the fluctuations into ρ, ω and ϕ mesons but also a continuum of heavier mesons. They predict appreciable shadowing even at Q^2 of a few GeV^2 as shown by Schildknecht in 1973 [23]. This model describes the data qualitatively well especially at low x, provided the theoretical curve is calculated at the mean Q^2 of the data points [15]. In the region above $x = 0.03$ the model deviates from the data as it does not attempt to describe the enhancement region.

A special hybrid model was developed by Kwieciński and Badełek to explain the EMC data at low x and Q^2. This model combines the VMD with the parton approach. It contains contributions from the low lying vector mesons and from pomeron exchange at higher masses [16]. In fig. 3 this model is compared to the data (solid line). It shows a similar qualitative agreement as the model of [15], but tends to overestimate the shadowing at small x and low Q^2 whereas it underestimates the effect at higher Q^2 for Sn.

Partonic models. Partonic models try to explain the EMC effect including the shadowing region with a modification of the parton distributions in the nucleon. Most of them rely on the Q^2 dependence calculated by Mueller and Qiu [17] and try to describe the x

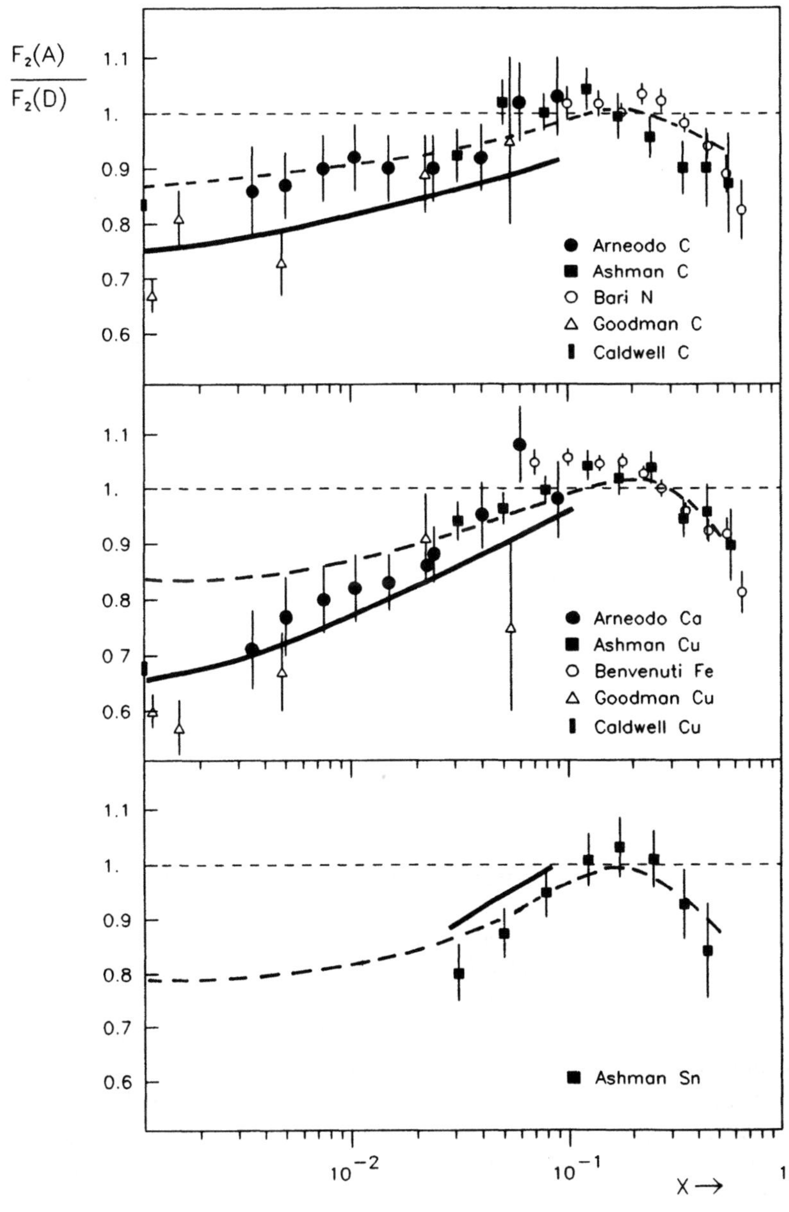

Fig. 3. Structure function ratios for C/D, Ca/D and Sn/D obtained by the EMC (Ashman [11], Arneodo [12]) compared with results from BCDMS (Bari [14], Benvenuti[8]) and FNAL (Goodman [10]). On the vertical axis the results from a photoproduction experiment are marked (Caldwell [13]). The error bars represent statistical and systematic errors added in quadrature. In addition model predictions from ref. 16 (solid lines) and from ref. 18 (dashed lines) are shown.

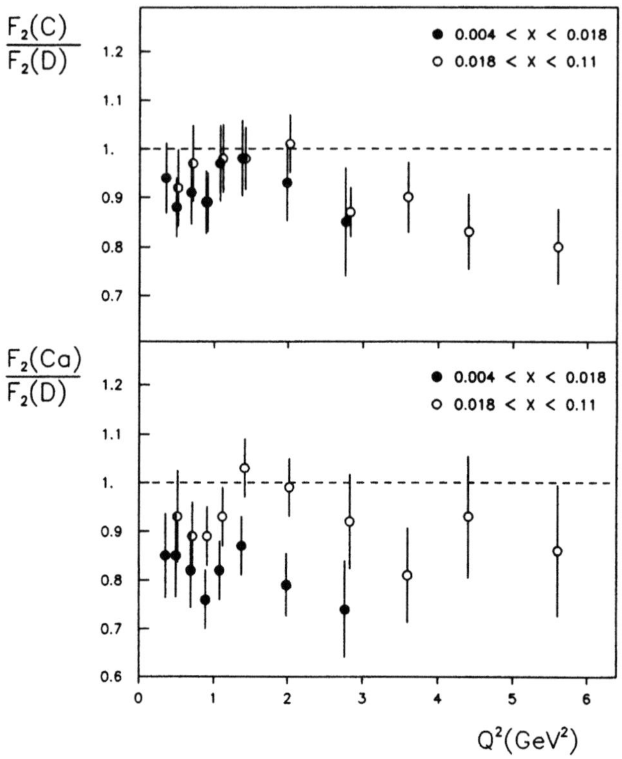

Fig. 4. Structure function ratios versus Q^2 for two x intervalls (from [12])

dependence at fixed Q^2. The Q^2 dependence is calculated based on QCD arguments and is found to be weak, consistent with the data.

There are several phenomenological models using a parametrisation of the x and A dependence of shadowing at fixed Q^2, in combination with a model for the EMC effect at higher x including the enhancement region.

As an example the model of Castorina and Donnachie [18] is compared to the data in fig.3 (dashed line). Their parametrisation is based on the idea that in nuclei the effective coupling of the Pomeron is suppressed. This causes a harder sea distribution. To describe the large x region they use the model for the EMC effect from [24]. A reasonable qualitative agreement with the data is obtained, although the enhancement region is not well described.

A somewhat different approach is taken in the aligned QCD jet model of Frankfurt and Strikman [22]. Jets of $q\bar{q}$ pairs are produced by the virtual γ and shadowing of the sea quark distribution is obtained via interactions of aligned asymmetric jets with the nucleus. This model shows good agreement with the data in the shadowing region, similarly to the VMD models, but it also does not include the enhancement region.

In general most models give a qualitative but not quantitative description of the data in the shadowing region at small x, and there are considerable differences between the models. However, the data are not precise enough to distinguish between the models. In addition, to describe the approach to the enhancement region, an extension of the models seems to be necessary to include the high x region.

4. New Results from the NMC

In 1986 the New Muon Collaboration (NMC) started a high precision measurement of the cross section ratios at small and medium x on a series of nuclei ranging from He to Pb: He, Li, Be, C, Al, Ca, Fe, Sn and Pb. The NMC uses the upgraded EMC spectrometer and extends the kinematic range of the measurements by an additional small angle trigger, thus covering an x range from 0.0035 to 0.7 and a Q^2 range from $0.7\,\text{GeV}^2$ to $90\,\text{GeV}^2$ for a beam energy of 200 GeV [25].

Systematic errors are minimised using a complementary target setup. The target system consists of two sets of target pairs which are alternatively moved into the muon beam. For the He/D measurements a target set contains two targets, ("upstream" and "downstream"); one of the target vessels is filled with liquid He, the other with liquid D_2. The sequence of target materials is interchanged in the second target set. By a lateral transport mechanism either set can be moved into the beam. Frequent exchange of the two sets (typically twice per hour) minimizes effects of a time dependent detector response.

Acceptance and flux normalisation have been the main sources for systematic errors in previous experiments. With the complementary target setup acceptance and flux normalisation cancel in the double ratio $N_u^{He} N_d^{He} / N_u^D N_d^D$. Therefore, the cross section ratio can be taken as

$$\frac{\sigma^{He}}{\sigma^D} = \kappa \sqrt{\frac{N_d^{He} N_u^{He}}{N_d^D N_u^D}} \tag{6}$$

where κ is the ratio of the number of nucleons per area in the upstream and downstream targets and N the event numbers after radiative corrections. Similar setups have been used for the measurements with solid targets. They consisted of several equally spaced slices distributed over the region previously occupied by the He targets to cover the same acceptance.

The preliminary results obtained for Ca/D and C/D are shown in fig.5. The data are averaged over Q^2 and the mean Q^2 varies between $0.8\,\text{GeV}^2$ and $41\,\text{GeV}^2$ over the x range shown. The remaining normalisation error due to uncertainties in the target densities is about 0.5%. Statistical and systematic errors stay below 1% in the x range from 0.02 to 0.25. The most important contributions to the remaining systematic errors come from the radiative corrections (small x), the finite vertex resolution (small x) and the momentum calibration of the spectrometers measuring the incoming and the scattered muon (large x) [26].

A strong shadowing signal is observed, larger for Ca than for C, with an approximately linear increase in $\log x$ towards small x. For Carbon there might be an indication of a saturation of shadowing at $x \sim 0.009$, whereas no saturation is seen for Calcium. A clear signal for an enhancement of 1–2% is observed for both targets in the region of x between 0.07 and 0.17. No Q^2 dependence is found within the errors.

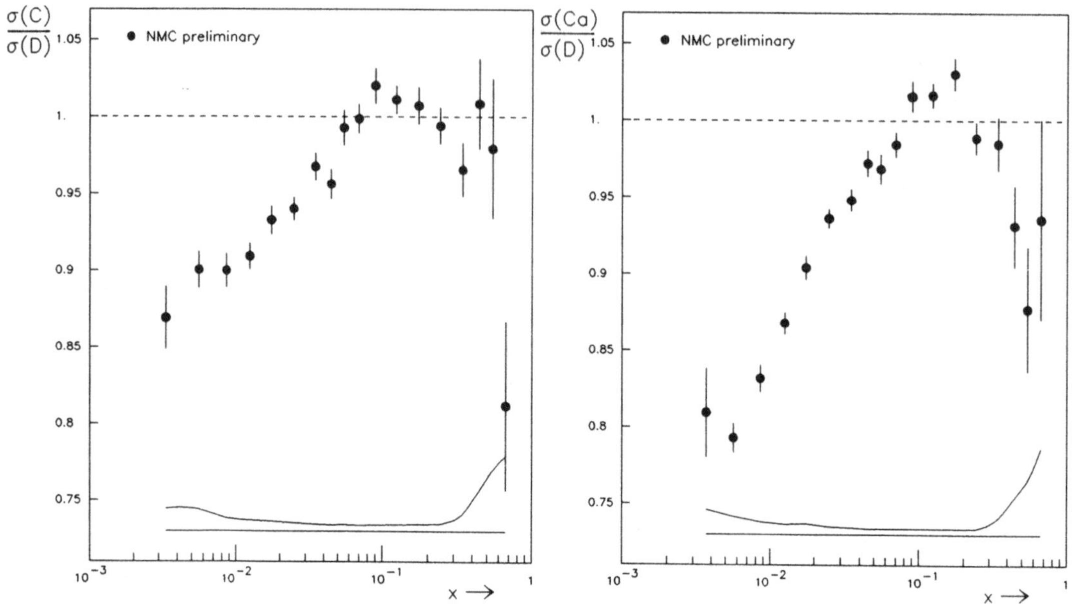

Fig. 5. Preliminary NMC results for the cross section ratios C/D and Ca/D. The error bars are statistical only, the size of systematic errors is given by bands at the bottom.

Figure 6 demonstrates the A dependence of the ratios measured by the NMC for He/D, C/D and Ca/D. The data are compared with results from EMC, SLAC and FNAL for the same target materials.

The precise results obtained in a wide x range by the NMC agree well with the previous measurements. A strong increase of shadowing with increasing A can be observed, whereas no A dependence is visible in the amount of the enhancement.

With these high precision data it is now possible to check various models. In fig. 7 the NMC results are compared with some of the recent models. At present none of them gives a satisfactory description of the data in the whole x range. It should be noted, however, that some of them need further refinement to be precise enough or have to be adapted to the Q^2 range of the new data.

As was shown by Weise during this workshop, it is possible to describe the shadowing in the NMC results with a generalized VMD model. For a good description in the whole measured range an extension of the model to the enhancement and large x region is needed.

To obtain a better understanding of the EMC effect in the whole kinematic range one needs precise measurements of the Q^2 and A dependence and definite model predictions to allow a quantitative comparison between models and data. In addition to the presented data the NMC has made in 1988 a high statistics measurement on carbon and tin at several beam energies between 90 GeV and 280 GeV. This will allow to do detailed

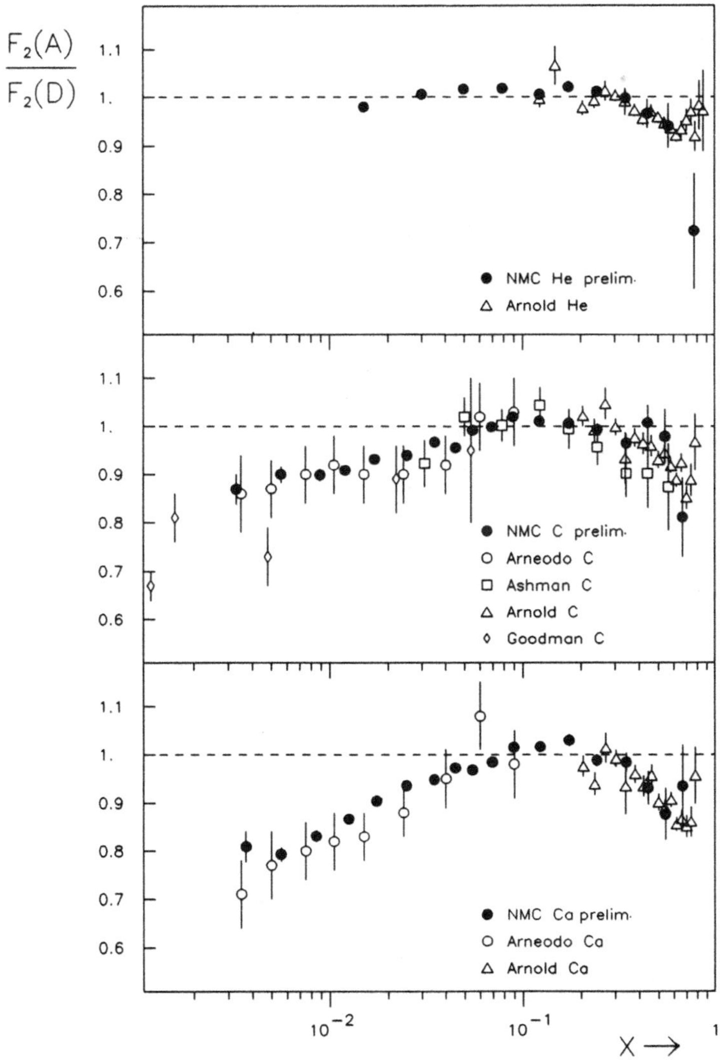

Fig. 6. **Comparison of the preliminary NMC results for He, C and Ca with previous measurements from SLAC (Arnold [7]), EMC (Ashman [11], Arneodo [12]) and FNAL (Goodman [10]).All errors are quadratic sums of statistical and systematic error, when systematic errors are available.**

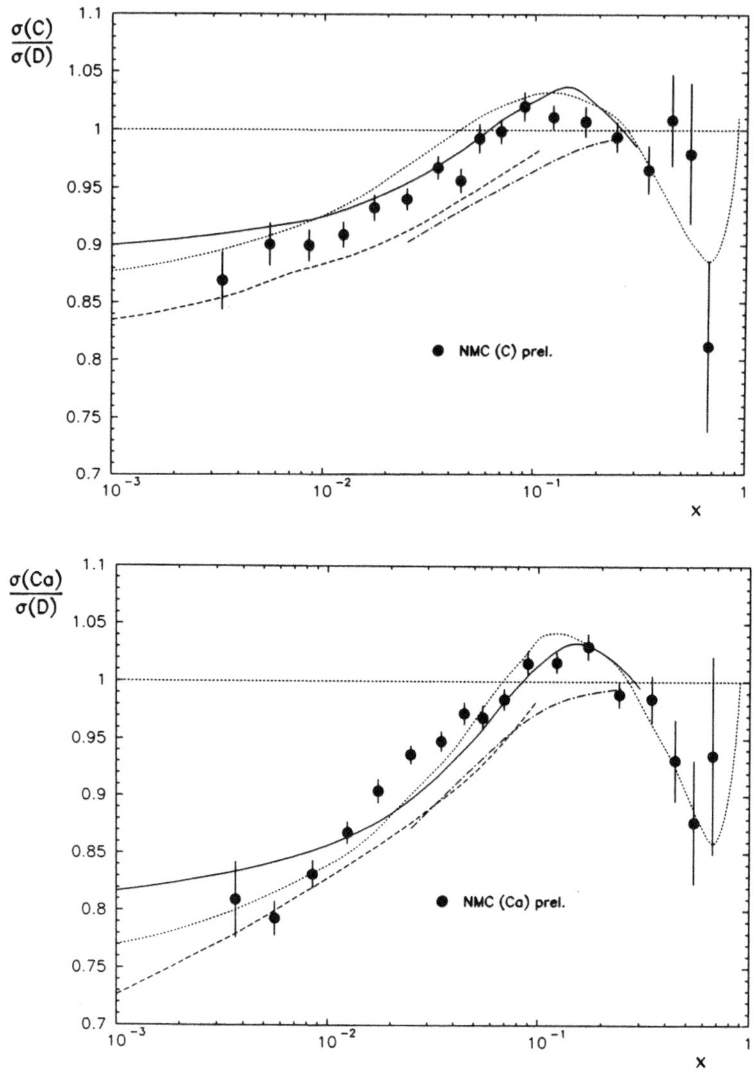

Fig. 7. Comparison of the cross section ratio obtained by the NMC with models from ref. 19 (solid line), ref. 16 (dashed line), ref. 21 (dotted line) and ref. 15 (dashed dotted line). The model from ref. 15 has been interpolated to the mean Q^2 of the NMC data.

comparisons of the Q^2 dependence of the cross sections ratios with model predictions and to study the change of the gluon distribution from carbon to tin via J/ψ production.

References

1. For a review see e.g. G. Grammer and J. Sullivan, in: Electromagnetic interactions at high energies, eds. A. Donnachie, G. Shaw (Plenum , New York 1978)
2. For a review see e.g. T. Sloan, G. Smadja, R. Voss: Phys.Rep. **162** 45 (1988)
3. J. Dasu et al.SLAC E140): Phys. Rev. Lett. **60** 2593 (1988)
4. G. Franz et al.: Z. Phys. C **10** 205 (1981);
   J. Bailey et al.: Nucl. Phys. B **151** 367 (1979)
5. J. Aubert et al.(EMC): Phys. Lett. B **123** 275 (1983)
6. L. Frankfurt, M. Strikman: Phys. Rep. **160** 235 (1988)
7. R. Arnold et al.(SLAC E139): Phys. Rev. Lett. **52** 727 (1984)
8. A. Benvenuti et al.(BCDMS): Phys. Lett. B **189** 473 (1987);
9. S. Stein et al.(SLAC): Phys. Rev. D bf12 1884 (1975)
10. M. Goodman et al.(FNAL): Phys. Rev. Lett. **47** 293 (1981)
11. J. Ashman et al.(EMC): Phys. Lett. B **202** 603 (1988)
12. M. Arneodo et al.(EMC): Phys. Lett. B **211** 493 (1988);
    M. Arneodo et al.(EMC): Nucl. Phys. B, to be published
13. D. Caldwell et al.: Phys. Rev. Lett. **42** 553 (1979)
14. G. Bari et al.(BCDMS): Phys. Lett. B **163** 282 (1985);
15. C. Bilchak, D. Schildknecht, J. Stroughair: Phys. Lett. B **214** 441 (1988)
16. J. Kwieciński, B. Badełek: Phys. Lett. B **208** 508 (1988)
17. J. Qiu: Nucl. Phys. B **291** 746 (1987);
    A. Mueller, J. Qiu: Nucl. Phys. B **268** 427 (1986)
18. P. Castorina, A. Donnachie: Phys. Lett. B **215** 589 (1988)
19. S. Brodsky, H. Lu: Phys. Rev. Lett. **64** 1342 (1990)
20. F. Close, R. Roberts: Phys. Lett. B **2213** 91 (1988)
21. W. Zhu, J.G. Shen: Phys. Lett. B **235** 170 (1990)
22. L. Frankfurt, M. Strikman: Nucl. Phys. B **316** 340 (1989)
23. D. Schildknecht: Nucl. Phys. B **66** 398 (1973)
24. F. Close, R. Roberts, G. Ross: Phys. Lett. B **168** 400 (1986)
25. D. Allasia et al.: CERN report CERN/SPSC 85-18
26. N. Novotny: PhD thesis, Heidelberg 1988;
    C. Scholz: PhD thesis, Heidelberg 1989;

The Electromagnetic Interaction of Rhos

Paolo Christillin

Dipartimento di Fisica, Università di Pisa
and I.N.F.N., Sezione di Pisa, PISA, Italy

Abstract: The necessity of a spontaneous symmetry breaking mechanism for electrostrong interactions is justified and its relevance in low energy hadronic phenomenology is examined.

1 Introduction

The present contribution is concerned with the electromagnetic interaction of hadrons at low energy. Photons , rhos, nucleons, pions and electrons only are explicitly considered. They constitute the effective degrees of freedom in our approach and are described by the corresponding fields.

Our treatment is therefore admittedly phenomenological, although the connection with a more fundamental description i.e. with quarks will be briefly touched upon in the following. It conflicts however in many respects with a whole body of currently accepted phenomenology. We will hence try to justify the necessity and the correctness of the present framework in as simple terms as possible. The material covered here is based on [1], presented in a preliminary form in [2], and now published as part of [3] where additional details can be found. It will however closely follow the oral presentation so as to allow the unbiased reader to realize whether the author "is just blowing up words" , according to the expression used on that occasion in his regards, or whether somebody else does not know very well what he is talking about.

2 The $\gamma - \rho$ Interaction

We outline here the well known difficulties connected with the electromagnetic interaction of rho mesons (isospin triplet of vector mesons).

If the γ is massless and if it has to remain so (the rho being massive) no direct $\gamma - \rho^0$ coupling is possible.

Proof: Obvious. As a matter of fact off-diagonal terms in the 2×2 $\gamma - \rho^0$ mass matrix entail $m \neq 0$ for both eigenstates. Hence, without counterterms, one would have a massive photon.

The reason to fiercely oppose such a coupling is that it is particularly misleading in suggesting non existing " hadronic component of the photon". In turn, such a catch phrase has been widely used to justify shadowing i.e. a non linear behaviour in the atomic number A of the total nuclear photoabsorption cross section, which is well known [4,5,6] to result from a coherence effect totally independent from such a direct $\gamma - \rho^0$ coupling.

Kroll-Lee and Zumino [7] have succeeded in overcoming the mass problem by supplementing the direct $\gamma - \rho^0$ coupling with another term which can be interpreted as a photon mass term for a transformed vector potential (see their App. B) so as to make the total interaction gauge invariant.

However this is not enough to solve the problem of the e.m. interaction of rhos. To see it let us consider $e^+e^- \to \rho^+\rho^-$. The electron Lagrangian is

$$\mathcal{L}_e = -e\bar{\psi}_e(\gamma^\mu \partial_\mu - m_e)\psi_e \tag{1}$$

the photon Lagrangian

$$\mathcal{L}_\gamma = -\frac{1}{4} F_{\mu\nu} F^{\mu\nu} \tag{2}$$

where $F_{\mu\nu} = \partial_\mu F_\nu - \partial_\nu F_\mu$, and suppose to take the standard abelian Proca one for the rho

$$\mathcal{L}^*_\rho = -\frac{1}{4}\vec{\rho}^*_{\mu\nu} . \vec{\rho}^{*\mu\nu} + \frac{1}{2} m^2_\rho \vec{\rho}_\mu . \vec{\rho}^\mu \tag{3}$$

where $\vec{\rho}^*_{\mu\nu} = \partial_\mu \vec{\rho}_\nu - \partial_\nu \vec{\rho}_\mu$, and the arrow denotes isospin.

Obviously the principle of minimal electromagnetic coupling $\partial_\mu \to \partial_\mu - ieA_\mu$ generates a gauge invariant interaction. The only allowed intermediate particle (see Fig.1a)) is however the photon and since for massive spin one particles the sum over polarizations ($\rho^{(\lambda)}$) is such that $\Sigma_\lambda \rho^{(\lambda)}_\mu \rho^{(\lambda)}_\nu = -g_{\mu\nu} + (k_\mu k_\nu)/m^2_\rho$, unlike for photons where $\Sigma_\lambda \epsilon^{(\lambda)}_\mu \epsilon^{(\lambda)}_\nu = -g_{\mu\nu}$, it is easy to get $\sigma(s) \sim const$, totally contradicted by experiment. One has therefore, in close analogy to the $e^+e^- \to W^+W^-$ case the unpleasant feature of a non well behaved (non-unitary) cross section. The reason for that is due to the "hard " rho mass [8] and to their non abelian formulation. Indeed, even disregarding the mentioned problem of the direct $\gamma - \rho^0$ coupling, it is easy to understand in elementary terms that such a prescription would be instrumental in reducing (as observed experimentally) the $e^+e^- \to \rho^+\rho^-$ cross section only by allowing for a trilinear $\rho^0\rho^+\rho^-$ coupling (since only in this case the photon propagator could be depressed by the additional $m^2_\rho/(s - m^2_\rho)$ term due to the ρ^0 propagator).

We will outline along these lines the solution of the problem which had been parenthetically (only rhos and photons) treated in [9]. We will add, as mentioned before, electrons nucleons and pions.

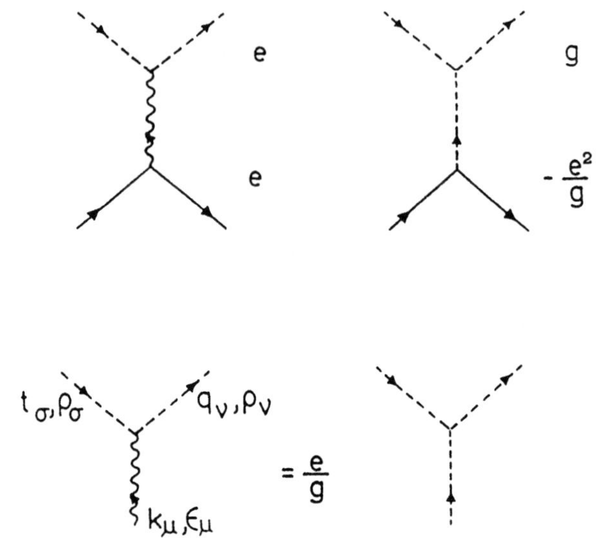

Fig.1a) 1b) $e^+e^- \rightarrow \rho^+\rho^-$. Solid line for electron, wavy for photon and dashed for ρ. Fig.1c) Vertex $V_{\sigma\mu\nu}$ for $\gamma\rho^+\rho^-$ and for $\rho^0\rho^+\rho^-$ (see eq.(12)).

The starting point is provided by the relation

$$Q = T_3 + Y/2 \tag{4}$$

between the conserved e.m. charge, isospin T (to which we associate an isospin triplet of vector mesons $\vec{V}_\mu$ with coupling constant g_0) and hypercharge Y (associated with a (isoscalar) field B_μ with coupling constant e_0). They will turn in the end to be strictly connected with rhos and with the photon.

We impose invariance of the theory under a local $SU(2)_I \times U(1)_Y$. The covariant derivative is thus

$$D_\mu = (\partial_\mu + ig_0 \vec{V}_\mu.\frac{\vec{\tau}}{2} - ie_0 \frac{Y}{2} B_\mu) \tag{5}$$

The first consequence of invariance under a local isospin transformation is to forbid the mass term in the spin 1 Lagrangian. Hence the vector boson cannot be identified with the ρ.

The additional consequences of local isospin invariance are twofold. As regards the free "ρ" Lagrangian $\vec{\rho}^{*}_{\mu\nu}$ must go into

$$\vec{V}_{\mu\nu} = \partial_\mu \vec{V}_\nu - \partial_\nu \vec{V}_\mu + ig[\vec{V}_\mu, \vec{V}_\nu] \tag{6}$$

i.e. there is an extra term due to the non abelian structure of the group [10]. Second all coupling constants are now connected realizing *universality* i.e. the same g_0 for VVV,$V\pi\pi$ and VNN interactions.

The "soft" mass generation mechanism and the simultaneous correct introduction of the photon closely parallel the electroweak case. The boson mass is generated in a standard way through the spontaneous symmetry breaking due to Higgs bosons. In the simplest way (consistent with the *global* $SU(2)$ symmetry of our problem) they are a scalar complex isodoublet i.e. four fields, three of which go into the longitudinal components of the now massive gauge bosons, the surviving fourth (of non-zero vacuum expectation value $< H >= \phi_0$) remaining as a physical state. Its interaction with $\vec{V}_\mu$ and B_μ yields a mass term

$$\frac{1}{4}\phi_0^2 g_0^2 [V_\mu^{(1)^2} + V_\mu^{(2)^2}] + \frac{1}{4}\phi_0^2 (g_0 V_\mu^3 - e_0 B_\mu)^2 \tag{7}$$

which can be diagonalized through a rotation in terms of the physical ρ_μ^0 and A_μ

$$V_\mu^3 = \frac{1}{\sqrt{1+\xi^2}}(\rho_\mu^0 + \xi A_\mu)$$

$$B_\mu = \frac{1}{\sqrt{1+\xi^2}}(A_\mu - \xi\rho_\mu^0) \tag{8}$$

where

$$\xi = \frac{e_0^2}{g_0^2} \qquad g = \frac{g_0}{\sqrt{1+\xi^2}} \qquad e = \frac{e_0}{\sqrt{1+\xi^2}}$$

Hence the physical photon is massless as due, and there is a mass splitting between charged and neutral ρ's

$$2\frac{m_{\rho^0} - m_{\rho^\pm}}{m_\rho} = (\frac{e}{g})^2 \tag{9}$$

which yields for $g^2/4\pi = 2.0$, $\Delta m = 1.06$ MeV.

This prediction, first obtained by 't Hooft [9] has to be compared with the weighted experimental average of 0.3$\pm$ 2.2 MeV [11]. More precise measurements would be welcome, although conclusions could be drawn only by having the background under control to the same accuracy.

By recalling that in this electrostrong coupling the free parameter of the theory is given by the Higgs mass m_H, we now examine two points where differences with respect to the standard treatments arise.

It has to be stressed first that *the previous diagonalization is possible at all k^2-values* i.e. also for virtual particles. It entails that in considering the propagators of the physical ρ_μ^0 and A_μ, the off-diagonal term $V_\mu^3 B_\mu$ is absent. To see the consequences in the interaction with matter fields let us first introduce the coupling with electrons

$$\mathcal{L}_{e,int} = -e_0\bar{\psi}_e\gamma^\mu B_\mu\psi_e = -e\bar{\psi}_e\gamma^\mu(A_\mu - \frac{e}{g}\rho_\mu^0)\psi_e \tag{10}$$

One has therefore the standard γee coupling, plus a *direct* $\rho^0 ee$ interaction of strength $-e^2/g$.

In all electromagnetic processes i.e. $e^+e^- \to e^+e^-$ or Bhabha scattering, propagators combine so to yield

$$e^2(\frac{1}{k^2} + \frac{e^2}{g^2}\frac{1}{k^2 - m_\rho^2}) \tag{11}$$

the momenta of both virtual particles being labelled by k.
When considering $e^+e^- \to \rho^+\rho^-$, *if and only if*,

$$\frac{V^{\gamma\rho^+\rho^-}_{\sigma\mu\nu}}{e} = \frac{V^{\rho^0\rho^+\rho^-}_{\sigma\mu\nu}}{g} = (2q_\mu - k_\mu)g_{\sigma\nu} - (q_\sigma + k_\sigma)g_{\mu\nu} - (q_\nu - 2k_\nu)g_{\sigma\mu} \tag{12}$$

as in Fig.(1c), the fields appearing at the vertices naturally combine to yield the propagators

$$e\frac{1}{k^2}e - \frac{e^2}{g}\frac{1}{k^2 - m_\rho^2}g \equiv \frac{e^2}{k^2}\frac{m_\rho^2}{k^2 - m_\rho^2} \tag{13}$$

as shown in Fig.(1b), i.e. the VMD recipe [12] and $\sigma(e^+e^- \to \rho^+\rho^-)$ can be easily shown to be cured from the unitarity problems mentioned above.

The above relation for the $\gamma\rho^+\rho^-$ and $\rho^0\rho^+\rho^-$ vertices results naturally in the present scheme because of the rotation through which the fields V_μ entering the starting Lagrangian (three linear couplings from $\vec{V}_{\mu\nu}.\vec{V}^{\mu\nu}$ are expressed as a function of the physical fields A_μ, ρ^0_μ). As a consequence of eq.(12) the $\gamma\rho\rho$ vertex has additional contributions with respect to the standard one coming from the principle of minimal e.m. coupling in the Proca Lagrangian

$$V^*_{\sigma\mu\nu}/e = (2q_\mu - k_\mu)g_{\sigma\nu} - q_\sigma g_{\mu\nu} - (q_\nu - k_\nu)g_{\sigma\mu} \tag{14}$$

The two prescriptions differ by

$$\Delta V_{\sigma\mu\nu} = e(-k_\sigma g_{\mu\nu} + k_\nu g_{\sigma\mu}) \tag{15}$$

which corresponds to a (k divergenceless) unit anomalous magnetic moment for the ρ.

In this connection it is worth pointing out that the naive quark model (e.g. $u\bar{d}$ for ρ^+) yields a total ρ magnetic moment $\mu_\rho = e/2m_q$, where $m_q = m_u \sim m_d \sim 350$ MeV. It is therefore clear that, in contrast with Proca's formulation where $\mu_\rho = e/2m_\rho$ [13], only a non abelian treatment of rhos, which intrinsically contains the "anomalous" part, agrees with quark model predictions.

The possibility of using e^+e^- as a further test of the present framework by predicting in a parameter free way (because of the fixed $\rho\rho\rho$ and $\rho\rho\rho\rho$ couplings, which generate at the same time three and four body forces) the 4π and 6π cross sections (eventually proceeding through the Higgs meson of unknown mass) are examined elsewhere [14].

Second there is the problem of the relation of the anomalous magnetic ρNN coupling κ^ρ

$$\mathcal{L}_{\rho NN} = g_{\rho NN}\bar{\psi}(\gamma^\mu + \frac{i\kappa^\rho}{2M}\sigma^{\mu\nu}k_\nu)\frac{\vec{\tau}}{2}\psi.\vec{\rho}_\mu \tag{16}$$

to the γNN Pauli isovector one κ^V

$$\mathcal{L}_{\gamma NN} = e\bar{\psi}(\frac{1+\tau_3}{2}\gamma_\mu + \frac{i\sigma_{\mu\nu}}{2M}k^\nu \frac{\kappa^S + \tau^3\kappa^V}{2})\psi A^\mu \tag{17}$$

Taking for simplicity only the vector part, $\mathcal{L}_{\rho NN}$ and $\mathcal{L}_{\gamma NN}$ are easily seen to derive from the VNN interaction by reexpressing the gauge fields appearing in the covariant derivative

$$\mathcal{L} = g_0\bar{\psi}\gamma^\mu(\frac{\rho_\mu}{\sqrt{2}}\tau^- + \frac{\rho_\mu^*}{\sqrt{2}}\tau^+ + A_\mu(\frac{1+\tau^3}{2})\frac{\xi}{\sqrt{1+\xi^2}} + \frac{\tau^3-\xi^2}{2}\frac{1}{\sqrt{1+\xi^2}}\rho_\mu^0)\psi \tag{18}$$

One gets therefore at the same time the usual coupling constant $g = g_{\rho NN}$ (apart from small $O(\frac{e^2}{g^2})$ isospin breaking effects) of the ρ to nucleons and a corresponding e for the e.m. field.

As before, for the *isovector part* A_μ and ρ_μ^0 appear in the right combination because they are stemming from the V_μ^3 coupling, which makes the propagators subtract, thus reproducing also in $eN \to eN$ the VMD result.

The direct coupling of the photon and of the ρ both to (e, e') and to (N, N') makes clear that, automatically, *such a combination is possible also for the induced tensor part only if the corresponding Pauli form factors are equal.*In particular at $t = 0$

$$\kappa^V = \kappa^\rho = 3.7 \tag{19}$$

This is at variance by a factor of two with the value obtained by Höhler and Pietarinen [15] from a dispersion relation analysis. More than that their predicted value at $s = m_\rho^2$ (expressed as a function of F_π) is roughly given by $G_2(m_\rho^2) \simeq i(m_\rho/\Gamma_\rho)G_2(0)$. Can this be understood at all? Without entering in too many details some points in their approach look questionable. For instance the graph $\rho \to \pi\pi \to \rho$ loop should not be included when the renormalized rho is used. In a similar way the graph $\gamma \to \pi\pi \to \rho$ would give again (beyond the tree approximation) a mass to the photon.

One should have therefore a sound theoretical model to handle the problem of the unitary addition of a background and a resonance in a coupled channel case, with the constraint that the photon must be massless. We will not attempt it here. The importance of unitary corrections [16] can be appreciated from the comparison of the measured $\frac{g^2_{\rho e^+e^-}}{4\pi} = 2.0 = \frac{g^2}{4\pi}$ with $\frac{g^2_{\rho\pi\pi}(m_\rho^2)}{4\pi} \simeq 2.8$ Because of its manifest lack of structure, the first process yields the universal g coupling constant in agreement with the measured $g_{\rho\pi\pi}$ and $g_{\rho NN}$ at $t = 0$. A rough (and non unitary) estimate of the isovector Pauli form factor can come from the old standard [17] (see however the criticisms in [18] exercise 15.2) $\pi\pi$ ρNN and γNN vertex corrections (denoted by G_2 and F_2 respectively). Let us recall first that (here $g = g_{\pi NN}$)

$$G_2(k^2) = g^2/2\pi^2 \int_0^1 (1-x)^2 dx \int_0^x \frac{M^2 dy}{M^2(1-x)^2 + m_\pi^2 x - k^2 y(x-y)} \tag{20}$$

It is straightforward to obtain to the leading order in $(m_\pi/M)^2$, $G_2(0) \equiv \kappa^\rho = g^2/4\pi^2 \simeq 4.6$ and the same value for $F_2(0)$ of course. Let us also mention that by performing the first integration one finds again to the leading order

$$\lim_{k^2 \to \infty} G_2(k^2) = (g^2/4\pi)M/\sqrt{k^2}$$

with a consequent $1/(\sqrt{k^2}(k^2 - m_\rho^2))$ behaviour for F_2.

It is also straightforward to calculate

$$ImG_2(m_\rho^2) = g^2/2\pi \int_0^1 (1-x)^2 dx \int_0^x M^2 \delta(M^2(1-x)^2 + m_\pi^2 x - m_\rho^2 y(x-y))\, dy \tag{21}$$

with a value $O(\kappa_\rho)$. This is in even stronger disagreement than the *t=0* value with the quoted results [15].

The following problems arise when trying to test the two alternatives. The only place where one can measure $g_{\rho NN}(m_\rho^2)$ is represented by ρ photoproduction. The process $\gamma p \to p\rho^0$ in the Born approximation, i.e. by considering only direct and crossed intermadiate nucleons, with the full e.m. and ρNN interaction eq.(16,17) be calculated with the use of REDUCE. The result can be checked with straightforward modifications against $\gamma p \to \gamma p$ i.e. against the Powell cross section [19]. Both cross sections grow linearly in s. This is obvious since intermediate nucleons which are sizeably off-shell have been completely mistreated. A form factor enters at each vertex with the cut-off as an essentially free parameter. Once form factors are allowed however, the photoproduction amplitude is, as well known, no longer gauge invariant nor ρ transverse. Both these properties have to be always obeyed by adding counterterms not uniquely determined by low energy theorems [20,21,22],which therefore do not guarantee at all the reconstruction of the amplitude. Therefore even the large possible differences we are talking about may be obscured by other theoretical uncertainties (in addition to the very poor dynamics intervening in such a study case).

In an analogous way in $\gamma n \to p\rho^-$ measurements of $d\sigma/dt$ exist [23], from which one might hope both to get further information on $G_2(0)$ (entering the seagull and the ρ-in-flight diagram) and to test the form of the $\gamma\rho\rho$ vertex. The same problems mentioned before make however this extraction dubious.

3 Conclusions and Outlook

In conclusion there is in our opinion a compelling evidence for a spontaneous symmetry breaking mechanism in the low energy electromagnetic interaction of hadrons.

The ensuing phenomenology differs in some places from the standard treatment although quantitative differences are hard to be pinned down. As regards the forthcoming intermediate energy program we list here our expectations and prejudices by considering both the nucleon and the nuclear case in the different energy regions.

3.1 Δ region

i) A correct unitary treatment of πN dynamics has allowed to conclude that the Δ is essentially an elementary object, with a large transition quadrupole moment, brought to an almost symmetric situation by the pion cloud [24] (small renormalized electric quadrupole coupling constant g_E).

The extension of the analysis to virtual photons has allowed to clarify the connection between g_E and the Coulomb coupling constant g_C, showing the incorrectness of existing experimental data. New data supplemented by a forthcoming theoretical analysis would probably let us understand better the interplay between the different degrees of freedom.

ii) In nuclei the Δ deformation plays practically no role. The most relevant gross feature is represented by the absence of shadowing, because of the transverse ($\mathbf{q}.\mathbf{k} \times \epsilon$) nature of coherent (γ, π^0) which might otherwise develop a coherent damping effect. The necessary quantitative study of coherent (γ, π^0) [25] along the nuclear table proves at the same time the inadequacy of the Δ - hole recipe, which stems from the inconsistent dynamical inputs for πN and (γ, π) (no crossed Born term in the former). In conclusion partial transitions can indeed discriminate quantitatively among different formulations.

3.2 $\pi\pi$ region

i) The reaction $\gamma N \to \Delta\pi$ has been recently reconsidered [26]. It has been shown that the pion-in-flight is irrelevant, the excitation mechanism being given (neglecting s channel resonances) just by the $\mathbf{S}.\epsilon$ contact term, although the role of the vertex form factor vs. "absorptive corrections" still needs clarification.

ii) As regards nuclei, a spreading of the total cross section follows from the cooperative effect of the Δ and pion self energies. No shadowing has been predicted in this energy region either [6], since the coherent part of this reaction mechanism ($\mathbf{q}.\epsilon$) is as much transverse as the single pion photoproduction which contributes a comparable amount. Therefore no coherent effect develops.

3.2 ρ region

i) The $\gamma p \to p\rho^0$ total cross section in the GeV region is of the order of 20 μb [27] i.e. a rather small fraction of the total (multiparticle) one, with a marked diffractive behaviour. On the contrary $\gamma n \to p\rho^-$ is rather flattish as a function of t and is, in turn, a small fraction (1/5 , 1/10 depending on the energy) of $\gamma p \to p\rho^0$.

In this transition region the theoretical situation looks in our opinion still unsatisfactory. Born terms, in sharp contrast with the situation met e.g. in the Δ region, have the intermediate nucleon heavily off-shell. Therefore, in addition to not explaining the diffractive behaviour of the differential cross section, an essentially unknown form factor, which affects differently direct and crossed terms, enters the amplitude introducing a non curable gauge non invariance. The introduction of possible diffractive mechanisms and/or of "absorptive corrections" makes the situation even more complicated.

In this context the possibility of testing the present scheme e.g. through Proca's vs. the spontaneous symmetry breaking $\gamma\rho\rho$ vertex, looks rather difficult.

ii) As regards nuclei let us restate that the appearence of shadowing has nothing to do with pseudo hadronic components of the photon, but that derives this time from a sizable presence (due to the quantum numbers of ρ^0) of the coherent process. Is it likely that something can be learned from the nuclear process, for instance through exclusive measurements? In consideration of how unsettled the situation is already on the nucleon we are unfortunately rather skeptical about it.

ACKNOWLEDGMENTS. I would like to thank prof. J.D.Bjorken for an interesting remark about the ρ magnetic moment.

References

1. P. Christillin, preprint IPNO/TH 87-27
2. P. Christillin, contribution to the Seillac Conf.,Nucl.Phys. **A 497** (1989) 567c
3. P. Christillin, Phys.Rep **190 C** (1990) 63
4. J.S.Bell, Phys.Rev.Lett.**13** (1964) 57
5. L.Stodolsky, Phys.Rev.Lett.**18** (1967) 135
6. P. Christillin, Nuovo Cimento **102A** (1989) 165
7. N.M.Kroll,T.D. Lee and B.Zumino, Phys. Rev. **157** (1967) 1376
8. L.B. Okun, Leptoni i Kwarki, Nauka, Moscow 1981, Chapter 19
9. G.'t Hooft, Nucl. Phys. **B35** (1971) 167
10. C.N. Yang and R.L. Mills, Phys. Rev. **96** (1954) 191
11. Review of Particle Properties, Phys. Lett. **170B**(1986)
12. J.J. Sakurai, Ann. Phys. (N.Y.) **11** (1960) 1
13. T.D. Lee and C.N. Yang, Phys. Rev. **128** (1962) 885
14. P. Christillin, preprint IFUP-TH 11/90
15. G. Höhler and E. Pietarinen, Nucl.Phys. **B95** (1975) 210
16. D.Lohse et al., Phys.Lett **234B** (1990) 235
    D.Lohse et al. , in preparation
17. P.Federbush,M.L.Goldberger and S.B.Treiman, Phys.Rev. **112** (1958) 642
18. M.D.Scadron, Advanced Quantum Theory, Springer Verlag, New York 1979
19. J.L.Powell, Phys.Rev. **75** (1949) 32
20. F.E. Low, Phys.Rev. **110** (1958) 974
21. S.L. Adler, Y. Dothan, Phys.Rev. **151** (1966) 1267
22. P. Christillin, S. Servadio, Nuovo Cimento **42A** (1977) 165
23. Aachen–Bonn–Hamburg–Heidelberg–München Collaboration Nucl.Phys. **B21** (1970) 23
24. P. Christillin, G.Dillon, J. of Phys. **G 16** (1989) 967
    P. Christillin, G.Dillon, J. of Phys. **G 16** (1990) 805
25. S.Boffi, L.Bracci, P. Christillin, Invited talk at the Trieste Meeting (1989), World Scientific Publ. Singapore, pag.571
26. L.Bracci, P. Christillin,G.Dillon preprint IFUP-TH 9/90
27. Aachen–Bonn–Hamburg–Heidelberg–München Collaboration Phys.Rev. **175** (1968) 1669

Vector Meson Dominance and Pointlike Coupling of the Photon in Soft and Hard Processes

E. Paul

Physikalisches Institut der Universität Bonn, Fed. Rep. Germany

1 Introduction

In this talk I will discuss hadronlike photon and pointlike (direct) photon coupling in a new kinematical range which has been studied recently on photoproduction of hadrons with real photons in the reaction

$$\gamma A \rightarrow \text{hadrons}$$

where A means proton or nucleus. Previous photoproduction experiments were mainly restricted to soft phenomena, i.e. to the production of hadrons at small 4-momentum transfers to the target particle. This is the kinematical range where Vector Meson Dominance (VMD) models have been introduced in photoproduction. Two CERN experiments labelled NA14 and WA69 have recently published results on photoproduction with the kinematical range extended to higher photon energies (of order 100 GeV) and larger momentum transfers, too. With the new data production mechanisms were studied over the so-called soft-hard transition region I will consider here. In the first part I will begin with reminding you to the usual picture of the interacting photon and will give a brief summary on the manifestation of Simple Vector Meson Dominance in established data. In the second part I will discuss the new results on photoproduction of hadrons from experiment WA69.

2 The Nature of the Interacting Photon

The physics of photoproduction is usually classified with respect to the nature of the photon when interacting with matter. We know very well the electromagnetic process of e^+e^- pair production: the free propagating photon can move into a virtual e^+e^- state with a coupling strength essentially determined by the electric charge of the electron (Fig. 1a). This virtual state may exist for a dissociation time propotional to $E\gamma$, the energy of the photon, divided by the squared e^+e^- mass. In the neigbourhood of a nucleus the e^+e^- pair can become real, while energy and momentum are conserved by a corresponding 4-momentum transfer to the nucleus.

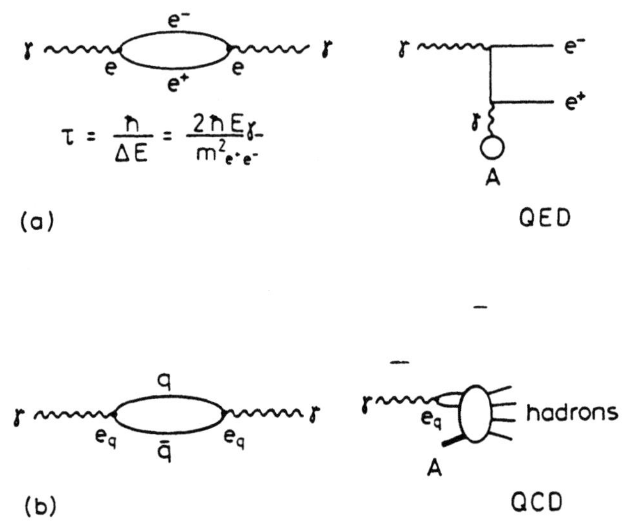

Fig. 1. Real photon interactions; (a) production of e^+e^- pairs, (b) production of hadrons

Photoproduction of hadrons can be described by a corresponding picture where the photon couples to a virtual $q\bar{q}$ pair (Fig. 1b). QCD-type ideas suggest that there are strong interactions between q and $\bar{q}$; in particular at low masses the $q\bar{q}$ may form a vector meson state so that the photon scattering can be linked to vector meson scattering (Vector Meson Dominance Model VMD, see Fig. 2a).

The mass of the virtual $q\bar{q}$ state might also be large due to a large opening angle. Then the hadron-like photon serves as source of free quarks and gluons ("resolved photon") which is as usual described by hadronic structure functions; in the hard regime these are calculated by QCD (Fig. 2b) [1]. The quark structure function may be simply a constant, i.e. the electromagnetic field couples pointlike to a $q\bar{q}$ pair which then enters into the strong interaction (Fig. 2c).

The contributions from the resolved photon are small with respect to the total photoproduction cross section since this is dominated by soft hadronlike-photon processes; however, when the momentum transfer is increased, they become important since they are harder. Substantial contributions are predicted first of all from two processes where the structure of the resolved photon is simply pointlike. In one process (Fig. 3a), called Photon Gluon Fusion the photon interacts with a gluon from a nucleon yielding a $q\bar{q}$ pair which then fragments into hadrons. The other process (Fig. 3b) is the QCD-analogon to

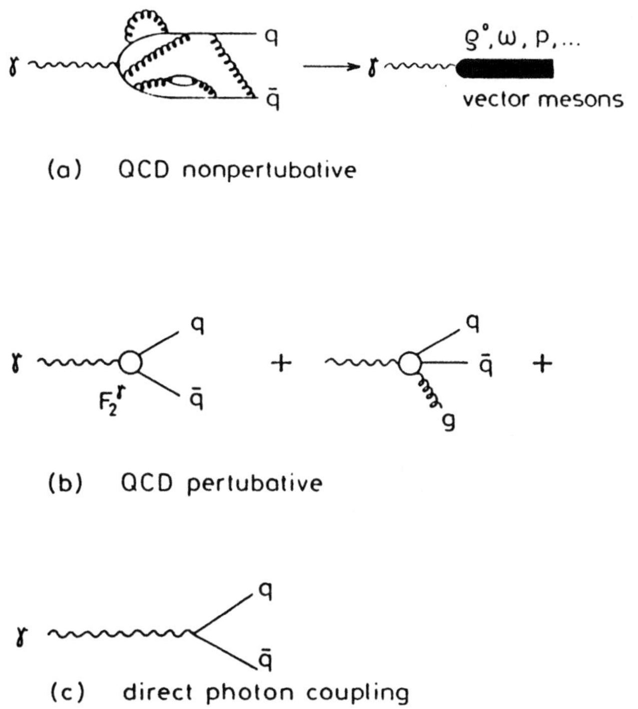

Fig. 2. Hadronlike-photon interactions; (a) VMD model, (b) resolved photon, (c) pointlike photon

the well-established QED-Compton process with a gluon instead of a photon in the outgoing state (QCD-Compton process).

Both processes are calculated in QCD, and for predicting the final state hadrons, connected with a fragmentation model. The main prediction is a simple event topology characterized by two clusters of hadrons in addition to the cluster formed by target spectators. In the kinematical range to be studied contributions from second order processes (with respect to α_s) are nonnegligible. Particularly interesting is the so-called Higher Twist process (Fig. 3c) which can be distinguished from the two lowest-order processes (see above), also called Minimum Twist, and from other contributions by a unique event topology [2]. The HT process is predicted to show one hadron cluster which is balanced in tranverse momentum by a single hadron produced promptly by a form factor-like coupling.

In recent analyses (NA14 and WA69) the hypothesis has been tested whether the data can be described by assuming a superposition of a VMD-component and a pointlike

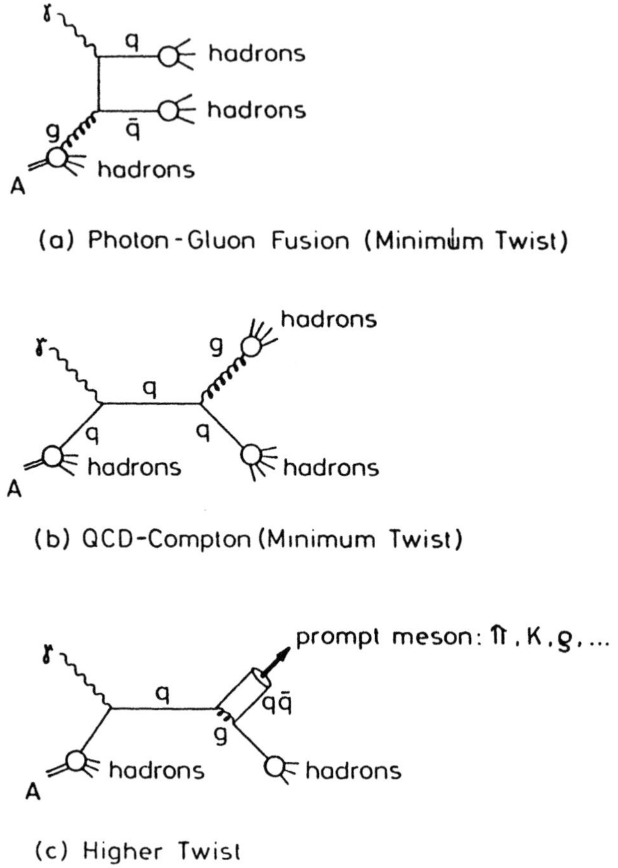

Fig. 3. Pointlike photon processes; (a) Photon-Gluon Fusion, (b) QCD-Compton, (c) Higher Twist

component (with and without higher order processes) throughout the soft-hard transition region. The results are discussed in Sect. 4.

3 Simple Vector Dominance and Established Experimental Data (brief overview)

Next we consider the hadronlike photon as it is described in terms of the simple VMD model. This model assumes that the intermediate hadronic states of the photon are entirely represented by the existing neutral vector meson states ρ^0, ω and ϕ according to

$$|\gamma\rangle = \sum_{V=\rho^0,\omega,\phi} \frac{4\pi}{\gamma_V^2} |V\rangle \,.$$

3.1 e$^+$e$^-$ $\rightarrow$ Vector Meson

The VMD hypothesis has been verified a long time ago on e$^+$e$^-$ annihilations at storage rings assuming the process sketched in Fig. 4. The coupling strengths, $4\pi/\gamma_V^2$, are related to the leptonic decay widths according to

$$\frac{4\pi}{\gamma_V^2} \cdot m_V = \frac{12}{\alpha^2} \cdot \Gamma(V \rightarrow e^+e^-)$$

The relative coupling strengths are predicted by the quark content of the vector mesons, i.e.

$$\frac{4\pi}{\gamma_V^2} \cdot m_V \propto \left(\sum_{q \in V} e_q \right)^2 \,.$$

The prediction yields for ρ^0, ω and ϕ (assuming an ideally mixed SU(3) nonet structure) a relative strenghts of 9:1:2. The coupling strengths determined from measurements of the annihilation process [3] are (8.8±2.6):1:(1.7±0.4) when normalising to ω. Prediction and experimental results are consistent. The measurement errors are large which is due to experimental difficulties to measure the vector meson signals [3].

3.2 γ A $\rightarrow$ V A

A second well-understood domain is the soft elastic photoproduction of vector mesons according to the reaction

$$\gamma p(A) \rightarrow V p(A) \,.$$

It is linked to elastic vector meson scattering in the framework of VMD (Fig. 5). Then the cross section of the elastic photoproduction process can be expressed by the total vector meson scattering cross section according to

$$\frac{d\sigma}{dt}(\gamma p \rightarrow V p) = \frac{\alpha\pi}{\gamma_V^2} \frac{d\sigma}{dt}(V p \rightarrow V p) \propto \exp(-bt)$$

$$\frac{d\sigma}{dt}(\gamma p \rightarrow V p)|_{t=0} = \frac{\alpha\pi}{16} \frac{1}{\gamma_V^2} \sigma_{\rm tot}^2(V p)$$

$$\sigma(\gamma p \rightarrow V p) = \frac{\alpha\pi}{16b} \frac{1}{\gamma_V^2} \sigma_{tot}^2(V p) \,.$$

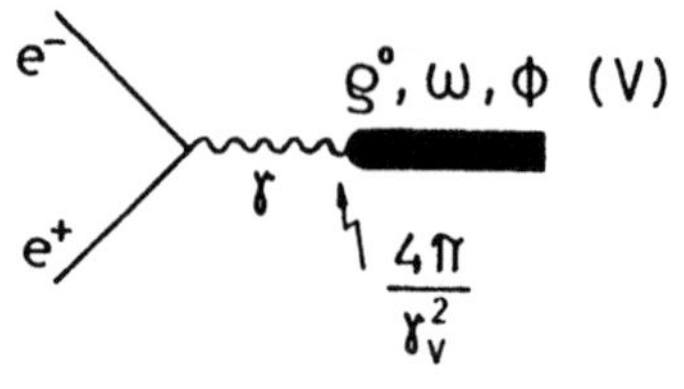

Fig. 4. Production of vector mesons in e^+e^- annihilation

The Vp cross sections are not directly measured. However, by assuming the validity of the additive quark model it can be derived from measured πp and K p cross sections since they contain the same quarks as the vector mesons. Figure 6 shows for illustration the measured elastic ϕ photoproduction in comparison to a VMD-quark model prediction derived from hadron beam measurements according to

$$\sigma(\phi p) = \sqrt{\sigma^2(K^+p) + \sigma^2(K^-p) - \sigma^2(\pi^-p)} \; .$$

The slopes agree very well. Relying on the estimate of the ϕp cross section (as intoduced) the normalisation provides another measurement of $4\pi/\gamma_V^2$ independently of $e^+e^- \rightarrow \phi$.

3.3 e(μ) A $\rightarrow$ Hadrons

Spacelike virtual photons are also described in the framework of the VMD model (Fig. 7). The photon emitted from a charged lepton can be in a vector meson state for a dissociation time given by the uncertainty principle

$$\tau = \frac{2\nu}{Q^2 + m_V^2} \; .$$

For real photons is $Q^2 = 0$ and we obtain

$$\tau = \frac{2E_\gamma}{m_V^2} \; .$$

A corresponding relation has been introduced already for e^+e^- annihilation (see Fig. 1a). Experimental results support this picture: Electroproduction data with electrons in the range of a few GeV are described by simple VMD predictions at low Q^2 as it is illustrated in Fig. 8 for single pion electroproduction [5].

The dissociation time can be decreased by increasing Q^2 so that the mesonlike photon state is finally suppressed and the photon interacts purely as bare state. At the same time a small interaction distance is approached and the scattering becomes sensitive to the substructure of the nucleons. Thinking on this line it was predicted to see Mott-type

Fig. 5. Elastic photoproduction of vector mesons

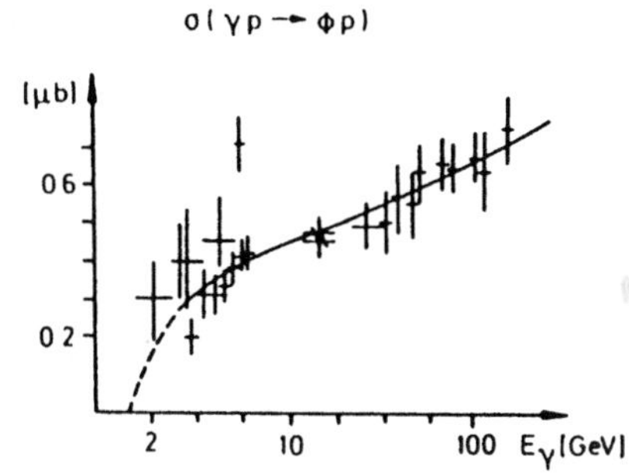

Fig. 6. Elastic ϕ photoproduction in comparison to the VMD prediction (see text)[4]

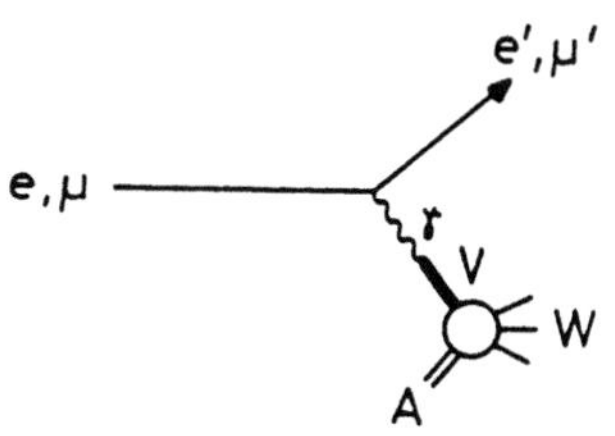

$$\nu = E_l - E_{l'}'$$
$$Q^2 = -(p_l - p_{l'}')^2$$
$$W^2 = (q + p_A)^2$$

Fig. 7. Electroproduction of hadrons

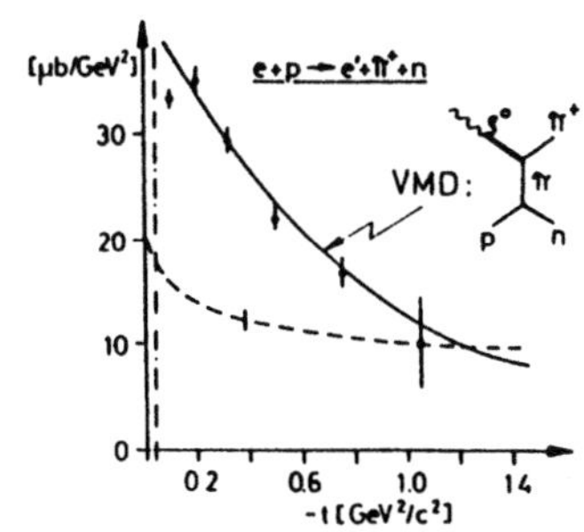

Fig. 8. π^+ electroproduction as function of t at $Q^2 = 0.26\,\mathrm{GeV}^2$ and $W = 2.2\,\mathrm{GeV}$; full line: VMD prediction; dashed line: π^+ photoproduction data at $W = 2.2\,\mathrm{GeV}$ [5]

scattering already at moderate energies and Q^2 values. This is observed in electroproduction data: as example is shown in Fig. 9 the ratio of the cross section of electroproduction of hadrons and of Mott scattering as function of Q^2 [5]. One sees that this ratio becomes approximately constant for Q^2 bigger $1\,\mathrm{GeV}^2$ already, i.e. the electroproduction of hadrons can be described by the scattering of relativistic spin-1/2 electrons by spinless pointlike nuclei.

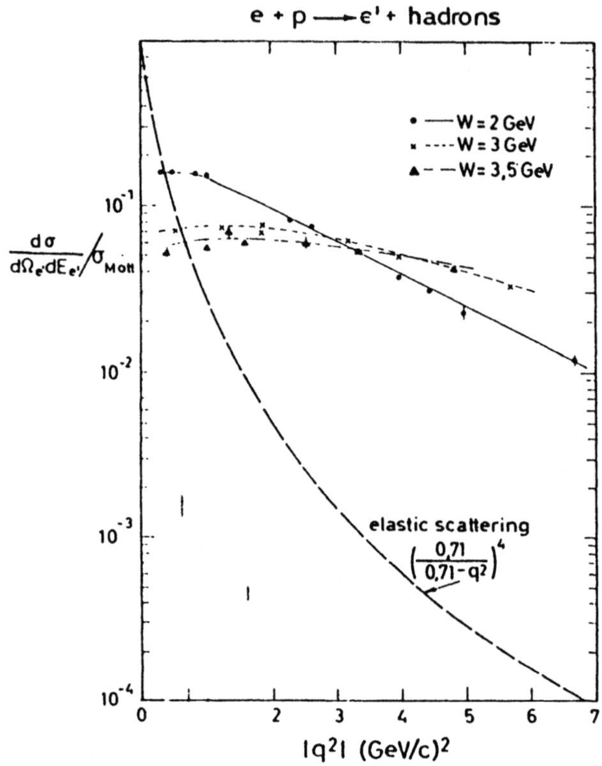

Fig. 9. Electroproduction of hadrons divided by the Mott cross section vs. Q^2 [5]

A more established tool to study virtual photons in the transition region to hard processes provides the shadowing effect on nuclei. I will not discuss this subject here since there are other talks at this workshop devoted to it (see e.g. W. Weise in these proceedings). Shadowing data also suggest that bare photon interactions become observable for Q^2 bigger 1 GeV^2 already [6]. The high luminosity accelerators ELSA and CEBAF might open possibilities for improved experiments to study the onset of hard processes in electroproduction in this kinematical range.

3.4 $\gamma\gamma \rightarrow$ Hadrons

More recently interesting results with spacelike virtual photons came from photon-photon interactions which were explored at e^+e^- storage rings (PETRA and PEP). The Q^2 values of the two virtual photons are determined from the kinematics of the scattered electron and positron (Fig. 10). Three kinematical ranges are distinguished (Fig. 10a–c): both photons are almost real ($Q^2 \approx 0$), an almost real and a highly virtual photon (large Q^2) and thirdly two highly virtual photons. For the first class(a) VMD predicts that the cross section is given by the scattering of two vector mesons in agreement with measurements [7]. The second class(b) was studied over a range which connects soft and hard processes. The situation is illustrated on $e^+e^- \rightarrow h^\pm X$ in Fig. 11 where a measurement of PLUTO at PETRA is shown: the inclusive p_t^2 distribution (with p_t being the transverse momentum w.r.t. the photon direction in the photon-photon c.m.s.) is described by $\rho\gamma$ scattering for p_t below 1 GeV whereas above a QPM calculation assuming pointlike coupling for both photons is needed to describe the data.

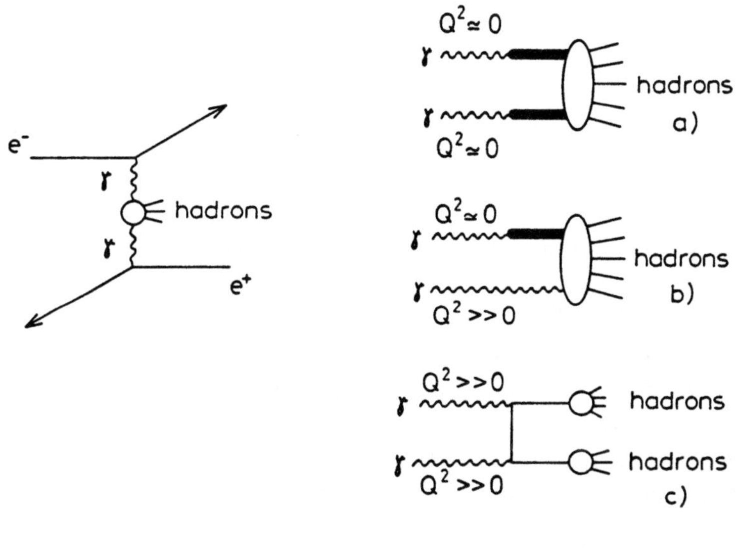

Fig. 10. Photon-photon interactions; (a)with two almost real photon, (b) with one almost real photon and one hghly virtual photon, (c) with two highly virtual photons

We note that here appears besides Q^2 another parameter which allows to vary the kinematical range: p_t. The latter is essential for real photon scattering ($Q^2 = 0$). It defines the spacial resolution according to

$$\lambda = \frac{\hbar}{p_t}\,.$$

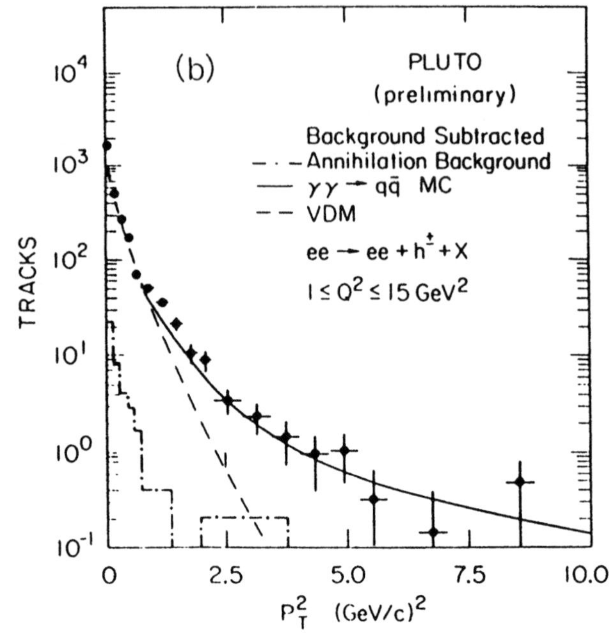

Fig. 11. Inclusive p_t^2 distribution compared to Monte-Carlo expectations for PLUTO single tag data [8]

My next subject will cover results on inelastic photoproduction of hadrons in a wide p_t range published recently.

4 New Results from Experiment WA69 on γ p $\rightarrow$ Hadrons

4.1 VMD-Model Predictions

The total photoproduction of hadrons covers besides the elastic vector meson production all inelastic channels. Following Fig. 12 the inelastic channels rise the question whether the VMD model is also valid here.

With essentially the same assumptions about the VMD model as for the elastic vector meson production, I discussed in Sect. 3, the following relations are derived:

$$\sigma(\gamma p \rightarrow X) = \sum_{V=\rho^0,\omega,\phi} \sigma(Vp \rightarrow X)$$

$$= \frac{1}{VMD}\left[\frac{2}{3}\left(\sigma(\pi^+ p \rightarrow X) + \sigma(\pi^- p \rightarrow X)\right) + \frac{1}{3}\left(\sigma(K^+ p \rightarrow X) + \sigma(K^- p \rightarrow X)\right)\right] ,$$

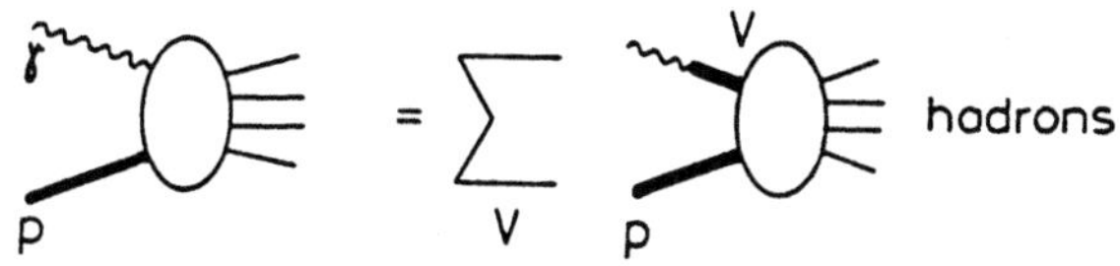

Fig. 12. Inelastic photoproduction of hadrons

where X is the hadronic final state.

The photon-vector meson coupling strenghts (see Sect. 3) yield, taking into account ρ^0, ω and ϕ, an effective value 1/VMD=1/200 where the uncertainty is about 30%. Assuming a hadron beam cross section of 20 mb we estimate for the VMD component a cross section of 100 μb, which is in the same order of magnitude as the total cross section for the photoproduction of hadrons.

The VMD component of inelastic photoproduction of hadrons is moreover a soft phenomenon. A relevant parameter to specify the kinematical range of the scattering process is the p_t of the produced hadron(s) defined perpendicular to the photon-beam direction in the photon-proton c.m.s. (Fig. 13b). The p_t slope of the VMD-component is, according to the model assumptions, determined by the slopes of the hadron beam data which are very steep so that even small contributions can be measured at high p_t provided their slope is less steep. The poinlike-photon processes Minimum Twist and, to a lower extent, Higher Twist (defined in Sect. 2) are such processes: VMD component and MT/HT are of comparable magnitude at p_t = 2GeV; above the bare component is expected to be the more dominant the more p_t is increased (Fig. 13a).

The CERN-experiments NA14 and WA69 carried out measurements and analyses in the p_t range up to 5 GeV. I concentrate on WA69 in the following.

4.2 Experiment WA69

The experiment WA69 has been carried out by the OMEGA-PHOTON collaboration at the CERN-SPS [10]. A proton beam (Fig. 14) was converted into an electron beam via π^0 production on a first target and $\gamma \rightarrow e^+e^-$ production on a second target. The electrons produced the final photon beam by bremsstrahlung on a third target. The photons were tagged in energy by measuring the single electrons before and after bremsstrahlung. The photon energy range was 70 to 170 GeV and the energy resolution was typically $\Delta(E_\gamma)/E_\gamma$ = 1%. The tagged photon beam was put on a hydrogen target inside the OMEGA spectrometer for photoproduction (Fig. 15). Secondary charged particles were bent in the homogeneous magnetic field and detected in a set of MWPC's and drift chambers downstream and beside the target. Charged particles leaving the OMEGA magnetic field passed the Ring Imaging Cerenkov counter and the Transition Radiation detector where pion/ kaon/ proton identification was provided. Photons from decays of π^0 and η were measured in a large lead-liquid scintillator calorimeter with another lead-scintillating fibres calorimeter in the central part.

For the results discussed here the data analysis was restricted to charged particles only. Studies including also the neutrals are in progress. Data were taken with tagged photons and with pion and kaon beams (at 80 and 140 GeV for computing the VMD component in two ranges of the photon energy).

4.3 Single Particle Spectra

We consider first the ratio of cross sections of photon beam data and VMD component computed from the hadron beam data (Sect. 4.1) in the x_F vs. p_t plane of single inclusive charged particles (Fig. 16a). Here x_F is the fractional longitudinal momentum carried away by the final state hadron , and p_t is defined due to Fig. 13b. One can see, there is a region where the ratio is approximately constant ($0 < x_F < 0.7$ and $p_t < 1.5$ GeV) its value is eqal to 1 when assuming 1/VMD= 215. This factor is, within 20% uncertainty consistent with predictions and measurements elsewhere (see Sect. 3) . The range where VMD is valid is limited to p_t values less than 1.5 GeV: this is particularly obvious when looking at the projection (Fig. 16b). The observed rise when p_t is increased can be explained by the onset of the pointlike photon processes (MT mainly) as we will discuss in a moment.

The apparent fall of the ratio at large x_F is caused by extra contributions to the hadron beam data which disturb somewhat the simple quark model relations applied: these are, over the whole p_t range, substantial contributions from diffraction dissociation processes where the beam hadron punches through in a dominant manner. Those processes have no counterpart in vector meson induced inclusive single hadron production. They can be effectively excluded by taking into account only those hadrons where beam meson and final state hadron are different in charge (see below).

The cross sections for photon beam data and VMD component as function of p_t are shown in Fig. 17 (same data sets as in Fig. 16 and 1/VMD=1/215). One can see again the significant difference in the slope of p_t.

The difference plots (Fig. 18) were considered next in order to quantify the extra component in the photon data and to compare to theoretical predictions for MT and HT processes. For the comparison to theory two approaches were followed:
(1) a prediction of Aurenche et al. including Minimum Twist (MT) and higher order graphs [1]. The hadrons in the final state were predicted assuming a simple fragmentation model ("independent fragmentation").
(2) A lowest order calculation of MT with a complete fragmentation into hadrons on the basis of a string model (LUCIFER-LUND program [11]). Also Higher Twist(HT) was predicted at this level.
The difference plots (Fig. 18) were compared to both sets of predictions for p_t bigger 2 GeV. The full line is the prediction of Aurenche et al., the dashed line is the same prediction, but including in addition an estimate for a contribution assuming an explicite photon structure function ("resolved photon component"). Data and both predictions agree within expected accuracies, i.e. it is not possible to gain any insight on this extra component. The triangles are the MT predictions of the LUCIFER-LUND program with statistical errors of the simulated events. The normalisation (adjusted on a phe-

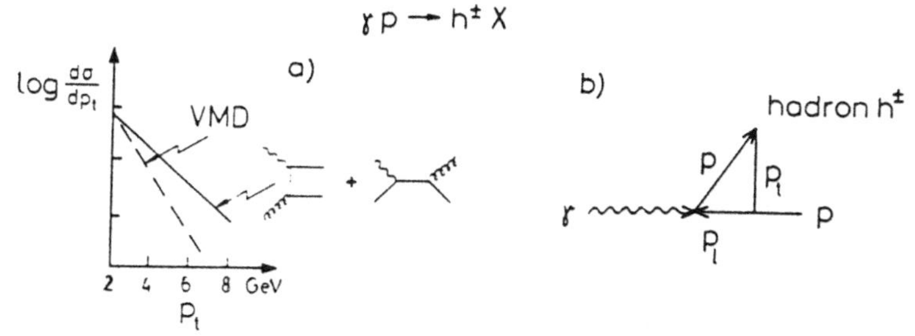

Fig. 13. (a)Predictions for VMD component and pointlike component as function of p_t of single charged hadrons[9]. (b) Definition of p_t

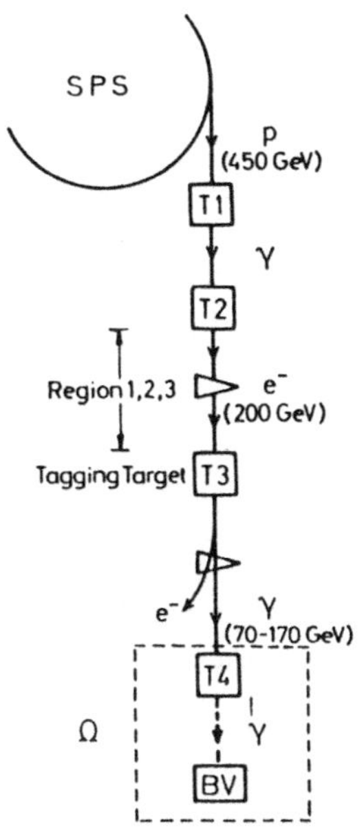

Fig. 14. Electron-photon beam line of experiment WA69

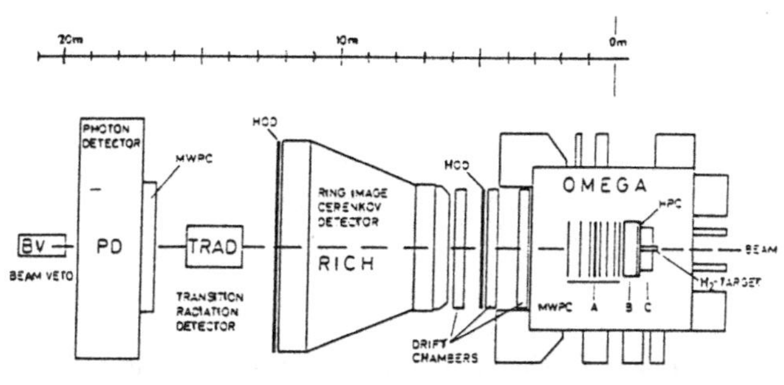

Fig. 15. OMEGA setup for WA69

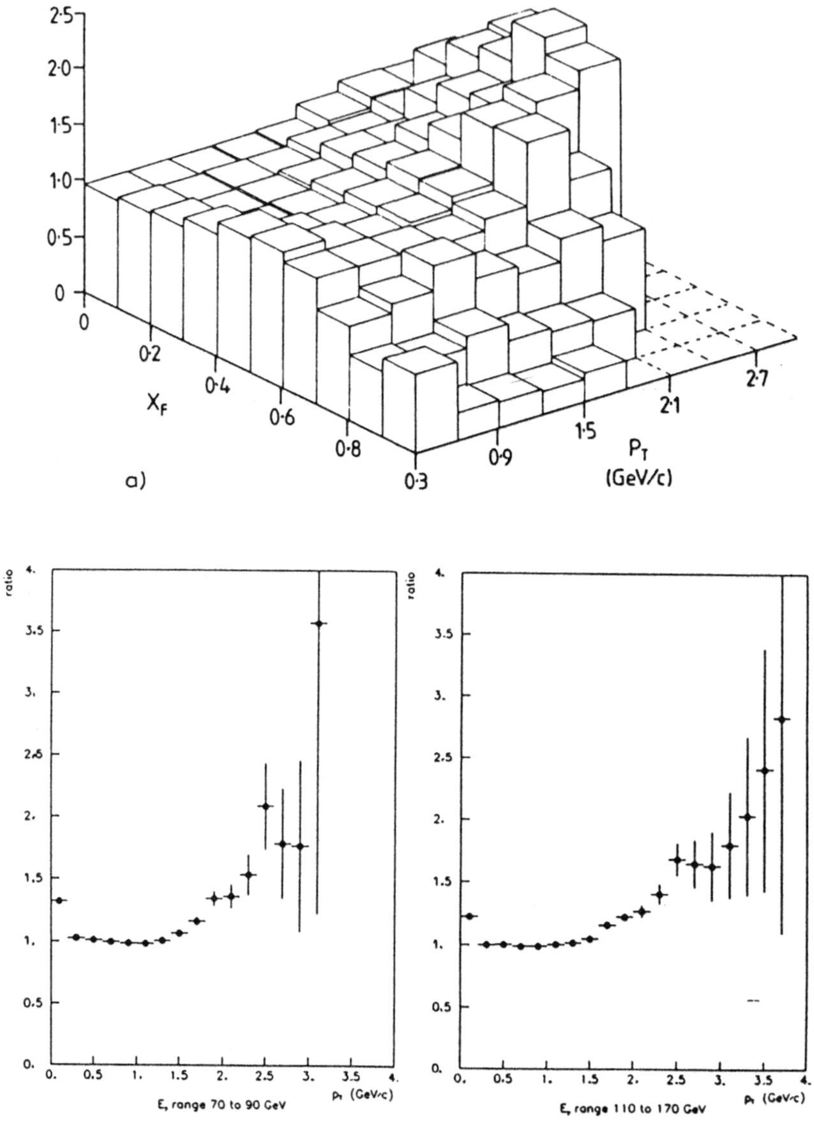

Fig. 16. Ratio of photon-beam data to normalized hadron-beam data for the photon energy range 110 to 170 GeV; (a) x_F vs. p_t, (b) p_t integrated over x_F from 0.0 to 0.7 [12]

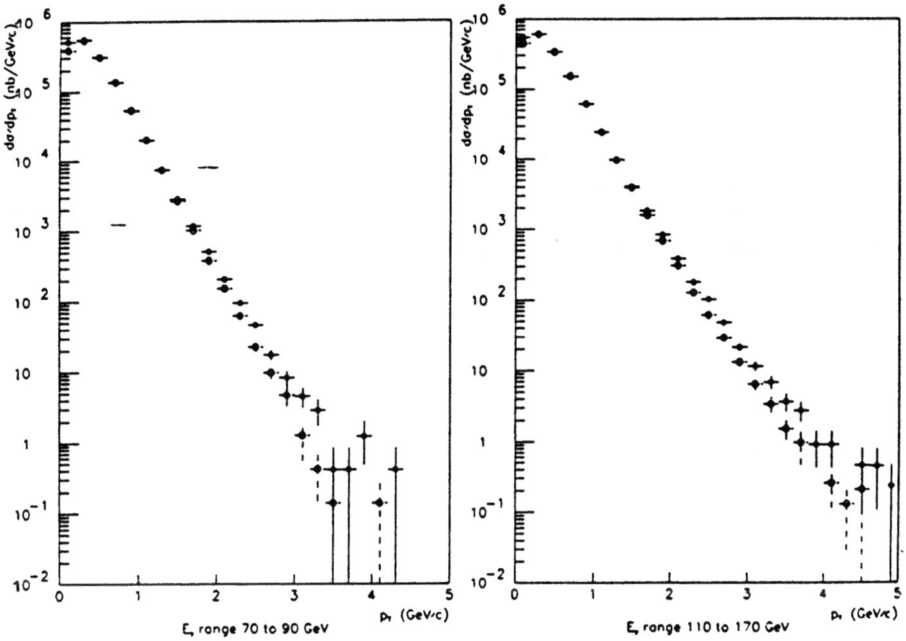

Fig. 17. Single inclusive cross section as function of p_t; full circles: photon-beam data, open circles: normalized hadron beam data

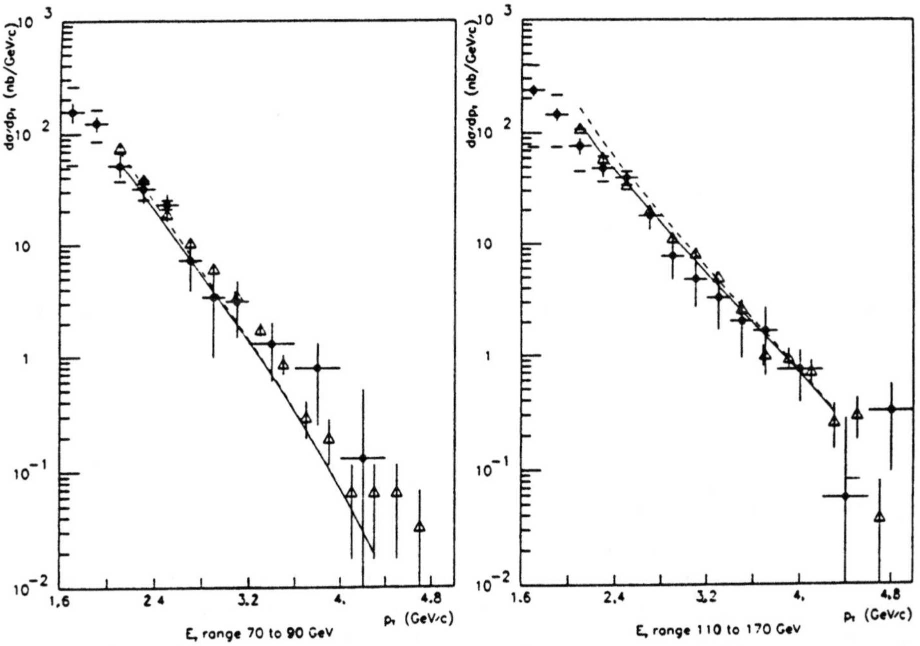

Fig. 18. Subtracted pt distributions integrated over x_F from 0.0 to 0.7 for p_t bigger 1.6 GeV; full circles: data points (exp. WA69), full and dashed lines: predictions by Aurenche et al. (see text); triangles: prediction by the LUCIFER-LUND program

nomenological basis [11]) has an overall uncertainty factor of about two. All predictions are roughly consistent with each other and with the data.

A similar result was obtained by experiment NA14 ([12], see talk by Wormser in this proc.). As an example is shown in Fig. 19 the subtracted NA14 data in comparison to the prediction of Aurenche et al. It was discussed to extract from this data the QCD Λ-parameter and the gluon fragmentation function measured via the QCD Compton process at large p_t; however the sensitivity is low [13].

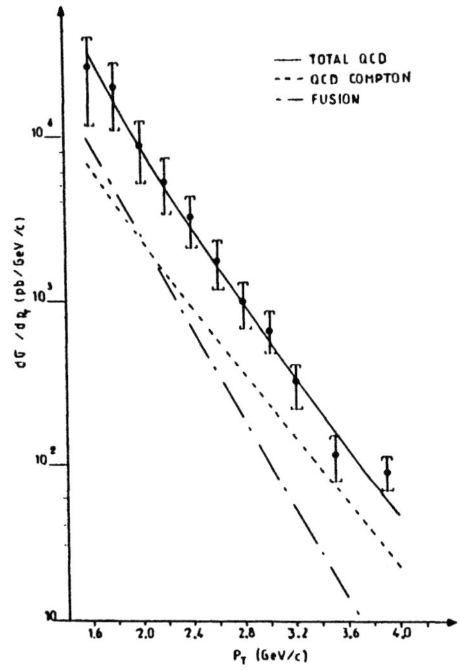

Fig. 19. Subtracted p_t distribution for inclusive π^0 photoproduction for p_l bigger 40 GeV (exp. NA14[12])

Next we consider the x_F distribution of single charged particles. It was studied in order to search for Higher Twist processes with prompt pions and kaons (Fig. 3c). The WA69 data for p_t bigger 2 GeV (Fig. 20a) were fitted to an incoherent sum of predicted shapes for the VMD component, MT and HT [14]. The shape of the VMD component was "predicted" from hadron beam data taken in the same experiment (see above) including in the x_F distribution only hadrons with a charge opposite to the beam meson (Fig. 20b). This shape was found to be not distorted by the diffraction dissociation effects visible in Fig. 16a; in this case the scattered meson and the beam meson have no common valence quark. Moreover, as a matter of fact, the shape of this distribution is very similar to that of pion beam induced π^0 spectra where one quark can be common (the latter corresponds to the case of vector meson induced production of charged mesons).

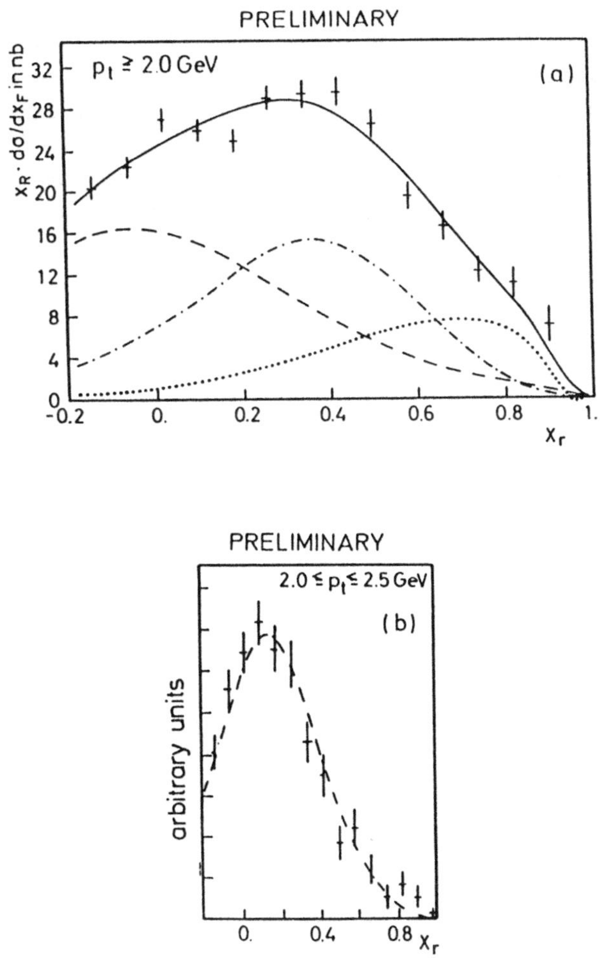

Fig. 20. (a) Single inclusive charged particle x_P distribution (exp.WA69); full line: total fit, dashed: MT, dashed-dotted: HT; (b) VMD component derived from hadron-beam data with a curve describing this data (identical to the dashed curve in (a)); $x_R^2 = x_P^2 + x_t^2$

The fit finds with respect to the predicted cross sections, typically 70% of the VMD component and 140% of MT [14]. The small HT component (predicted [2] to be less than 10% at p_t bigger 2 GeV) is a 3 standard deviation effect; the cross section is 3 to 4 times larger than predicted [2]. Attempts have been made to improve the separation between MT and HT by taking into account that the prompt meson of HT is isolated from other final state hadrons. The fits carried out on subsamples of the data where further cuts (on topological parameters) were introduced confirm the overall fit result.

4.4 Analysis of the Azimuthal Energy Flow

So far we have considered single particle spectra. In experiment WA69 the study was extended to another observable which takes into account the complete charged particle topology of the events: the azimuthal energy flow [15]. MT and HT events are predicted to have a special angular distribution of the hadrons w.r.t. to the photon direction in the overall c.m.s. As an example a Monte Carlo event of the Photon Gluon Process is shown in Fig. 21. The nine tracks of the event are projected into the event plane (defined as the plane where the sum of p_t^2 has a maximum). The track momenta in this plane are given by the lengths in the proper directions w.r.t. to the photon direction with the transverse momentum in this plane: $p_{t,in}$.

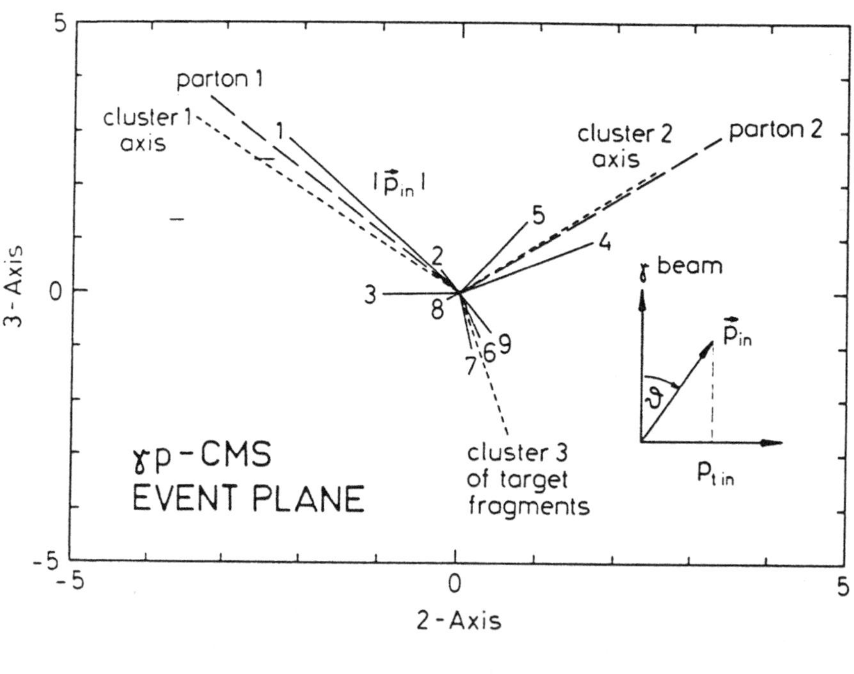

Fig. 21. Monte-Carlo event of Photon-Gluon-Fusion; full lines: projected momenta of charged particles,dashed: projected parton (quark) momenta; dotted: cluster axes determined by a cluster finding algorithm

The characteristic feature is the lack of momentum in the forward direction. The forward dip structure was analysed in terms of the azimuthal energy flow i.e. the cross section as function of the azimuthal angle θ.

The data were subdivided into bins of sum of $p^2_{t,in}$ shown in Fig. 22a–b. The difference plot (Fig. 22b) is well described by QCD predictions of the LUCIFER-LUND program for $\sum p^2_{t,in} \gtrsim 4\,\mathrm{GeV}^2$. We look next at the comparison of photon beam data and hadron beam data (superimposed and normalized to represent the VMD component). The data are shown for some bins of sum of $p^2_{t,in}$ in Fig. 23a–d, and as subtracted plots in Fig. 24a–c. A track enters into the plot at its angle θ with a weight of the normalized energy $2E/\sqrt{s}$. The distribution were finally converted to cross sections by normalizing to the total measured luminosity. Energy flows for photon and hadron beam data coincide for small values of the sum of $p^2_{t,in}$ (Fig. 23d) where one cannot expect to see pointlike photon processes. A difference is obvious at larger p_t values (Fig. 23a–c). The difference plots for the same large p_t data (Fig. 24) show a significant structure, which is well described by the QCD prediction for MT (and HT).

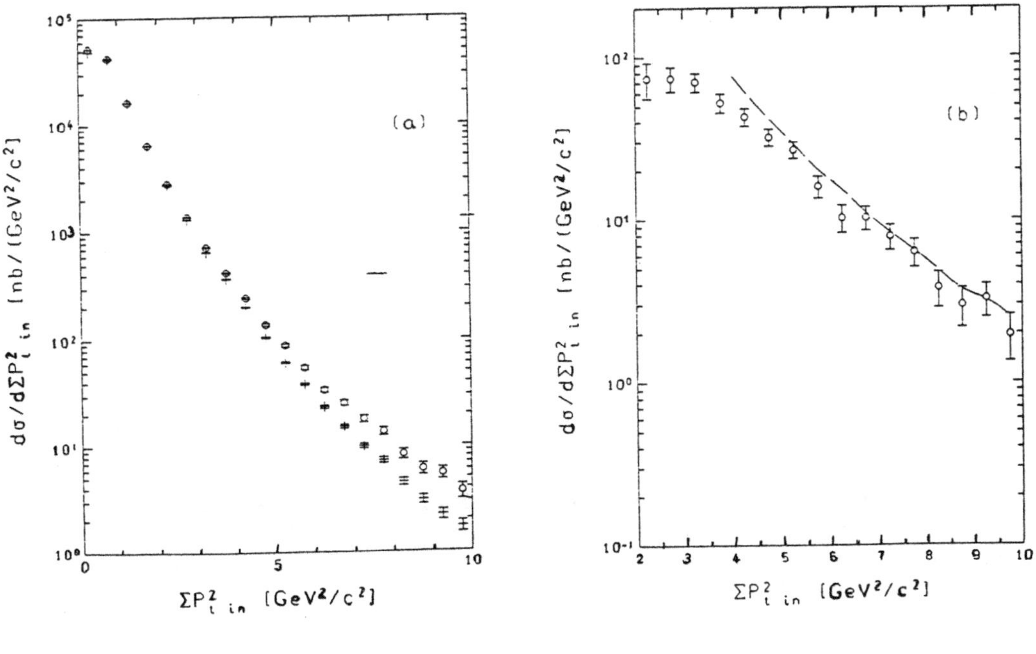

Fig. 22. (a) Distribution of $\sum p^2_{t,in}$ (exp. WA69); squares:photon-beam data, crosses: normalized hadron beam data; (b) Subtracted distribution compared to QCD prediction (LUCIFER-LUND)

Next we look at the asymmetry of energy flow plots. For a single event, in general, the energy flow will be different w.r.t. the photon beam, i.e. for positive and negative angles θ since p_l is shared between three parts including the target fragments. This sharing is also predicted. For testing this feature the energy flows of the two forward clusters were ordered so that the cluster with the bigger energy enters at positive and the smaller one at negative angles. Results are shown for two bins of sum of $p^2_{t,in}$ in Figs. 25 and 26. The prediction agrees well with the data also in further details. It can be nicely seen that the p_t balance of the two forward cluster (and not a kinematical artifact from our recipe to

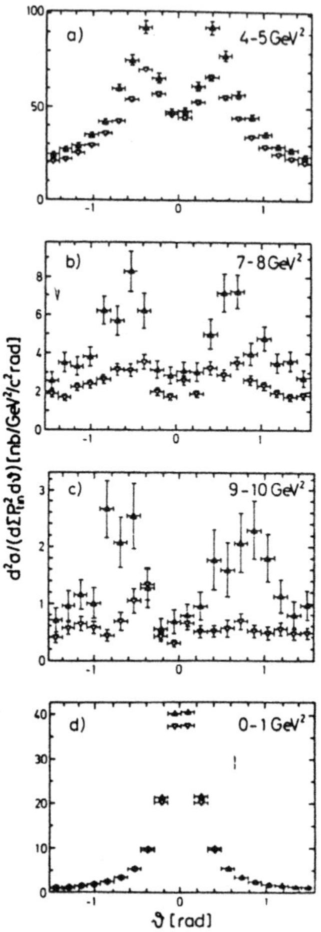

Fig. 23. Energy flow distributions for bins of $\sum p_{t,in}^2$ indicated; full triangles: photon data, open triangles: normalized hadron-beam data (exp. WA69)

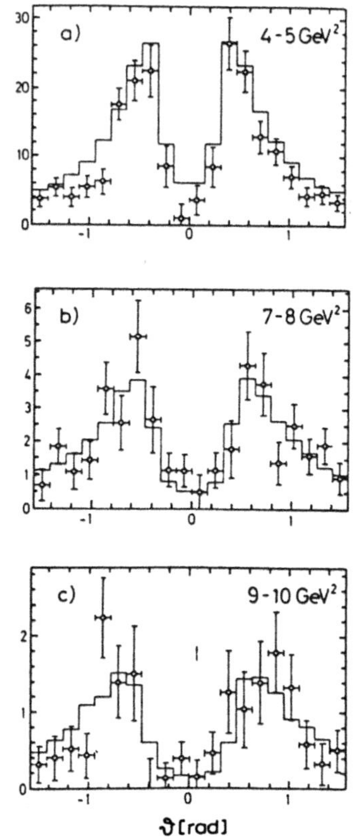

Fig. 24. Subtracted energy flow data corresponding to Fig.22a–c compared to QCD predictions (histogram of LUCIFER-LUND events)

reject the VMD component by a cut on $\sum p_{t,in}^2$) causes the dip structure; the same p_t for both clusters requires a larger angle θ for these clusters with lower energies and vice versa (as shown)

One can conclude that subtracted energy flow data are in accordance with the existence of the MT processes, and that this result amplifies what was concluded from single particle spectra.

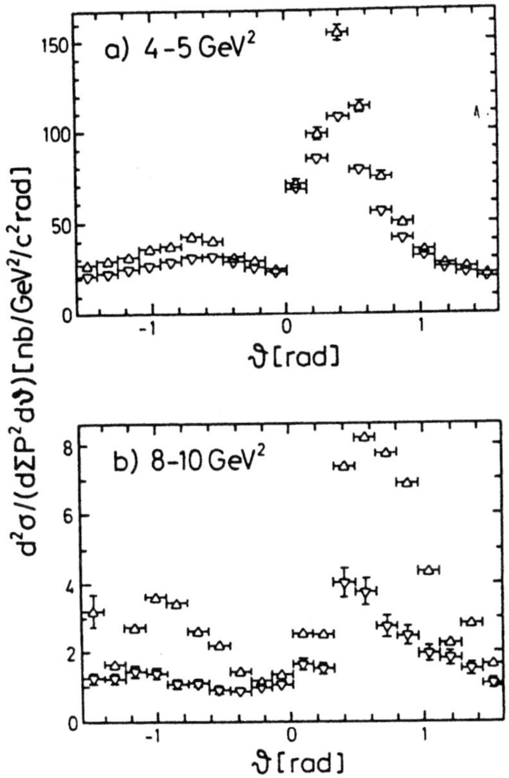

Fig. 25. Asymmetric energy flow data for two bins of $\sum p_{t,\mathrm{in}}^2$ (see text)

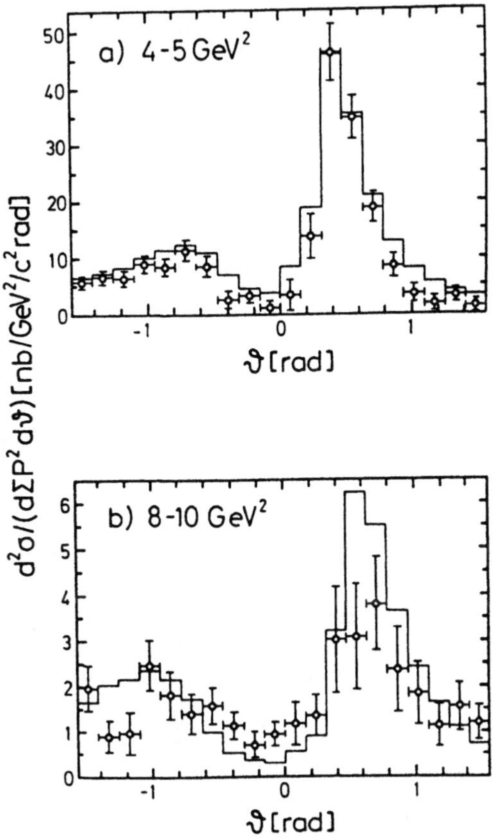

Fig. 26. Subtracted distributions to Fig. 24 compared to QCD (histogram of LUCIFER-LUND events)

5 Summary and Conclusions

Photoproduction of hadrons at photon energies up to a few hundred GeV is described on the level of gross features over a wide kinematical range from soft to hard processes by assuming that it is built up by a Vector Meson Dominance component plus a hard component of pointlike photon interactions. The accuracy level is about 20%. The pointlike component is measurable for transverse momenta of single hadrons bigger than 2 GeV or for a sum of the tranverse momenta squared of the charged particles bigger than 3 GeV^2. This means also that high energies are essential for this physics with real photons. Corresponding studies with spacelike virtual photons should be possible already at energies in the GeV range.

Acknowledgements

I would like to thank my collegues B. Diekmann and K. Heinloth for helpful discussions and careful reading of the manuscript. Special thanks go to M. Schumacher and his collegues in Göttingen for organizing this workshop.

References

1. P. Aurenche et al.: Nucl. Phys. B **286** 553 (1987) and references contained therein
2. E.L. Berger: Z.Phys. C **4** 63 (1980)
   M. Benayoun et al.: Nucl. Phys. B **282** 653 (1987)
   R.J. Fitch at al., Z. Phys. C **31** 51 (1986)
3. E. Paul: Proc. of the 1981 Photon-Lepton Conf.in Bonn, p. 301
4. B. Diekmann: Phys. Rep. **159** 100 (1988)
5. K. Heinloth: Springer Tracts of Mod. Physics 103 (1972)
6. E. Paul: Research Program of CEBAF, summer study 1986, p. 407
7. Ch. Berger and W.Wagner, Phys. Rep. **146** 2 (1987)
8. See e.g. N. Wermes, Lecture Notes on Physics Vol.191 (Springer Verlag 1983).
9. J.F. Owens: Phys. Rev. D **21** 64 (1980)
10. R.J. Apsimon et al.: Z. Phys. C **43** 63 (1989)
11. G. Ingelman, A. Weigend: DESY-87-018, Febr. 1987
12. E. Auge et al.: Phys. Lett. **168B** 163 (1986)
13. D.Treille: Proc. of a NATO advanced Research Workshop on QCD Hard Processes, 1987 in St. Croix, Ed. by B. Cox, NATO ASI Series B: Physics Vol.197, p. 193.
14. R.J. Apsimon et al.: Z. Phys. C, to be published
15. R.J. Apsimon et al.: Z. Phys. C **46** 35 (1990)

PROBING THE SHORT RANGE BEHAVIOR OF NUCLEI WITH HIGH P_T PHOTO- AND ELECTRO-NUCLEAR REACTIONS

J.M. Laget
DPhN/SEPN CEN-SACLAY
F91191 Gif-sur-Yvette CEDEX, France.

1. INTRODUCTION

The main goal of Nuclear Physics is the study of the structure and the dynamics of complex systems made of strongly interacting hadrons. As for any complex systems, the problem is to know:

- What are the relevant degrees of freedom;
- What picture is the best suited to the scale of the probe.

The extensive study of nuclei with hadronic probes, as well as with electromagnetic probes, has led us to an economical and consistent picture of nuclei made of nucleons, which interact by exchanging mesons. It relies on the description of baryons as a small core, of radius 0.5 fm, surrounded by a meson cloud which is polarized by the vicinity of the other baryons. We have also reached a good understanding of intermediate energy reactions, induced in few body systems, in terms of complex combinations of elementary processes, which can be independently single out by a judiciable choice of the kinematics and the probe. Up to now, there is no need to explicitely introduce quark degrees of freedom. Of course, there is no doubt that ultimately nuclei are made of quarks. But, in this picture, they are hidden in the various meson-baryon coupling constants and form-factors. At which scale will they start to show up is still an open question.

The internal structure of hadrons has been revealed by reactions induced at very high energies. Here, asymptotic freedom allows to compute pertubatively the interaction between quarks, and to understand the short range structure of hadrons. On the contrary, at lower energies, and larger distances, the non pertubative nature of QCD prevents to use the traditional methods of field theories. Inspired by QCD, phenomenological models have been taylored to reproduce the static properties of hadrons, but they encounter some problems when dealing with their dynamical properties or their interactions. A consistent theory of the hadron structure, valid at short as well as long distances, is still missing.

To day, Intermediate Energy Nuclear Physics is precisely concerned with the non pertubative low energy limit of QCD. The relevant distances are small with respect to the usual nuclear scale, but large with respect to the characteristic distance between quarks. Four main topics are relevant:

- The study of nuclei at short distance (smaller than the nucleon radius) will allow us to understand the interplay between the hadron substructure and the nuclear structure.

- The use of the nucleus as a laboratory will provide us not only with a way to study the interaction of unstable particules (baryonic or mesonic resonances for instance) or the hadronisation of quarks in the nuclear medium, but also with a filter to select some part of the wave function of hadrons, or hadronic systems, when they propagate in the nuclear medium.

- The creation of new flavors (strangeness, charm) in nuclei will provide us with a powerfull tag to follow the propagation of the corresponding quark, and to study new degrees of freedom of nuclear matter.

- The study of spin observables will allow us to reveal interferences between non pertubative and pertubative QCD amplitudes.

High p_T reactions are a powerful tool to undertake the study of these four research axes. Since they probe distances comparable to $\hbar c/p_T$, they offer us with a way to study the short range behavior of hadronic matter. When p_T is large enough it is possible to factorize hard processes, where the photon couple directly to a single quark, and soft processes, which have to be described by non pertubative QCD. In some sense, these reactions look like quasi-free, nucleon or electron, scattering on nuclei (which have led to the determination of the nuclear wave functions), and they are expected to provide us with constraints on the wave functions of hadrons and the response functions of the nucleon as well as the nucleus.

In this talk I will restrict myself to two of these topics, which are related: the study of the short range behavior of the nucleus, and the use of the nucleus as a filter. I will put a special a emphasis on photon induced reactions, since this is the main topics of this worshop, but I will also deal with hadron induced reactions. I will start explaining how to disentangle and determine the various components of the nuclear wave function. I will review the evidences of hard scattering processes in reactions induced by real photons as well as by hadrons, on free nucleons, with a special emphasis on spin observables. I will try to foresee what remains to be done with a free nucleon target, and what new perspectives will be opened by the study of the same reactions in the nuclear environment. I will conclude with a discussion on the relevant energy ranges.

2. THE NUCLEUS AT SHORT DISTANCES.

2.1. The nuclear wave function.

The extensive study of nuclei with the (e,e'p) reaction has provided us with a good knowledge of the nucleon momentum distribution up to 300 MeV/c for heavy nuclei [1], and 500 MeV/c for few body systems [2]. Modern nucleon-nucleon potentials, as the Paris one [3] or the Argonne V14 one [4], lead to a fair agreement to existing data, provided that Final State Interaction (FSI) and Meson Exchange currents (MEC) are fully taken into account. Their long range part is described by the exchange of one meson, and their intermediate range part by the exchange of two mesons. But, below 0.8 fm, their short range part is parametrized in a phenomenological way.

Hybrid models have been developped to overcome this difficulty: the short range part is described by the exchange of one gluon between quarks. The work of Yamauchi et al. [5] is the most advanced attempt in that direction. The potential is of the form:

$$V(r) = V_Q(r)\ f(r) + V_M(r)\ (1 - f(r)\ )$$

where $V_Q(r)$ is the local approximation of the one gluon exchange potential beetwen quarks, $V_M(r)$ is the one and two meson exchange potential (which coincides with the Paris one at distances larger than 0.8 fm), and f(r) is a cut-off function. The wave function is expanded on a cluster basis, and the dynamical equations are solved in the framework of the Resonating Group Method (RGM). The antisymmetry between quarks and the orthogonality between each components are fully taken into account.

For instance the deuteron wave function takes the form:

$$\varphi_D = \chi_{NN} + \chi_{\Delta\Delta} + \chi_{CC}$$

where χ_{NN} and $\chi_{\Delta\Delta}$ are respectively the projections on clusters which have the same quantum numbers as the nucleon and the Δ. The hidden color component χ_{CC} is made of colored clusters, and is orthogonal to the nucleon and the Δ components. Although there are several equivalent ways to expand the two nucleon wave function, this is the most natural expansion, since the nucleon component dominates. This appears clearly in Fig. 1: the nucleon component χ_{NN} is almost the same as the Paris one, and the small components $\chi_{\Delta\Delta}$ and χ_{CC} are of the same order of magnitude and contribute mainly at short distances.

The most direct way to disentangle each component is to project the wave function on given asymptotic states. Obsiously, (e,e'N) and (e,e'NN) reactions single out the χ_{NN} component: the charge operator is scalar and couples it directly to the asymptotic NN state. The (e,e'Δ) or the (e,e'$\Delta\Delta$) reactions single out the Δ component of the nuclear wave function. But care must be taken that Δ can also be created by the probe. Since this Δ creation mechanism occurs mainly in the transverse part of the current [2], its coulomb part, which couples directly to the charge of the Δ existing in the nuclear wave function, has to be selected. This requires either a tranverse-longitudinal separation, via the Rosenbluth method [2], or the determination of the spin observables [6] which are proportional to the interferences between the transverse and the longitudinal components of the electromagnetic current.

The determination of the nucleon and the resonance sectors of the nuclear wave function will be a major part of the research program of a high intensity CW electron accelerator in the energy range 4-6 GeV. The physics case is well documented [2,7] and implies to carry out transverse-longitudinal separation, to perform coincidence experiments out of the plane and to determine the various spin observables. Since the high momentum components of the wave function will be probed, high p_T events will have to be detected.

The study of the hidden color component is not straightforward, since its symmetry does not correspond to the symmetry of any asymptoptic physical state. It requires quark rearrangement mechanisms in order to connect colored clusters and colorless particles, as depicted in Fig. 2 for instance.

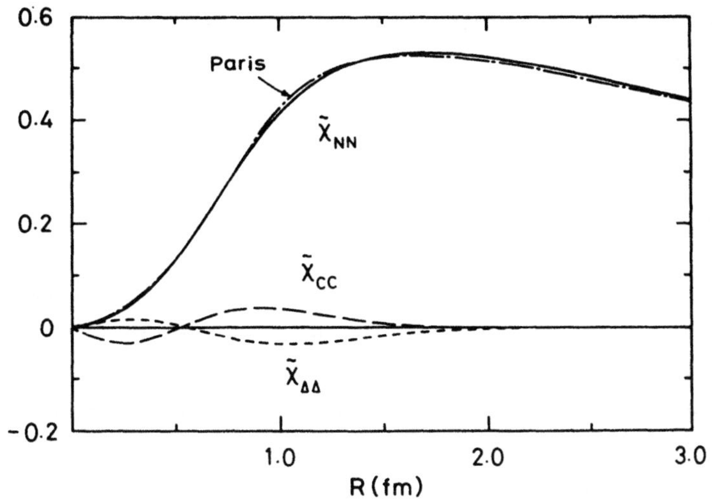

Figure 1

The various components of the deuteron wave function as computed in the hybrid model of Yamauchi et al.[5]

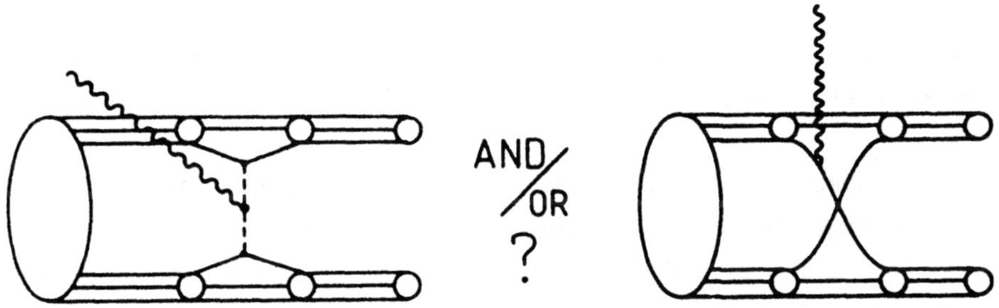

Figure 2

The two extreme couplings of a photon to a nuclear system.

They involve a hard interaction between the incoming photon and a quark. What are the evidences of such hard processes in the photon-nucleon interaction? What is the relevant energy range?

2.2. Hard mechanisms in photon induced reactions.

While below $E_\gamma = 1$ GeV the total absorption cross-section of a photon by a nucleon is dominated by the creation of the various baryonic resonances, above 1 GeV it is flat and almost constant when the energy increases. This is the domain of the Vector Dominance Model (VDM), the exchange of Regge trajectories, and hard scatterings of the photon constituents with the quarks of the nucleon.

Fig.3 clearly exhibits the main components of the cross section of the simplest channel, the $\gamma p \to n\pi^+$ reaction [8]. At forward angles ($t \simeq 0$), it is dominated by the exchange, in the t channel, of the pion Regge trajectory, and falls down very quickly when t increases. At backward angles (t large, $u \simeq 0$), it is dominated by the exchange, in the u channel, of the nucleon Regge trajectory, and falls down very quickly when u increases (or t decreases). Around 90°, where both t and u are large, the angular distribution exhibits a plateau, whose the variation with the squared total energy s is:

$$\frac{d\sigma}{d\Omega}(90°) \propto s^{-8}$$

This behavior is a signature of a hard interaction between the photon and the quarks inside the nucleon (Fig.4). Why ?

The basic assumption [9] is that it is possible to factorize the soft and the hard scatterings which occur between the constituents of the photon and the hadrons. For instance, the repeated soft interactions between the quarks insure the cohesion of the target nucleon, and govern the hadronisation of the quarks in the final state, or the coupling of the photon to a quark pair (free or interacting as a vector meson). In these soft mechanisms the typical length is:

$$c\Delta\tau \simeq R\, E/m$$

where R is the average distance of propagation of a free quark in the rest frame of a hadron (comparable to the radius of the core of this hadron) and E/m is the Lorentz dilatation factor.

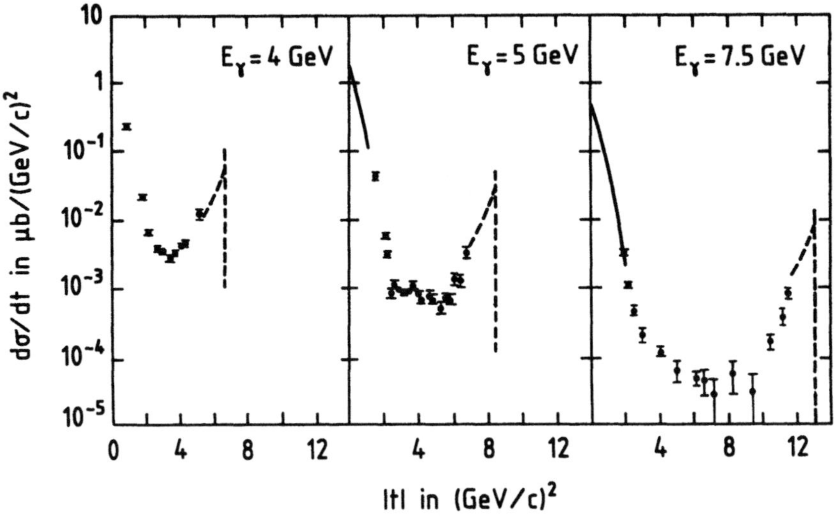

Figure 3

The cross section of the reaction $\gamma p \to n\pi^+$ is plotted against the momentum transfer $|t| = |(k_\gamma - k_\pi)^2|$, for selected values of the incoming photon energy E_γ. Adapted from ref.[8].

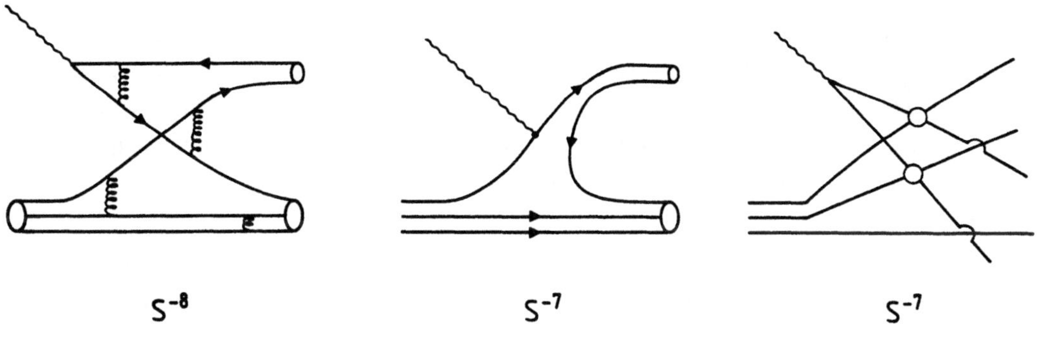

Figure 4

The various diagram, for hard scattering between the photons and the nucleon constituents : VDM hard scattering (left), point like photon coupling (center), VDM multiple scattering Landshoff mechanism (right), when the blob stand for iterated gluon exchange. In each case the characteristic power law is indicated.

On the contrary a single hard scattering between two quarks leads to the emission of fast particles with large transverse momentum p_T. The corresponding interaction length is of the order of:

$$c\Delta T \simeq \hbar c/p_T$$

The factorisation between hard and soft processes implies that:

$$c\Delta T \ll c\Delta\tau$$

In other words, the struck quark must escape quickly the volume of interaction and hadronize far away. To set the order of magnitude, if p_T = 1 GeV/c and R ≃ 0.5 fm, then cΔτ ≃ 0.2 fm and cΔT ≃ 4 fm for pions or 0.7 fm for protons. Above p_T = 1 Gev/c the factorization procedure is expected to be good.

The cross section is therefore made of two parts. The soft mechanisms are highly non pertubative, involve the hadron wave functions and set the norm of the cross section. The hard mechanisms can be treated in a pertubative way (high p_T), and govern the variation of the cross section with the energy. For instance, if the total energy $\sqrt{s}$ is larger than the mass of each hadron involved in the reaction ($\sqrt{s} \gg m$), the only scale in the problem is the energy $\sqrt{s}$ and dimentional scaling arguments lead to:

$$\frac{d\sigma}{dt} \propto \frac{1}{s^2}|M|^2 = \frac{1}{s^8}\, f(t/s)$$

This power law comes also from the high energy limit of the Feynman diagram in Fig. 4, by counting the powers of q^2 of the propagators of the far off-shell quarks. In Fig. 3 both p_T (or t) and $\sqrt{s}$ are large, since $\sqrt{s}$ = 2.1 GeV when E_γ = 2 GeV, $\sqrt{s}$ = 2.9 GeV when E_γ = 4 GeV and $\sqrt{s}$ = 3.9 GeV when E_γ = 7.5 GeV. It is also worthwhile to note that such a pertubative analysis leads also to a fair account of the cross section [10] of hadron pair production in $\gamma\gamma$ interaction in the same kinematical range ($p_T \simeq$ 1 Gev/c and $2 \le \sqrt{s} \le 3$ GeV).

Further evidences of hard mechanisms come from the analysis of the $\gamma p \to p\omega$ and $\gamma p \to \Delta^{++}\pi^-$ reactions [11], of which the cross sections are plotted on Fig. 5 in the energy range $3 \le E_\gamma \le 5$ GeV. In both cases the step decrease of the cross section, when u increases, is due to the exchange of the nucleon, or the Δ, Regge trajectory for the $\gamma p \to p\omega$, or the $\gamma p \to \Delta^{++}\pi^-$, reactions respectively. When u increases, and above u ≃ 0.5 $(\text{GeV/c})^2$, the cross section is flatter, depends weakly on the momentum transfer u, and scales as s^{-8}. The data can be reproduced assuming that two components dominate the cross section:

$$\frac{d\sigma}{du} = \frac{d\sigma}{du}\bigg|_{Regge} \left(\frac{s}{s_0}\right)^{2\alpha(u)-2} + \frac{d\sigma}{du}\bigg|_{Hard} \left(\frac{s}{s_0}\right)^{-8}$$

where the powers are characteristic of the corresponding Regge trajectory $(2\alpha(u)-2)$ or the hard mechanism (s^{-8}). The data do not require any phase between the two amplitudes and the values of the Regge and the hard cross sections are plotted on Fig. 5. It is remarkable that the hard cross section is the same in the two reactions and does not depend on the transfer u.

To close this list, it worthwhile to note that significant departures from the prediction of the Vector Dominance Model ocur, for high values of the momentum transfer t, in the photoproduction of vector mesons on nucleon targets (see ref. [10] for an extensive discussion). The $\gamma p \to p\phi$ channel is particularly appealing, since the ϕ meson is identified by its decay $\phi \to K^+K^-$: besides the P state of the kaon pair, the study of its S state reveals also interesting interference patterns. The study of these channels may prove to be useful to determine the strange sector of the wave function of hadrons.

2.3 High p_T reactions in nuclei.

To study short range quark rearrangement mechanisms in nuclei, one must select kinematics where at least two nucleons are active and where hard scatterings have already been observed in reactions induced on free nucleons.

A possible example would be the comparison between the reactions $\gamma D \to \Delta^{++}\pi^- n$ and $\gamma p \to \Delta^{++}\pi^-$. The value of the transverse momentum of the emitted Δ^{++} must be large in order to sign a hard scattering and the momentum of the neutron must be large enough in order to suppress the quasi-free mechanism (where the neutron is spectator). Of course many other reactions can be studied as for instance the simplest, the $\gamma D \to pn$ reaction [12] when the proton is emitted at 90°, or the $\gamma D \to \Delta^{++}\Delta^-$ and the $\gamma D \to pn\omega$ reactions for high values of p_T.

However, detailed and extensive theoretical studies remain to be done in order to unravel those quark rearrangement mechanisms from more trivial nuclear effects, as Final State Interactions for instance. As at lower energies [2], Gauge Invariance of the electromagnetic current will be a powerfull way to link the various reaction mechanisms and the various short range parts of the nuclear wave function.

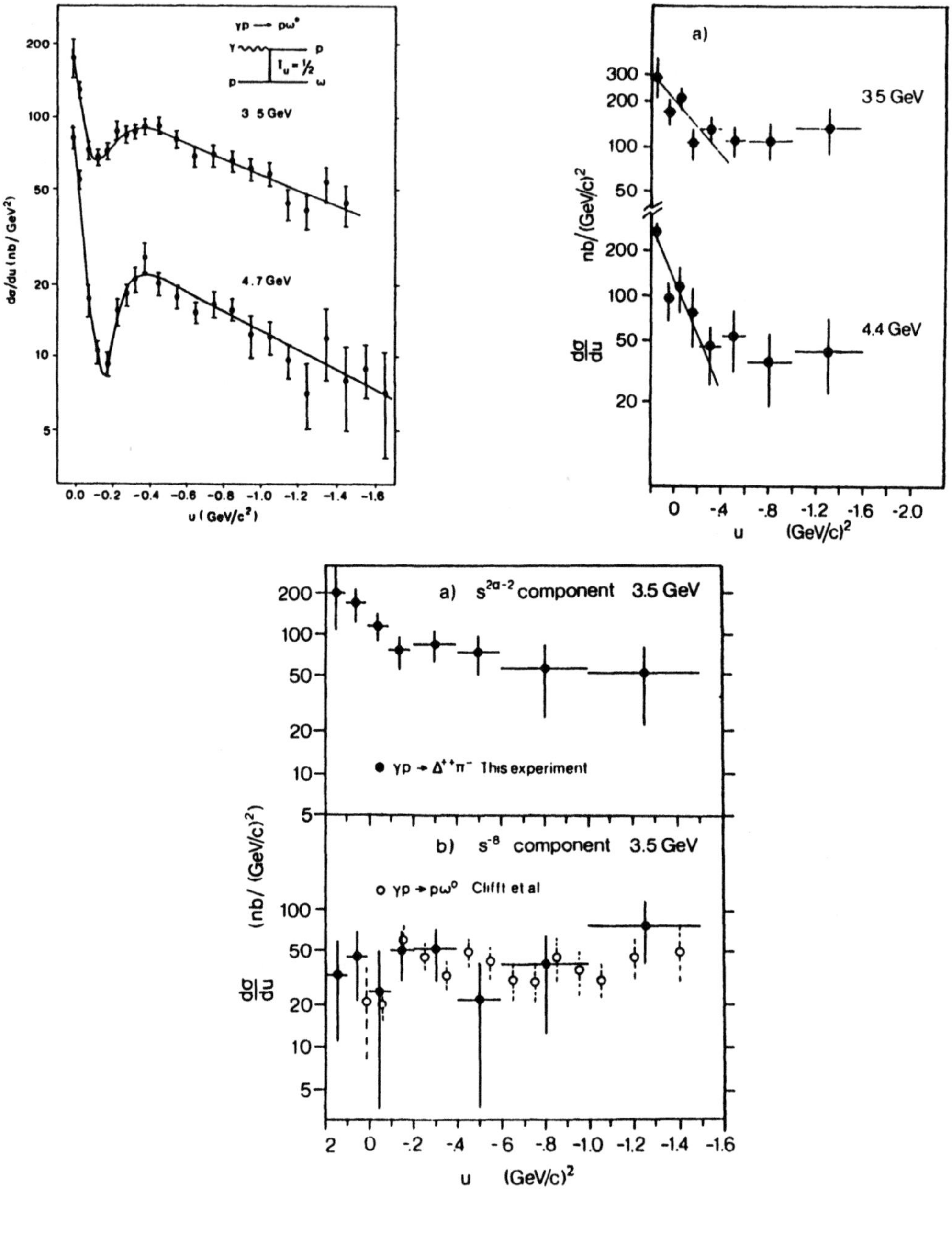

Figure 5

The cross section of the $\gamma p \rightarrow p\omega^0$ (top left) and the $\gamma p \rightarrow \Delta^{++}\pi^-$ (top right) reaction cross section are plotted against the momentum transfer $|u| = |(k_\gamma - k_\omega)^2|$ or $|u| = |(k_\gamma - k_\Delta)^2|$. The two components of the reaction cross section are plotted on the bottom part of the figure. Adapted from ref.[11].

On the experimental side, a CW electron beam is necessary to perform these experiments. The reactions $\gamma p \to \Delta^{++}\pi^{-}$ and $\gamma p \to p\omega$ were studied [11] with a tagged photon beam of 10^6 γ/s (due to the low duty factor, 5%, of the Daresbury machine). A continuous electron beam allows to increase this figure by more than one order of magnitude and to reach a few 10^7 γ/s. The experiments on a free nucleon target can be performed with a higher accuracy and at higher values of p_T. A systematic study of high p_T reactions induced by photons on nuclear targets can be undertaken.

To summarize, an intense tagged photon beam, in the energy range $2 \le E_\gamma \le 8$ GeV, will open up a new domain of research and allow to study the short range properties of nuclear matter, by means of high p_T quark rearrangement mechanisms.

3. THE NUCLEUS AS A FILTER.

3.1 Beyond Pertubative QCD.

Pertubative QCD (PQCD) and dimentional counting rules have led to a good understanding of an impressive number of reactions induced between hadrons at high energy and high momentum transfer [13]: hadron form factors, pair production in $\gamma\gamma$ scattering, compton scattering, or hadron scattering. However, deviations systematically occur.

For instance, at 90° the nucleon-nucleon scattering cross section $d\sigma/dt$ behaves as s^{-10}, as predicted by PQCD, but exhibits oscillations around this mean value (Fig.6). Ralston and Pire [14] call this pattern "color oscillations" They explain it as an interference between the hard one gluon exchange pertubative amplitude and the resummation of the triple gluon exchange higher order graph (Landshoff mechanism). While the former occurs at short distances, the latter involves the full size of the nucleon.

3.2 Spin observables.

Since PQCD is chirality conserving, the leading-twist contributions to scattering amplitudes are helicity conserving [15]: one cannot flip the helicity of a single quark, although one can reverse the helicity of a gluon. It predicts vanishing polarizations or asymmetries, and definite values for spin transfer observables (in which the helicity of at least two quarks are reversed). Experiments completely rule out these simple

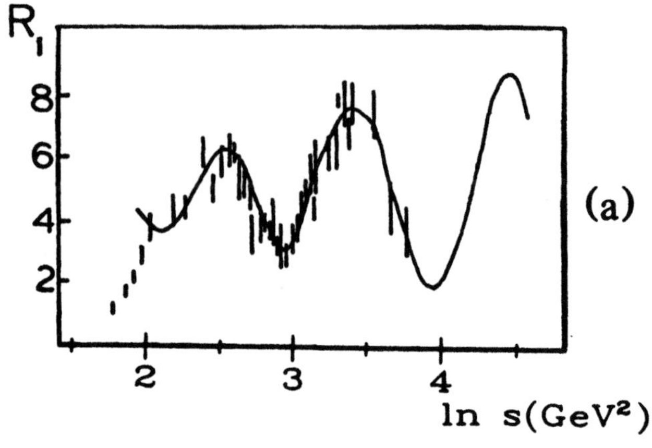

Figure 6

The values of the product $S^{10}\frac{d\sigma}{dt}$, for the proton proton elastic scattering at 90°, are plotted against the total c.m. square energy s. Adapted from ref.[14].

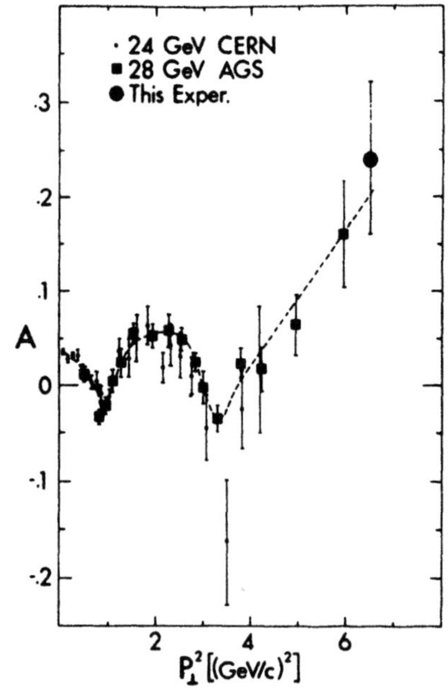

Figure 7

The values of the analysing power of the reaction $\vec{p}p \rightarrow pp$ are plotted against the transverse momentum p_T of the scattered proton. Adapted from ref.[17].

predictions [16], and show evidences for interferences between the leading-twist PQCD amplitude and higher-twist or non-pertubative QCD amplitudes.

Two examples of helicity nonconserving observables are the analysing power in pp scattering and the density-matrix element $\rho_{1,-1}$ in the $\pi^- p \to \rho^- p$ reaction. The analysing power [17], measured in pp scattering at $\sqrt{s}$ = 7.35 GeV, exhibits an oscillatory pattern when p_T varies from 0 to 2 GeV/c, and rises quickly up to A = 0.24 ± 0.08 when p_T= 2.54 GeV/c (Fig. 7), in contrast with the PQCD prediction. The non vanishing value [18] $\rho_{1,-1}$ = 0.32 ± 0.10 measured at 90° for incident pion momentum 9.9 GeV/c ($\sqrt{s}$ = 4.4 GeV, p_T $\propto$ 3 GeV/c) contradicts also PQCD. The analysis of these data led G. Farrar[19] to conclude that the amplitude of higher twist helicity nonconserving mechanisms should not exceed about 1/10 to 1/3 of the leading-twist amplitude in that energy and momentum range. They are revealed in spin observables, but do not drive the mean behavior of the unpolarized cross sections.

The high degree of polarization of the Λ emitted in the $pp \to K^+ p\Lambda$ reaction (Fig.8) is also a puzzle in the framework of PQCD. Fujita and Susuki [20] explain it as the result of the interference between the one gluon exchange hard process and a non pertubative multigluon exchange mechanism, which they parametrize by an effective pole in the t channel. This picture leads to a good accounting of the experimental data, polarization as well as unpolarized cross section, in a wide range of energies when p_T is about 1 GeV/c.

The variations [21] with the incoming proton momentum of the spin transfer coefficient A_{NN}, in proton-proton elastic scattering at 90°, are plotted on Fig.9. While PQCD predicts a constant value around A_{NN} = 0.3, the experimental values exhibit large variations and reach a maximum, A_{NN}= 0.6, when p_{Lab} = 11.75 GeV/c ($\sqrt{s}$ = 4.9 GeV, p_T= 2.2 GeV). To overcome this difficulty Brodsky and de Teramond [21] have added two resonant contributions to the PQCD amplitude. They are also able to reproduce the oscillatory pattern of the unpolarized cross-section (Fig.6). It turn out that the first resonance lies close to the pΔ threshold (p_{Lab} = 1.26 Gev/c, $\sqrt{s}$ = 2.17 Gev) and the second lies close the charm threshold (p_{Lab}=12 GeV/c, $\sqrt{s}$ =4.93 GeV). Whether they are real resonances or whether they simulate a threshold effect is still an open question.

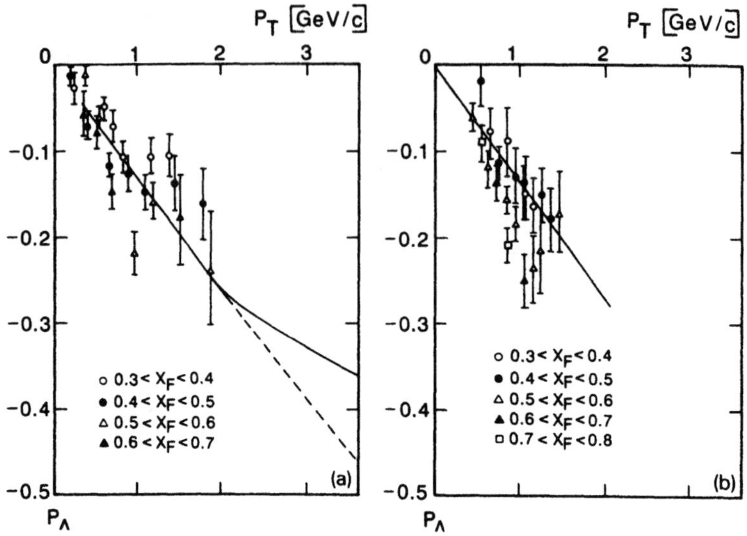

Figure 8

The values of the polarization of the Λ, emitted in the reaction $pp \rightarrow K^+ \Lambda p$, are plotted against its transverse momentum p_T, for two extreme values of the incoming energy : T_p = 300 GeV (left) and T_p = 12 GeV (right). The curve is the prediction of Fujita et al.[20]. Adapted from ref.[20].

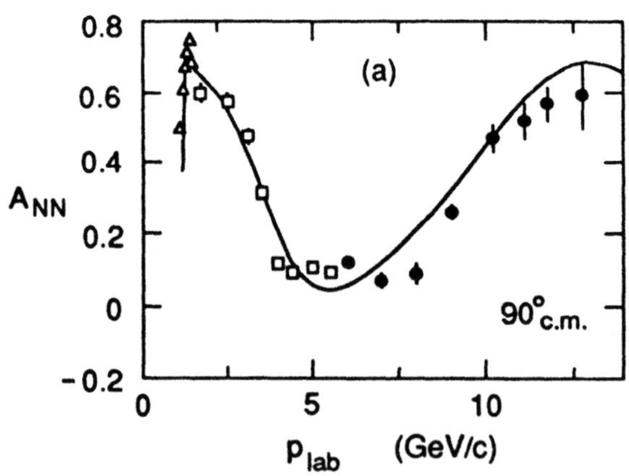

Figure 9

The values[21] of the spin transfer coefficient A_{NN}, in the $\vec{p}\vec{p} \rightarrow pp$ elastic scattering at θ_p = 90°, are plotted against the incoming proton momentum. The curve is the result of the model of Brodsky and de Teramond[22]. Adapted from ref.[22].

Although these non-pertubative mechanisms have been added in an ad-hoc way to fit the spin observables, they do not dominate the unpolarized cross-sections, which still obey a scaling behavior. It is very likely that the large values of the measured spin observables is the hint of possible interferences between a hard scattering dominant mechanism, which occurs at short distances in the interacting hadrons, and a non-pertubative mechanism or a higher-twist contribution, which involve the full size of the hadrons.

How can the nuclear medium help us in disentangling these two components of the reaction amplitudes?

3.3 The nucleus as a filter.

In exclusive processes (as elastic scattering for instance), high p_T mechanisms select the part of the hadronic wave function which has a small extension. On the one hand, high p_T mechanisms implies a small impact parameter between the quarks. On the other hand, the outgoing hadrons must emerge intact: during the interaction the momentum transfer is shared between all the quarks and the size of the hadrons must be comparable to the volume of interaction. They eventually evolve to their normal size when they reach their asymptotic states. If such an interaction occurs in a nucleus, as for instance in quasi elastic nucleon-nucleon scattering, such a small sized configuration of the nucleon is expected to travel almost freely in the nucleus. On the contrary, its normal sized configuration should be attenuated by multiple scatterings on the other nucleons. The nucleus damps the non pertubative mechanisms, which involve the full size of the nucleons, but filters the pertubative mechanisms, which occur at short distances. This picture is at the basis of color transparency [14,23,24], and leads to the paradox that PQCD would be more valid in a nucleus than in reactions involving free nucleons!

It has been confirmed by a recent experiment [25] where the cross-section of quasi-free proton-proton scattering at 90° has been measured in various nuclei (Fig.10). Up to p_{lab} = 10 GeV/c, the transparency (ratio of the pA quasi-free to the pp elastic scattering cross-section scaled by the number of protons) increases from a small value compatible with a normal nuclear attenuation (as computed in a Glauber formalism) to a value five times higher. However, it drops down to a small value compatible again with a normal attenuation at p_{lab} = 12 GeV/c. It turns out that this variation of the transparency occurs close to the charm threshold, where also the spin correlation coefficient A_{NN} is large (Fig.9). A more striking feature [27] is

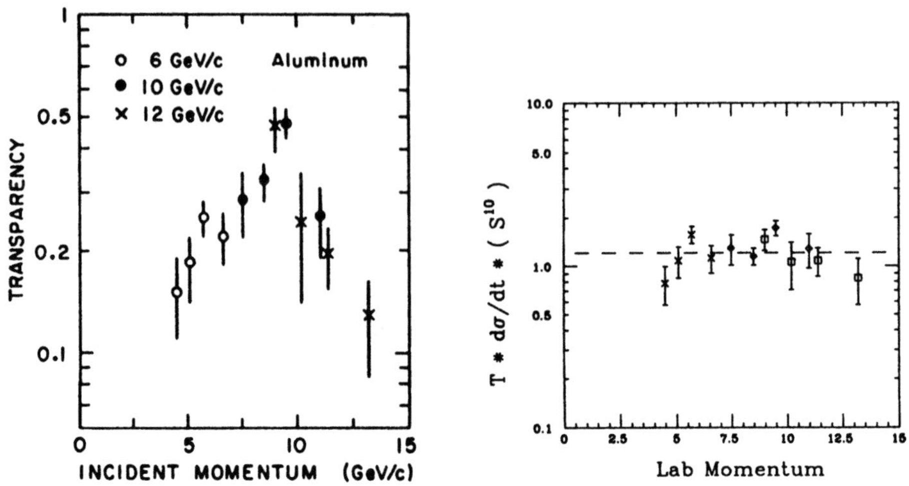

Figure 10

On the left part of the figure, the values[25] of the transparency coefficient (ratio $T = \sigma(pA) / Z\sigma(pp)$), are plotted against the value of the incoming proton laboratory momentum (adapted from ref.[26]). On the right part of the figure, the values of the product of the transparency T and the nucleon-nucleon cross section, scaled by S^{10} as in fig. 6, are plotted against the incoming proton momentum (adapted from ref.[27]).

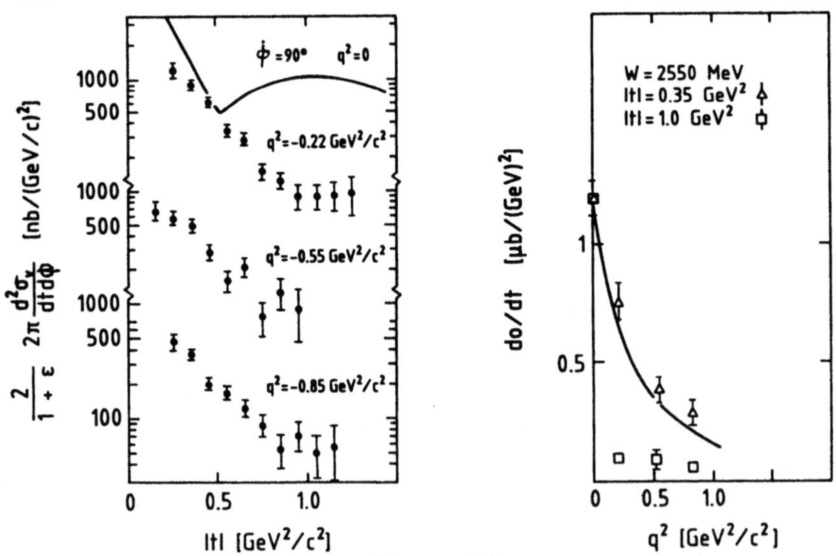

Figure 11

On the left part of the figure, the value[29] of the cross section of the reaction $p(e,e'\pi^0)p$ is plotted against the momentum transfer t, for different values of the squared mass q^2 of the exchanged virtual photon. On the right part, the variation of the cross section is plotted against q^2, for two values of the transfer t : a low value $t = 0.35\ (\mathrm{GeV}/c)^2$ and a large value $t = 1\ (\mathrm{GeV}/c)^2$. Adapted from ref.[29].

that the product of the elementary pp cross section by the transparency (Fig.10) behaves as s^{-10}. The oscillation of the nuclear transparency compensates the oscillation of the free pp cross section. In other words the oscillations of the elementary pp cross section (Fig.6), have disapeared: the nucleus filters away much of the stuctures beyond PQCD, which involve the full size of the nucleon.

Clearly further tests of this idea are needed. Quasi-free electron scattering could be a complementary way to study color transparency, since the electron nucleon elastic scattering amplitude is free of interferences with resonant or Landshoff effects. The rise of the color transparency T toward unity should be free of oscillations. Spin observables, in hadron as well as in electron scattering, could also be very useful, since they allow to study in more details how the nuclear medium could affect the interferences between pertubative and non pertubative mechanisms. In this respect, self analysing reactions, as $(e,e'\rho)$ or $(e,e'\Lambda)$ induced by polarized electrons, are paticularly promising.

What is the relevant energy and momentum range?. In hadron induced reactions, these interferences between soft and hard processes occur in the range of momentum transfer $1 \le p_T \le 3$ GeV/c. In the $(e,e'X)$ quasi-free reaction this corresponds to a range of values of the virtual photon squared mass $1 \le Q^2 \le 9$ $(GeV/c)^2$ and energy $0.5 \le \omega = Q^2/2m \le 4.5$ GeV. This range of energy and momentum transfer is achievable with a 8 to 10 GeV machine.

4. FROM REAL TO VIRTUAL PHOTONS.

Most of the evidences of hard processes, which I reviewed, come from hadron or real photon induced reactions. Except for the case of electroproduction of vector mesons [28], the experiments performed with virtual photons are very scarce. However, they open two new axes of research. On the one hand, they allow to go far from the domain of validity of the Vector Dominance Model (VDM) and to reach the domain where the photon behaves as a point like object. On the other hand, they allow to determine the coulomb part of the cross section, which vanishes at the photon point.

Due to quantum fluctuations, a photon can convert, for a short time, into a vector meson (ρ, ω, ϕ) which subsequently scatters on a nucleon. The characteristic propagation length of this meson is:

$$c\Delta t = \frac{2\hbar c\omega}{Q^2 + m_v^2}$$

where m_v is its mass, and ω and Q^2 are respectively the energy and the squared mass of the virtual photon. To set the order of magnitude, for a real photon of energy E_γ = 4 GeV (ω = 4 GeV and Q^2 = 0) $c\Delta t$ = 3.2 fm, while for a virtual photon of squared mass Q^2 = 1 $(\text{GeV}/c)^2$ and ω = 4.5 GeV (same $\sqrt{s}$ as a real photon of E_γ = 4 GeV) the vector meson lifetime drops down to $c\Delta t$ = 1.2 fm. At high energy, a virtual photon is more point like than a real photon: above Q^2 = 1 $(\text{GeV}/c)^2$, the coupling of a virtual photon with a nucleon is expected to differ from the coupling of a real photon.

Such a point like coupling of the virtual photon seems to occur for smaller values of its squared mass, provided that the momentum transfer (t or p_T) is high enough. This is apparent in Fig. 11 where the variation, with the momentum transfer $t = (k_\gamma - k_\pi)^2$ for selected values of Q^2, of the cross section [29] of the $p(e,e'\pi^0)p$ reaction is compared to the cross section of the $p(\gamma,\pi^0)$ reaction, at the same value of the total c.m. energy $\sqrt{s}$ = 2.55 GeV. Below t = 0.7 $(\text{GeV}/c)^2$, the cross section decreases exponentially when t increases and decreases as $(Q^2 + m_\rho^2)^{-2}$ when Q^2 increases, as expected from VDM. Above t = 0.7 $(\text{GeV}/c)^2$, the virtual photon cross section is flat and is almost constant when either t or Q^2 vary. This is the hint of the onset of hard mechanisms of a point like object. Clearly the restricted domain of momentum (t and Q^2) and energy $\sqrt{s}$, covered by this experiment, must be extended to confirm this picture and get a better understanding of pion electroproduction on a nucleon. Nevertheless, this is a good example of an exclusive reaction induced by electrons: besides the energy ω and the squared mass Q^2 of the exchanged virtual photon (which are the only kinematical variables accessible in inclusive electron scattering), the transverse momentum p_T (or t) of the detected hadron is an important variable. Its variations open up the possibility to reveal and map out different mechanisms.

Virtual photons allow also to determine the coulomb part of the cross section, where the physics is different than in the transverse part (which looks like the real photon absorption cross section for small values of t). For instance, PQCD predicts a vanishing coulomb cross section: indeed the measured ratio $R = \sigma_L/\sigma_T$, between the longitudinal and transverse part of the cross section, vanishes [30] for large values of Q^2. However it is significantly different from zero when Q^2 decreases, in the range $Q^2 \leq 5$ $(\text{GeV}/c)^2$. This is a signature of higher order corrections to PQCD, as higher twists for instance.

Traditionally the separation between the longitudinal and the transverse parts of the cross section is performed with the Rosenbluth method. It requires a high energy of the incoming electron for the measurement at the most forward angle. Furthermore, to pin down the small contribution of the coulomb amplitude to the cross section, a perfect control of the accuracy, not only stastistical but also systematical, is mandatory. An alternative way [6], to overcome these difficulty, would be to measure the various spin transfer observables which depend on the interference between the coulomb and the transverse amplitudes. The two advantages of this method are: i) such a measurement does not require to be performed at forward angle of the scattered electron: the energy of the incoming electron beam can be lower than in the Rosenbluth method; ii) only relative measurements (in which only the spin of the electron or of the target is flipped) are necessary: the sources of systematical errors are minimized.

For achieving this goal, the energy of the incoming electron would not be much more higher than the energy of a real photon beam (see section 2.3). A continuous beam of polarized electrons of energy $E_- = 8$ to 10 GeV will allow to study exclusive electron induced reactions in the range of c.m. energy $2 \leq \sqrt{s} \leq 3$ GeV (for the γp system), of squared mass of the virtual photon $0 \leq Q^2 \leq 4$ $(\mathrm{GeV/c})^2$ and of transfer momentum $0 \leq t \leq 10$ $(\mathrm{GeV/c})^2$ (or equivalently $0 \leq p_T \leq 3$ GeV/c). For instance, if $\sqrt{s} = 3$ Gev, $Q^2 = 4$ $(\mathrm{GeV/c})^2$ and $E_- = 8$ GeV, then $\omega = 6$ GeV and the electron scattering angle is $\theta = 29°$: the energy of the scattered electron, $E' = E_- - \omega = 2$ GeV, is large enough to prevent large radiative corrections.

Higher electron energies could be justified by the study of two topics which lie outside the scope of this talk: the study of charm production in nuclei, and the study of quark hadronisation mechanisms in nuclei. In both cases, it must be demonstrated how an electron beam will complement the various available hadron beams, and to what extent the small electromagnetic cross section will allow to undertake a systematic research programme. For instance the charm threshold is 8.2 GeV, for the γp system. At $E_\gamma = 13$ GeV the charm photoproduction cross section is only of the order of 10 nb, while at $E_\gamma = 300$ GeV it is close to 600 nb. It could be likely that the study of charm production in nuclei would better be carried out with very high energy electron beams or hadron beams (protons or antiprotons).

5. CONCLUSION.

Three ranges of momentum and energy are relevant to study the behavior of hadronic matter at intermediate and short distances.

In the range of c.m.energy $\sqrt{s} \leq 2$ GeV (in the γp system) and momentum transfer $p_T \leq 1$ GeV/c, nuclear reactions induced by electrons and hadrons allow the study of the properties of the nucleons, the Δ, the lightest hyperons (Λ and Σ), and their resonances in nuclei as well as in free space. They also allow the study of the nucleon and the resonance components of the nuclear wave function. Few Gev polarized proton beams (as the 3 GeV intense beam of Saturne [31] for instance) and polarized electron C.W. beams (as the 4 GeV intense beam of CEBAF [32] for instance) cover this kinematical region.

In the range of momentum transfer $1 \leq p_T \leq 3$ GeV/c and total c.m. energy $2 \leq \sqrt{s} \leq 4$ GeV, leading twist pertubative mechanisms dominate the cross section, although spin observables reveal large interferences with smaller contributions due to higher twist or non pertubative mechanisms. A consistent body of experiments shows that high p_T reactions, induced either by real photons or hadron beams, are driven by hard scatterings, when *both* the transfered momentum and energy are large enough to be far from the domain of the Vector Dominance Model or Regge pole exchange.

A more systematic study, with a special emphasis on spin observables, of these hard mechanisms in the free nucleon will lead to a better understanding of the interplay between pertubative and non pertubative processes, and lead to strong constraints on the wave functions of hadrons. High p_T reactions provide us with the way of disentangling the hard interaction of the probe with the quarks of the target, which can be treated in a pertubative way, and the soft interactions between the quarks inside the probe or the target, which insure their cohesion and determine their wave functions.

Their study in the nuclear medium enlarges the dynamical domain in two related ways. On the one hand, quark rearrangement mechanisms allow to determine the short range part of the nuclear wave function, and to single out its hidden color component. On the other hand, the nucleus acts as a filter: if the momentum transfer is large enough, the nuclear medium damps the non pertubative mechanisms, which involve the full size of the hadrons, and selects the hard mechanisms, which occur at short distances inside the interacting hadrons.

This is the domain of hadron and electron beams of intermediate energy. On fixed targets, an intense 15 to 20 GeV polarized proton beam, or an intense 8 to 10 GeV polarized electron beam, will cover the relevant range of momentum and energy transfer. Colliding beams could also be an alternative solution: for instance, 3+3 GeV/nucleon beams of nucleons or light ions, or an electron beam colliding against a light ion beam. Several possibilities are open and must be investigated.

At higher energies ($\sqrt{s}$ > 5 GeV) and momentum transfer (p_T > 4 GeV/c), one enter far into the pertubative QCD regime, where the relevant topics are the study of the logarithmic evolution of the nuclear structure functions and the study of the structure of multi-jet events.

While Europe is well equipped in the highest energy range (HERA or CERN for instance), only Saturne covers, with hadronic probes, the lower part of the energy domain. The two major projects, which will allow to study hadronic matter at low and intermediate energies, will be located in North-America. The 4 GeV electron machine CEBAF [32] is already under construction in USA, and the funding of the 30 GeV proton machine KAON [33] is actively discussed in view of a construction on the TRIUMF site in Canada.

Therefore there is room for a major project in Europe. This could be a 8 to 10 GeV electron machine, with an intense polarized continuous beam. In a step by step construction scheme, it could usefully compete with CEBAF and supercede the PEGASYS [34] project (higher duty factor and external beams).

On the theoretical side, most of the evidences of hard processes which I reviewed rely on the dimentional analysis of pertubative QCD, and a phenomenological treatment of the non pertubative mechanisms. Although such an approach has been very useful to set the order of magnitude and define the relevant range of momentum and energies, a full quantitative theoretical treatment is still badly needed. This is a formidable challenge, which will become more exciting when new experimental data will be gathered.

REFERENCES.

1. S. Frullani and J. Mougey, Advances in Nuclear Physics 14 (1984) 1.
2. J.M. Laget, Nucl. Phys. A497 (1989) 391c;
   J.M. Laget, New Vistas in Electronuclear Physics. E. Tomusiak, H. Caplan and E. Dressler, Eds., Plenum Press, NY (1986) p. 361.

3. M. Lacombe et al., Phys.Lett. B101 (1981) 139.
4. R.B. Wiringa, R.A. Smith and T.L. Ainsworth, Phys.Rev. C39 (1984) 1207.
5. Y .Yamauchi and M. Wakamatsu, Nucl. Phys. A457 (1986) 621.
6. J.M. Laget, Physics with Polarized Beams on Polarized Targets. J. Sowinsky and S.Vigdor, Eds., World Scientific (to be published).
7. J. Arvieux et al., Rapport du Groupe de Reflexion "Physique et Aires Experimentales", October 1989, unpublished.
8. R.L. Anderson et al., Phys. Rev. Lett. 30 (1973) 627.
9. P.D.B. Collins and A.D. Martin, Hadron Interactions, Adam Hilger Ltd, Bristol (1984). See chapter 3.
10. Ch. Berger et al., Phys. Lett. B137 (1984);
S.J. Brodsky and G.P. Lepage, Phys. Rev. D24 (1981) 1808.
11. J.B. Dainton, Lecture Notes in Physics 234 (1985) 80;
R.W. Clifft et al., Phys. Lett. 72B (1977) 144;
D.P. Barber et al., Phys. Lett. 98B (1981) 135;
H.J. Behrends et al., Nucl. Phys. B144 (1978) 22.
D. Fries et al, Nucl. Phys. B143 (1978) 408.
12. J. Napolitano et al., Phys. Rev. Lett. 61 (1988) 2530.
13. S.J. Brodsky, Short Distance Phenomena in Nuclear Physics, D.H. Boal and R.M. Woloshyn, Eds., Plenum Press (1983) 141.
14. J.P. Ralston and B.Pire, Phys. Rev. Lett. 61 (1988) 1823.
15. S.J. Brodsky and G.P. Lepage Phys. Rev. D24 (1981) 2848.
16. E. Leader, Intersection Between Particle and Nuclear Physics, D.F. Geesaman, Ed., AIP Conference Proceedings 150, NY (1986), p. 257.
17. P.R. Cameron et al., Phys.Rev.D32 (1985) 3070.
18. S. Heppelman et al.,Phys.Rev. Lett. 55 (1985) 1824.
19. G. Farrar, Phys. Rev. Lett. 56 (1986) 1643.
20. T. Fujita and N. Suzuki, Nucl. Phys. A503 (1989) 899.
21. G.R. Court et al., Phys. Rev. Lett. 57 (1986) 507, and references therein.
22. S.J. Brodsky and G.F. de Teramond, Phys.Rev.Lett. 60 (1988) 1924.
23. G. Bertsch et al., Phys. Rev. Lett. 47 (1981) 297;
A.H. Mueller, XVII Rencontre de Moriond (1982).
24. G.R. Farrar et al., Phys. Rev. Lett. 61 (1988) 686; 62 (1989) 1095.
25. A.S. Carroll et al., Phys. Rev. Lett. 61 (1988) 1698.
26. S.J. Brodsky, Electronuclear Physics with Internal Targets, R.G. Arnold Ed., World Scientific (1989), p. 3.
27. S. Heppelmann, Nucl. Phys. B (Proc. Suppl.) 12 (1990) 159.
28. J.J. Aubert et al., Phys. Lett. B161 (1985) 203; B133 (1983) 370;
D.G. Cassel et al., Phys. Rev. D24 (1981) 2787.

29. F.W. Brasse et al.,Phys. Lett. B58 (1975) 467.
H. Blume et al., Z. Phys. C16 (1983) 283.
30. A. Bodek et al., Nucl. Phys. B (Proc. Suppl.) 7 (1989) 141.
31. J.M. Laget, 5emes Journees d'Etude Saturne, M. Bordry and P. Radvanyi, Eds.. Postal address: LNS, CEN-Saclay, F91191 Gif-Sur-Yvette, France.
32. H.A. Grunder, Perspectives in Nuclear Physics at Intermediate Energies, S. Boffi et al., Eds., World Scientific (1989), p. 200.
33. H.W. Fearing, Perspectives in Nuclear Physics at Intermediate Energies, S. Boffi et al., Eds., World Scientific (1989), p. 223.
34. K. van Bibber, Electronuclear Physics with Internal Targets, R.G. Arnold Ed., World Scientific (1989), p. 189.

POINT-LIKE INTERACTIONS of PHOTONS IN HIGH P_T PHOTOPRODUCTION

Guy WORMSER
Laboratoire de l'Accélérateur Linéaire, IN2P3-CNRS
et Université de Paris-Sud, F-91405 Orsay Cedex, France

ABSTRACT

High p_T photoproduction offers a unique opportunity to observe the interplay between VDM-like and point-like interactions of the photon. The relative strength of these two phenomena strongly depends of the selected final state: high p_T photon, high p_T hadron or J/ψ . The results obtained by the NA14 experiment at the CERN-SPS concerning the inclusive cross-section for such reactions as well as detailed properties of the associated final state are reported here.

INTRODUCTION

It is well known that low p_T photoproduction processes can be well described by the VDM formalism, that is the assumption that the incoming photon is equivalent to the superposition of ρ, ω and ϕ vector mesons. However, at higher p_T , the point-like interaction, that is the direct coupling between the photons and the constituent quarks, will become dominant. It is therefore very interesting to study the transition between these two regimes.

This experimental program was carried out by the NA14 collaboration, at the CERN SPS, between 1980 and 1984. The photon beam was an intense (10^7 γ per burst) tagged wide band photon beam, with a mean energy of 90 GeV. The large acceptance spectrometer[1] allowed the study of a variety of final states in great detail. The experimental setup is displayed in Fig. 1, where can be seen the three electromagnetic calorimeters used for γ and π° reconstruction and the two magnets and the set of MWPCs used for charged tracks analysis.

1. High p_T prompt photon photoproduction

This reaction, the elastic photon-quark scattering, was suggested 20 years ago by Bjorken and Paschos to measure the quark charges[2]. It is an excellent way to exhibit point-like interactions of the photon. since important features are indeed met :

- The VDM contribution in the p_T range between 2 and 3 Gev/c is quite small, of the order of 1%, as inferred from the prompt photon hadroproduction results in that kinematical domain[3].

- The Born term, the γ -quark elastic scattering or QED Compton graph, is readily calculable and it only involves the nucleon structure function. The leading logarithm QCD corrections to this term have been calculated and are found to be large at low p_T and decreasing with p_T .[4]

- The next-to-leading order terms have also been calculated and turned out to be small, giving a good confidence in the convergence of the pertubative computation.

On the experimental side, the extraction of a prompt photon signal is a difficult task because of the large background of non-prompt photons coming from π° and η decays. In NA14, this contamination was measured in two ways: firstly by a MonteCarlo program based upon a careful simulation of the electromagnetic calorimeters, and secondly by an experimental measure fo the yield of prompt photon candidates obtained in a π^- beam where as mentioned above, no prompt photons are expected.

The results obtained[1] are summarized in Fig. 2, where the measured prompt photon cross-section, expressed in units of the QED Compton cross-section, is plotted against p_T . In the absence of the QCD corrections, the measured value should be 1, regardless of p_T . Were the quark charge integers, the result should be at least 2.65. The solid line is the QCD prediction computed beyond the leading order terms. The data exclude the interger charge model and are in good agreement with the QCD prediction.

We have also been able to perform a topological isolation of the QED Compton term, thanks to its specific γ -jet topology[5]. Fig. 3 represents the distance between the high p_T photon and all the other tracks in the event. The distance is defined as $d^2=(\Delta\phi)^2+(\Delta\eta)^2$, where ϕ and η are the azimuth angle and the pseudorapidity. The dotted line represents the distance distribution for events containing a reconstructed high p_T π° , which is, to a very good approximation, the distribution for non prompt photons. A clear difference is observed, showing the presence of isolated prompt photons, with all other tracks at large distance from it, that is the expected topology from QED Compton scattering. The solid line represents the fit to the data of the sum of a QED Compton term , simulated by Monte-Carlo, and the non prompt photon term. The fraction of events produced by the QED Compton term was thus measured to be 0.34±0.06 for p_T above 2.5 GeV/c and 0.05±0.02 below, in good agreement with inclusive measurements.

The predicted charge asymmetry in the final state of QED Compton events, due to the electromagnetic coupling favoring u quarks over d quarks by a factor 16, is also observed. Fig. 4 shows the charge asymmetry as a function of the distance defined above. At large distance, i.e inside the hadronic jet recoiling against the photon, the measured positive charge asymmetry is in good agreement with the QED Compton fraction mentioned above, since u quarks lead to an asymmetry of 33% after hadronisation.

In summary, the measurement of the prompt photon signal shows unambiguously the point-like interaction of the photon beam. All the observed properties of the events (rate,topology, charge asymmetry) have been found compatible with the expectations from the QED Compton scattering, once pertubative QCD corrections have been applied.

2. High p_T π° photoproduction[6]

In this reaction, the VDM contribution to the total cross-section is large and should be dominant at low p_T . It has been estimated using the experimental measurements of π° *hadroproduction* with the assumption that ρ-production is well described by the mean of π^+ and π^- production. The systematic uncertainty on the VDM component is estimated to be of 10%. The point-like interaction proceeds at the Born term level, through QCD Compton scattering (γ q$\rightarrow$ qg) and γ -gluon fusion (γ g$\rightarrow$ $q\bar{q}$). These terms are directly proportional to α_s and therefore this reaction constitutes a good laboratory to test pertubative QCD. Higher order terms are predicted beyond leading logarithm approximation[7].

Fig. 5 displays the π° cross section as a function of p_T . The dotted curve represents the VDM prediction which fails to explain the data by a factor 2. The solid line is the sum of the VDM and point-like contributions and reproduces well the data. This good agreement is achieved when :

- the VDM and QCD contributions are simply added
- the next-to-leading log terms are taken into account
- the gluon fragmentation is as hard as a quark's one

The QCD cross-section is of the same order of magnitude as the VDM term on the whole kinematical domain except in the low p_T , low p_l region (Fig. 6). Similar results have been obtained with charged pions in the final state[8].

With such an high signal to background ratio,topological differences between γ and π^- beam can be looked for. Striking difference are observed on Fig. 8, where the distance is defined as in the previous paragraph. Clearly,γ data look more jet-like and this is a good signature of QCD terms. This topological difference explains the absence of double counting between the VDM and QCD terms observed in the inclusive measurements.

The observed charge asymmetry (Fig. 8) can only be produced by the QCD Compton term, when the high p_T triggering particle comes from the gluon jet,hence confirming the hard fragmentation of the gluon in this Q^2 range.

3. J/ψ photoproduction[9]

This reaction is interesting because it can be interpreted using two different approaches : a VDM approach, where the photon couples to the J/ψ before interacting with the nucleus and the J/ψ is then scattered through the nuclear matter[10], or a QCD approach where the J/ψ is produced by the recombination of a $c\bar{c}$ pair produced by γ -gluon fusion. The p_T dependance of the cross section is shown in Fig. 9, where the three components : coherent elastic, incoherent elastic and inelastic can be easily distinguished.

For incoherent elastic J/ψ scattering, both approaches are in fact equivalent and predict the same cross-section, in good agreement with the experimental measurement. The A dependance of the cross section will bring information upon the real mechanism at work[11].

For the inelastic scattering, the pertubative QCD computaion predicts the correct shape for the z and p_T distribution but is off in rate by a large factor. It is interesting to note that a pure gluon jet sample can be obtained in inelastic J/ψ photoproduction and could therfore be used for detailed quark-gluon fragmentation comparison.

4. Conclusion

High p_T photoproduction offers a unique opportunity to observe the interplay between the point-like and VDM-like interactions of the photon. The relative strength of these two phenomena strongly depends of the selected final state.

In high p_T γ photoproduction, the VDM contribution is negligible and the elastic photon-quark scattering (QED Compton scattering) has been observed. Its cross-section is in good agreement with prediction once the pertubative QCD corrections have been applied. In high p_T π° production, VDM and point-like terms have the same order of magnitude but the final state topology is very different in the two cases. Finally, in elastic incoherent J/ψ photoproduction, both VDM and point-like approaches can be used to interpret the data in a satisfactory manner.

References

1. P. Astbury et al., Phys. Lett. 152B, 5,6(1985)419.
2. Bjorken and Paschos, Phys. Rev. 185 (1969) 1975.
3. High momentum transfer reactions, Session 8, in the Proceedings of the XXIV International Conference on High Energy Physics, pp. 713, Munich,1988, Springer-Verlag (Kotthaus and Kuhn editors)
4. Aurenche et al., Z. Phys. C24 (1984) 309.
5. Augé et al., Phys. Lett. 182B (1986)409.
6. Aurenche et al., Phys. Lett. 135B (1984) 164.
7. Augé et al., Phys. Lett. 168B (1986) 163.
8. Augé et al., Phys. Lett. 174B (1986) 458.
9. Barate et al., Z. Phys. C33 (1987) 505.
10. Sakurai and Schildknecht, Phys. Lett. B40 (1972),121.
    Margolis, Phys. Rev. D17 (1978) 1310.
11. Sokoloff et al., Phys. Rev. Lett. 57(1986) 3003.

Figure captions

Fig. 1. The NA14 set-up

Fig. 2. Ratio of the experimental cross section by the QED Compton one. The solid curve is the the QCD prediction.

Fig. 3. Track-high p_T γ distance distribution. The histogram represents the data for p_T >2.5 GeV/c. The dotted curve is the distribution obtained for high p_T π° . The solid line is the fit to the data using the π° data and the QED Compton Monte Carlo.

Fig. 4. Charge asymmetry in the final state associated with a prompt photon candidate as a function of the distance between the track and the photon.

Fig. 5. p_T distribution of reconstructed π° . The solid curve is the sum of the QCD prediction and the VDM contribution (dotted curve) ;

Fig. 6. Cross section for π° photoproduction versus longitudinal momentum for three p_T bins. The dotted curves are the VDM prediction, the solid curves are the sum of VDM and QCD predictions.

Fig. 7. Track-high p_T π° distance for p_T >1.7 GeV/c, produced by γ beam (solid line) and π^- beam (dotted line).

Fig. 8. Charge asymmetry with a high p_T π° in the final state measured using identified kaons and protons, as a function of the track-π° distance.

Fig. 9. p_T dependance of the J/ψ photoproduction cross section.

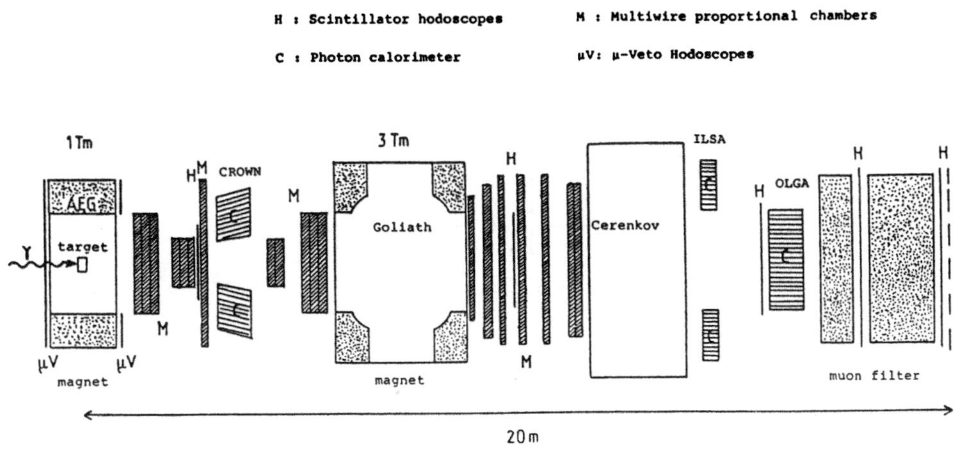

Figure 1

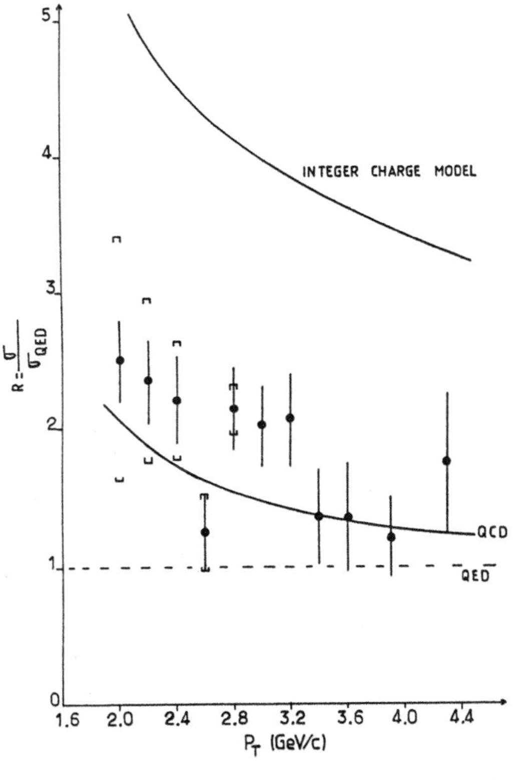

Figure 2

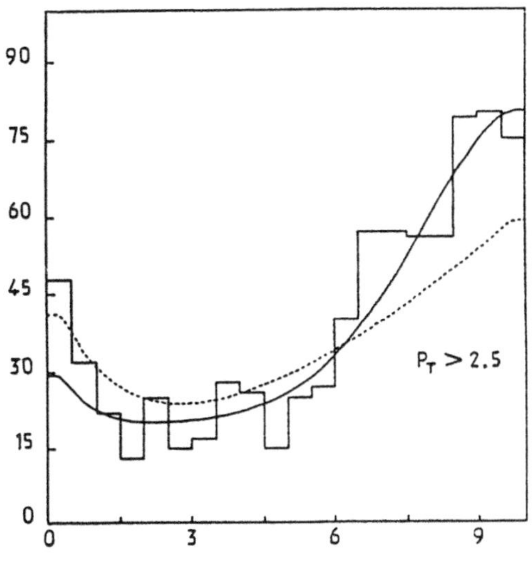

Figure 3

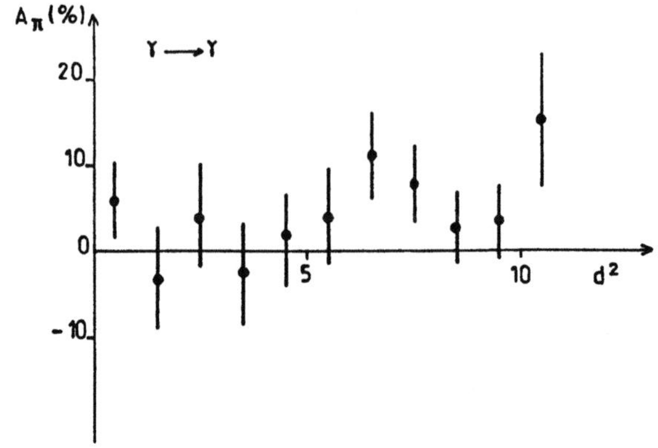

Figure 4

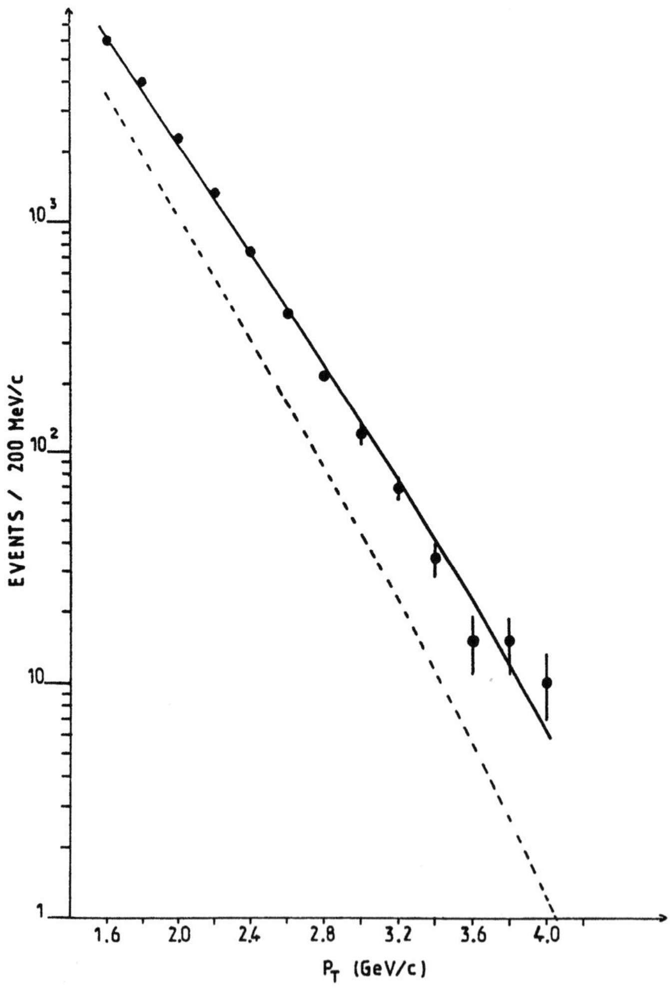

Figure 5

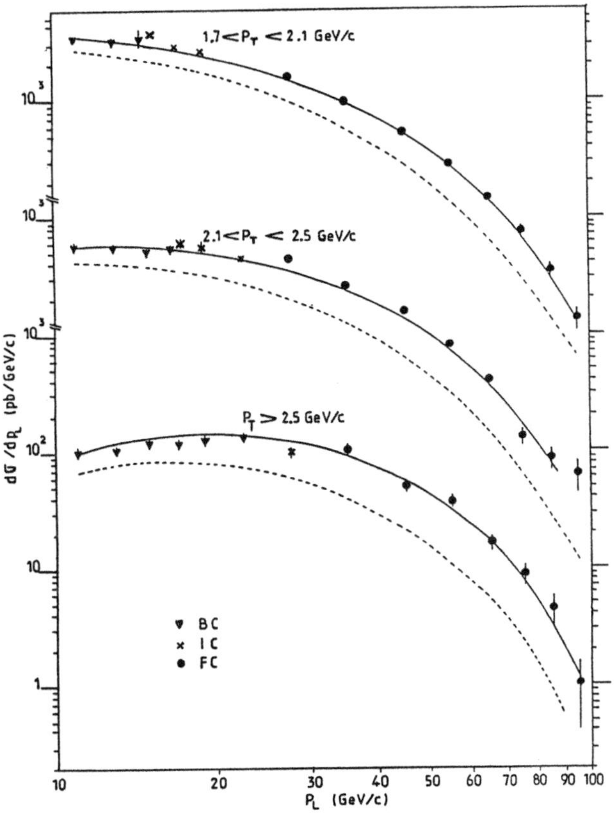

Figure 6

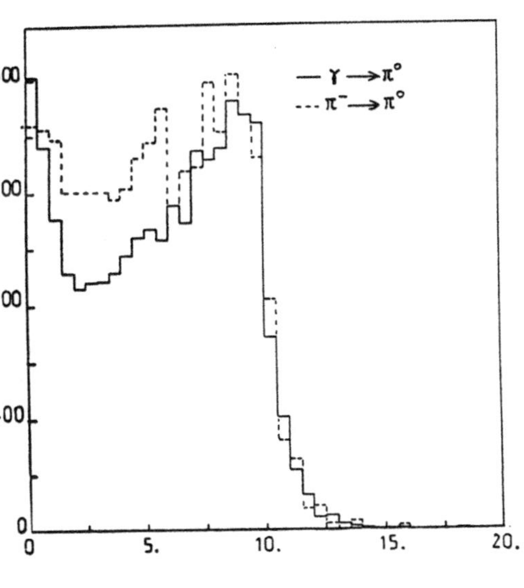

Figure 7

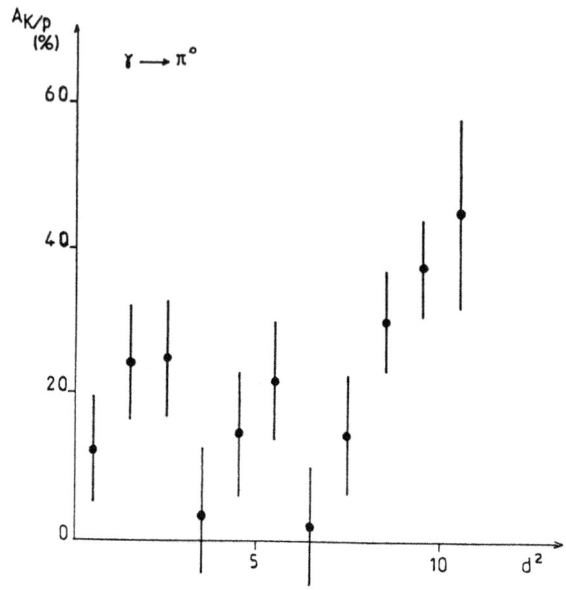

Figure 8

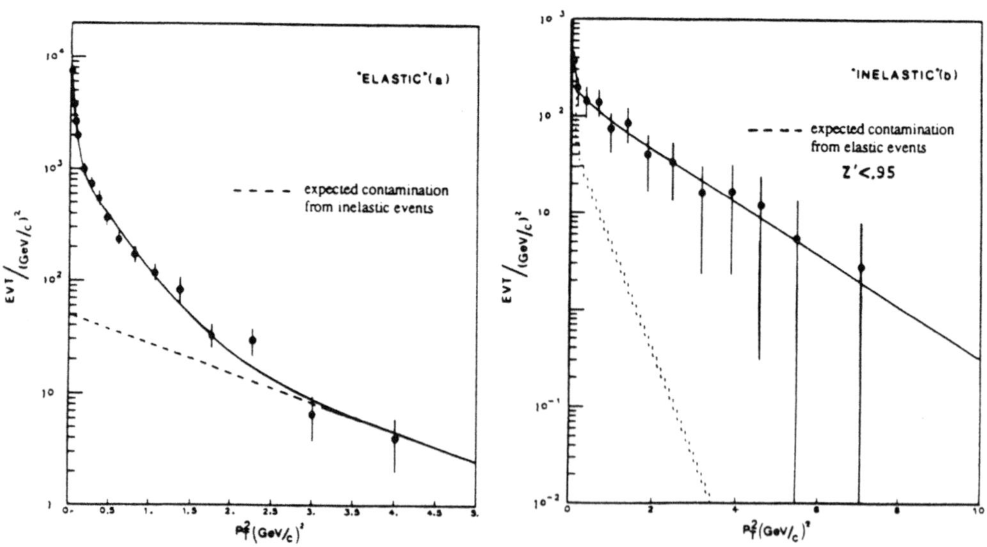

Figure 9

II.

EQUIVALENT PHOTONS

The Coulomb Field of Relativistic Nuclei: An Intense Source of (Almost) Real Photons

G. Baur
Institut für Kernphysik (Theorie), Forschungszentrum Jülich GmbH.
Postfach 1913, D–5170 Jülich

1. Introduction

The time–varying electromagnetic field of a fast nucleus can be looked upon as an intense continuous spectrum of photons. The basic idea of this equivalent photon (or Weizsäcker–Williams) approximation has been put forward by E. Fermi as early as 1924 (see Ref. 1), and it has been refined[2] and applied in many ways (for a recent review see e.g. Ref. 3).

In the domain of nuclear physics, the Coulomb excitation method has been used to determine electromagnetic matrix–elements, yielding valuable nuclear structure information. At low velocities, multiphoton processes are also very important and useful. Other applications are the use of Coulomb dissociation for the study of the structure of exotic nuclei[4-7], and for the (indirect) study of radiative capture processes for nuclear astrophysics[8].

In the low frequency approximation, the equivalent photon spectrum is given by

$$n(\omega) = \frac{2}{\pi} Z^2 \alpha \ln \frac{0.681 \gamma v}{\omega R} \tag{1a}$$

where $\gamma = (1-(\frac{v}{c})^2)^{-1/2}$, Z is the charge of the nucleus and R the minimum impact parameter. Beyond the adiabatic cut–off

$$E_\gamma^{max} \simeq \frac{\hbar c}{R} (\beta\gamma) \tag{1b}$$

the equivalent photon spectrum decreases exponentially. In terms of the corresponding photo–induced cross–section $\sigma_\gamma(\omega)$, the cross section σ_c for the equivalent photon process is given by

$$\sigma_c = \int \frac{d\omega}{\omega} n(\omega)\, \sigma_\gamma(\omega) \tag{2}$$

The flux of equivalent photons can therefore be calculated by multiplying the equivalent photon spectrum with the particle flux. E.g. at SIS/ESR in Darmstadt a ~ 1 GeV/A U–beam with ~ 10^{11} ions/s will be available[9]. Around $\omega = 10$ MeV, one obtains a flux of equivalent photons of the order of $\sim \frac{2}{\pi} \frac{Z^2\alpha}{\omega} \cdot 10^{11}\ s^{-1} \simeq 4 \cdot 10^{11}\ MeV^{-1}\ s^{-1}$. This could be compared to a bremsstrahlung photon beam intensity of ~ $10^9\ MeV^{-1}\ s^{-1}$ at around $\omega = 3$ MeV, as obtained at the bremsstrahlung facility at the Stuttgart dynamitron[10]. Of course, there are strong background problems inherent in the equivalent photon approach, which have to be solved experimentally. The method has already been of great value in the study of excitation and decay of giant resonance states in nuclei.[11-14]

In general, the equivalent photon spectrum depends on the multipolarity $\pi\lambda$. For $\gamma \to \infty$, the equivalent photon numbers $n_{\pi\lambda}(\omega)$ tend to a common limit, given by the Weizsäcker–Williams approximation. Since $n_{M1} \simeq (\frac{v}{c})^2\ n_{E1}$, one would have a new method to distinguish between E1 and M1 excitations, by determining the dependence of the electromagnetic excitation cross section on the projectile velocity.

In high energy physics, the equivalent photon method is widespread and has been especially useful for the investigation of the interaction of unstable particles (like Λ^0,π^0,...) with photons. This is commonly known as the Primakoff process. As a recent example, the interaction of 185 GeV/c polarized protons will the equivalent photons of a Pb target has been studied in the nucleon resonance (Δ^+(1323) and N*(1520)) region.[15] (Actually, in this case the Primakoff process is suggested to be used in order to measure the polarization of protons and of antiprotons at high energies).

In the present paper I want to concentrate on the following points: in chapter 2, the photo–nuclear interactions in the nucleon resonance region and beyond are discussed, in view of the possibilities at present and future relativistic heavy ion (RHI) facilities. There are already rather simple experiments which exploit the hard component of the equivalent photon spectra present at these collisions. In chapter 3, photon–photon interactions at RHI colliders are studied. In the invariant mass region around some GeV, which will be well accessible at RHIC, vector dominance studies should be possible with improved statistics, as compared to present results from e^+e^- colliders. At the high energy end of the invariant mass spectrum, effects due to strong absorption will become relevant, this is in contrast to the e^+e^- case, where a very detailed formalism to describe γ–γ processes exists (see e.g. Ref. 16). Conclusions are given in chapter 4.

2. Applications to photonuclear reactions

a) Nucleon resonance region

In order to achieve equivalent photon energies corresponding to the nucleon resonance region, the ion beam should have a Lorentz–factor of the order of $\gamma = 100$ (see eq. 1b). This is available presently at the CERN heavy ion facility. In Ref. 17, cross sections for electromagnetic excitation of nucleon resonances were calculated for heavy ion collisions with γ–values up to 200. They were found to be quite large. The effects of the hard part (above the nuclear giant resonance and quasideuteron region) of the equivalent photon spectrum were clearly identified in projectile fragmentation processes using plastic nuclear track detectors.[18-19] In the fragmentation of 200 GeV/A ^{32}S nuclei, electromagnetically produced fragments with Z_F as low as 8 were experimentally found, with a cross section of (116 ± 24) mb (on a Pb target). The electromagnetic process could be quite clearly separated from the dependence of the cross section on the target A and Z numbers. Whereas the total electromagnetic cross sections were readily calculated theoretically[17], the cascade process is more complicated. This process determines the differential cross sections (e.g. which fragments are produced).

I do not know how well the equivalent photon method would be able to investigate the nucleon resonance region and how it could compete with the investigation with real γ–sources. Certainly, dedicated detection systems will have to be used which can cope with the extreme forward angles and which have a good enough energy resolution. Fig. 2 of Ref. 15, where the invariant–mass spectrum of the π^0–p system in the Primakoff process $p+Pb \to \pi^0+p+Pb$ is shown, with a clear peak structure due to Δ^+ and N(1520) resonance excitation, certainly shows the potential of the method.

b) Photonuclear cross sections beyond the nucleon resonances

Experimentally, photonuclear reactions have been studied[20] up to an energy of $E_\gamma = 200$ GeV. There are plans[21] to fill the planned LHC (Large Hadron Collider) at CERN with relativistic heavy ions. In this case, the colliding nuclei would be subject to interactions with photons of energies up to $E_\gamma^{max} \simeq \gamma_p \frac{\hbar c}{R} \simeq 500$ TeV, where $\gamma_p = 2\gamma^2-1$ is the Lorentz factor, as seen from the rest frame of one nucleus. The corresponding energy at RHIC ($\gamma = 100$) would be $E_\gamma^{max} =$ 300 GeV, which exceeds already the experimentally studied range.

If we assume a constant value of $\sigma_{\gamma N} = 100\ \mu b$, and a photo–nucleus cross section proportional to A^α (with $\alpha < 1$, to take shadowing effects into account) we find a total electromagnetic cross section of[22]

$$\sigma(E_\gamma > 2\text{ GeV}) \cong 0.46\ \mu b\ A^\alpha Z^2 \ln^2(E_\gamma^{max}/2\text{ GeV}). \qquad (3)$$

We obtain a cross section (for both colliding nuclei) of $\sigma = 19$b, for the system Pb+Pb at $\gamma = 4000$. For the system U+U at $\gamma = 100$, one finds $\sigma = 2.6$ b. These processes contribute to a fragmentation of the colliding beams, and therefore to a decrease of the collider luminosity with time[22].

On the other hand, such (proton–heavy ion, or heavy ion–heavy ion) colliders could be useful for a study of photonuclear cross sections at energies beyond those presently available. The cross sections (see eq. 3) are large, and it seems a problem of the development of suitable detectors. Let me recall here that there is considerable discussion at the moment about the characteristics of photo–nucleus collisions at high energies. E.g. it is conjectured that[23] the observed excess of cosmic–ray mesons in the γ–ray induced showers from the direction of Cygnus X–3 could be explained by postulating a dramatic increase in the photonuclear cross section at energies above 10^4 GeV. This energy is well accessible at LHC.

The hadron structure of high–energy photons was recently studied in Ref. 24. It is argued there that the abundant gluon content of very high–energy photons could lead to the possibility of inclusive cross sections in excess of the $\sim$ 100 μb γp–cross section observed with existing photon beams. The results are model dependent; however, they have in common a rise of the γp cross section far above the 100 μb level at energies $E_\gamma \gtrsim 200$ GeV (see Fig. 3 of Ref. 24). In that reference it is discussed that one could study γp interactions with "almost real" photons radiated by the electrons at HERA up to an energy of $\sqrt{s}(\gamma p) \simeq 300$ GeV. (This corresponds to a photon energy in the "lab–system" of about 45 TeV, a factor of ten less compared to LHC, cross sections would be enhanced by the coherency factor Z^2 ($\lesssim 10^4$).)

3. Photon–Photon Collisions at Relativistic Heavy Ion Colliders

γ–γ physics has been studied extensively at e^+e^- colliders. It has proven to be a very rich field, contributing essentially to our understanding of QCD. For recent reviews see Refs. 25, 26, 27. Due to coherency, a Z^4 factor ($\lesssim 10^8$) appears in the cross section;[28] [29] therefore, relativistic heavy ions offer a new possibility for future accurate and extended studies. The cross section σ_C can be written in terms of the corresponding γ–γ cross section $\sigma_{\gamma\gamma}$ as

$$\sigma_C = \left(\frac{Z^2\alpha}{\pi}\right)^2 \int dW^2\, \sigma_{\gamma\gamma}(W^2)\, I(W^2) \qquad (4a)$$

where W is the invariant mass of the γ–γ system. In the low frequency approximation, $I(W^2)$ is given analytically as[28] [29]

$$I(W^2) = \frac{16}{3W^2}\left[\ln\frac{2\gamma}{W\cdot R}\right]^3 \tag{4b}$$

Especially at the high energy end, numerical evaluations are necessary to improve on eq. 4b. In contrast to the e^+e^-–case (see e.g. the review by Budnev et al.[16]) nuclear absorption effects have to be considered for the case of the collision of two strongly interacting particles.[3) 30) 31)] In contrast to the e^+e^-–case, only collisions of quasireal photons can be studied with RHI collisions.

The production of narrow C = +1 resonances R with a γ–γ partial width $\Gamma(R\to\gamma\gamma)$, mass m_R and spin J_R is given by

$$\sigma_c = 8(Z^2\alpha)^2 \frac{\Gamma(R\to\gamma\gamma)}{m_R}(2J_R+1)\, I(m_R^2) \tag{5}$$

In the low invariant mass region, QED processes like $\gamma+\gamma \to e^++e^-$ (or $\mu^++\mu^-$) are most prominent (see Ref. 3, and further references given there). Quite in analogy to the production of $q\bar{q}$–mesons, positronium states can also be produced with the γ–γ mechanism. E.g. in a U–U–collision with γ=100, one finds

$$\sigma_c\ (n=1 \text{ parapositronium}) = 10 \text{ mb}, \tag{6}$$

where $\Gamma((e^+e^-)_{n=1},\ {}^1S_0\to\gamma\gamma) = \frac{m_e c^2 \alpha^5}{2}$ was used.

Maximum invariant masses are of the order of (see eqs. 4a, b)

$$W^{max} \simeq \frac{2\gamma}{R}. \tag{7}$$

With $R \simeq 7$ fm and $\gamma = 100$ one has $W^{max} \simeq 6$ GeV. Thus a collider like RHIC seems to be especially promising to study low energy hadronic processes, which can shed light on low energy QCD. At colliders with even higher energies, like LHC ($\gamma \simeq 4000$) or even SSC ($\gamma \simeq 8000$), there are specially favourable scenarios[29) 32)] for producing medium mass (~ 100 GeV) Higgs particles, or particles beyond the Standard model (like SUSY–particles). γ–γ–interactions in the GeV–region are especially interesting for QCD studies. C = +1 mesons, like π^0, η, η' ... can be produced in γ–γ collisions, and their γ–γ–widths and decay modes can be studied. The non–appearance in γ–γ collisions of glueball candidates, with their lack of electric charges coupling to the photon, can strengthen their identification as such particles. This is discussed in detail in the recent reviews.[25) 26)] If the suppression of strong background events can be made effectively, it seems very promising to make use of the high γ–γ–luminosity at RHI colliders like RHIC for very detailed studies.

The production of vector meson pairs

$$\gamma + \gamma \rightarrow V + V' \; ; \; V,V' = \rho, \omega, \phi \tag{8}$$

is of special interest for the vector dominance model. It seems appropriate to summarize here some of the points of Refs. 25, 26. An unexpectedly huge cross section for the reaction $\gamma\gamma \rightarrow \rho^0\rho^0$ near threshold has been observed. Explanations in terms of resonance formation has been suggested. A $q\bar{q}$–resonance decaying into $\rho\rho$ has isospin I = 0, which would lead to a branching ratio of $\sigma(\rho^0\rho^0) : \rho(\rho^+\rho^-) = 1{:}2$. Since no significant $\rho^+\rho^-$ signal could be observed, an explanation in terms of such a simple resonance seems ruled out. Other possibilities could be two resonances, which interfere differently in the two channels. A natural explanation could also be given in terms of a $qq\bar{q}\bar{q}$–configuration. Crucial tests for the existence of four–quark resonances should be provided by refined spin–parity analyses of the final states. The $\gamma\gamma$–production of $\rho\omega$, $\omega\omega$ and $K^{*0}\overline{K^{*0}}$ final states is more complex as to be accounted for by any single model.

The total cross section for hadron production by two photons is complicated. In addition to the VDM mechanism for hadron production there is also the mechanism by direct quark exchange. (This is analogous to the lepton pair production mechanism in QED). Mixing between those mechanisms complicates the picture even more. In Ch. 3.3 of Ref. 26 it is stated that "... it seems premature to judge the quality of the models on the basis of the available data". RHI colliders could fill the gap.

With the probability $\frac{\alpha\pi}{\gamma_V^2}$ for a photon to turn into a vector meson, one obtains the VDM relation

$$\sigma(\gamma\gamma \rightarrow \text{hadrons}) = \sum_{VV'} \frac{\alpha\pi}{\gamma_V^2} \frac{\alpha\pi}{\gamma_{V'}^2} \sigma(VV' \rightarrow \text{hadrons}) \tag{9}$$

where

$$\frac{\alpha\pi}{\gamma_\rho^2} = 2.8 \cdot 10^{-3} \; , \; \frac{\alpha\pi}{\gamma_\omega^2} = 0.30 \cdot 10^{-3} \; , \; \frac{\alpha\pi}{\gamma_\phi^2} = 0.36 \cdot 10^{-3} \; . \tag{10}$$

All measurements approach a constant cross section at large W, somewhat higher than predicted,

$$\sigma_{\gamma\gamma}(W \rightarrow \infty) \approx 300 \text{ nb} \qquad (\text{exp.}). \tag{11}$$

Assuming a W–independent total hadron production cross section σ_0, one can easily obtain an estimate of the total hadronic cross section to be expected at RHI colliders.

$$\sigma_c = \frac{8}{3\pi^2} Z^4 \alpha^2 \sigma_0 \left(\ln \frac{2\gamma c}{W_0 R}\right)^4 \qquad (12)$$

where W_0 denotes a lower cut–off in invariant mass. (Actually, this formula underestimates the total yield, since the low frequency approximation, eq. 4b has been used.) With $Z = 92$, $\gamma = 100$, $\sigma_0 = 300$ nb, $R = 7.5$ fm and $W_0 = 1$ GeV one finds $\sigma_c = 2.5$ mb. With a luminosity of $5 \cdot 10^{26}$ $cm^{-2}s^{-1}$ (see Ref. 33) one obtains a rate of $\sim 1\ s^{-1}$; in a 10 hour run this would be 36000 events.

4. Conclusion

Quasireal photons from (heavy) nuclei have been and will continue to be useful to study interactions of photons with nuclei and nucleons. They complement and extend the study with real photons. They also allow the study of the interaction of more exotic short–lived particles (like secondary exotic nuclear beams (^{11}Li, ... or Λ^0, ...) with photons. Possible nuclear interaction effects have to (and can) be carefully considered. Especially the characteristically different dependence on momentum transfer allows a clear separation of the two processes.

The hard part of the equivalent photon spectrum, which appears in the 200 GeV/A heavy ion interactions at CERN, has been clearly seen in projectile fragmentation processes. The electromagnetic excitation of Δ^+ and N^* in a 185 GeV/c proton beam has been beautifully resolved recently. Further possibilities will open up at relativistic heavy ion colliders; they have been studied in the present work. They were seen to provide high fluxes of (quasireal) photons with energies exceeding the present experimental limits. RHI colliders were also shown to give high luminosities for γ–γ–collisions. The invariant mass region up to several GeV is well in reach at colliders like RHIC. In this region, the hadronic structure of the photon and their interactions can be studied in detail. It was the purpose of this work to provide a theoretical framework for the calculation of photon fluxes in nuclear collisions and to study the corresponding scenarios.

References:

1. J.D. Jackson, Classical Electrodynamics (Wiley, NY, 1975)
2. A. Winther and K. Alder, Nucl. Phys. A319 (1979) 518
3. G. Baur, proceedings of CBPF International Workshop on Relativistic Aspects of Nuclear Physics, Rio de Janeiro, Brazil, August 28–30, 1989, to be published by World Scientific
4. P.G. Hansen and B. Jonson, Europhys. Lett. 4 (1987) 409
5. C.A. Bertulani and G. Baur, Nucl. Phys. A480 (1988) 615
6. G. Baur, Proceedings of Int. Symp. on Heavy Ion Physics and Nuclear Astrophysical Problems, Tokyo, July 21–23, 1988, ed. by S. Kubono, M. Ishihara, and T. Nomua, World Scientific, p. 225
7. T. Kobayashi et al., Phys. Rev. Lett. 60 (1988) 2599, and Phys. Lett. B 232 (1989) 51
8. H. Rebel, workshop on nuclear reaction cross sections of astrophysical interest, unpubl. report, Kernforschungszentrum Karlsruhe, Feb. 1985
   G. Baur, C.A. Bertulani, and H. Rebel, Nucl. Phys. A459 (1986) 188
   G. Baur, M. Weber, Nucl. Phys. A504 (1989) 352
9. Proceedings of the workshop on experiments and experimental facilities at SIS/ESR, GSI–87–7, April 1987, p.3
10. B. Kasten et al., Phys. Rev. Lett. 63 (1989) 609
11. F.E. Bertrand and J.R. Beene, invited talk at the 1989 International Nuclear Physics Conference, Sao Paulo, Brazil, to be published in the proceedings
12. H. Emling, Proc. of the 8th High Energy Heavy Ion Study, Berkeley, November 16–20, 1987, J.W. Harris, G.J. Wozniak (eds.) p. 503, 1987
13. G. Baur, C.A. Bertulani, Phys. Lett. 174B (1986) 23, Phys. Rev. C34 (1986) 1654
14. J. Barrette et al., E814 Collaboration, BNL–43561, November 1989, submitted to Phys. Rev. C, P. Braun–Munzinger et al., Proposal E814 submitted to the AGS Program Committee (1985)
15. D.C. Carey et al., Phys. Rev. Lett. 64 (1990) 357
16. V.M. Budnev et al., Phys. Rep. 15 (1975) 181
17. C.A. Bertulani and G. Baur, Nucl. Phys. A458 (1986) 725, and Phys. Rep. 163 (1988) 299
18. C. Brechtmann and W. Heinrich, Z. Phys. A330 (1988) 407, Z. Phys. A331 (1988) 463
19. P.B. Price, Ren Guoxiao, and W.T. Williams, Phys. Rev. Lett. 61 (1988) 2193
20. D.O. Caldwell et al., Phys. Rev. Lett. 40 (1978) 1222
21. See e.g. D. Brandt, LHC/Note No. 87
22. G. Baur and C.A. Bertulani, Nucl. Phys. A505 (1989) 835
23. J. Szabelski, J. Wdowczyk, and A.W. Wolfendale, J. Phys. G: Nucl. Part. Phys. 15 (1989) 1893
24. M. Drees and F. Halzen, Phys. Rev. Lett. 61 (1988) 275
25. Ch. Berger, W. Wagner, Phys. Rep. 146 (1987) 1

26. H. Kolanoski and P.M. Zerwas, Two–photon physics in high–energy electron–positron physics, ed. A. Ali and P. Söding (World Scientific, Singapore, 1988)
27. S. Cooper, Ann. Rev. Nucl. Part. Sci. 38 (1988) 705
28. G. Baur, C.A. Bertulani, Z. Phys. A330 (1988) 77
29. E. Papageorgiu, Phys. Rev. D40 (1989) 92
30. G. Baur, 6th Nordic Meeting on Nuclear Physics at Haugesund, Norway, August 11–15, 1989, to be published in Physica Scripta
31. G. Baur, and L. Gonzaga Ferreira Filho, to be published
32. M. Grabiak, B. Müller, W. Greiner, G. Soff and P. Koch, J. of Phys. G15 (1989) L25, J. Rau, B. Müller, W. Greiner and G. Soff, J. Phys. G16 (1990) 211
    M. Drees, J. Ellis, and D. Zeppenfeld, Phys. Lett. B223 (1989) 454
33. N.P. Samios, Nucl. Phys. A498 (1989) 323c

III.

COMPTON SCATTERING, MESON SCATTERING AND MESON PHOTOPRODUCTION

PHOTON SCATTERING AS A TOOL FOR STUDY OF NUCLEON AND NUCLEUS STRUCTURE

A.I.L'vov, V.A.Petrun'kin
P.N.Lebedev Physical Institute
Moscow 117924, USSR

1. Introduction

Photon scattering on nucleons and nuclei at intermediate energies is a two-photon process proving to be very useful tool for study of structure of the particles. As compared with one-photon processes such as electron scattering it provides new options because the relevant s-channel intermediate states involve really or virtually the whole excitation spectrum of the target. By virtue of the photon scattering one can study the particle polarizability which cannot be reduced to static properties, like a size, measured in elastic (e,e')-scattering.

Although the photon scattering approximately reproduces the resonance picture observed in photoabsorption it cannot be expressed in terms of the latter, e.g. via s-channel dispersion relations, because of specific two-photon interactions such as t-channel exchanges by various mesons decaying into $\gamma\gamma$ or meson exchange currents displaying in different manner in one- and two-photon reactions with nuclei. The study of these interactions by means of γN- and γA-scattering is of independent interest.

γA-scattering at energies above the giant dipole resonance allows principally to obtain trustworthy information about properties of nucleons and isobars in nuclear media since the free propagation of photons through the nucleus both before and after the scattering simplifies an interpreting data.

Experimental investigations of the photon scattering are seriously impeded as compared with (e,e')-processes due to the lower cross section and worse quality of γ-beams (the intensity, $\Delta E/E$, focusing, etc.) and γ-detectors (the energy and space resolution, efficiency). That is why, in spite of a long history of studying of γN and γA-scattering, up to now at energies $\omega \gtrsim 50$ MeV there are very scanty data, a part of which is in conflict with others. That concerns even the γp-scattering where the data are least incomplete. As for investigations of the photon scattering on nuclei above the giant dipole resonance they make only first steps. Nevertheless at present we possess some experimental results about the polarizability of free and bound nucleons /1-3/, the γA-scattering above the pion threshold /4-5/, the $\gamma\pi$-scattering /6-7/ and some other points giving a rich food for theoretical speculations /8-13/.

In the present talk we discuss in what extent the elementary, for two-photon nuclear physics, amplitudes of γN-scattering are known, what information they contain about the nucleon structure, and what base they provide for understanding the photon scattering on nuclei.

2. Relation between scattering and absorption

The first question we want to discuss is whether the γN-scattering amplitude is determined by photoabsorption. Principally, in the framework of the local quantum field theory the affirmative answer is followed from the next expression for the scattering amplitude in terms of the e.m. current /14/ :

$$T = \varepsilon_\nu'^* \varepsilon_\mu \langle N' | \int \{ i T [J_\mu(x) J_\nu(0)] + S_{\mu\nu}(x) \delta^4(x) \} e^{ik'x} d^4x | N \rangle$$

$$= \varepsilon_\nu'^* \varepsilon_\mu \sum_n (2\pi)^3 \delta^3(\vec{p}+\vec{k}-\vec{p}_n) \frac{\langle N' | J_\mu(0) | n \rangle \langle n | J_\nu(0) | N \rangle}{E_n - E_N - \omega - i0} \qquad (1)$$

$$+ \text{ cross } + \text{ Schwinger term,}$$

where the Schwinger term $S_{\mu\nu}(x)\delta^4(x)$ originates from the interaction terms in the Hamiltonian of the second order in the e.m. potential ; it equals $ix_\mu \delta(x_0)[J_0(x), J_\nu(0)]$ and compensates a non-covariant and non-gauge-invariant part of the T-production. However, a direct using of the Eq. (1) to predict the γN-scattering amplitude on the base of our knowledge of photoabsorption meets with difficulties. Firstly, there are high-lying intermediate states in the sum (1). They have poor-known excitation amplitudes (e.g. for $\pi\pi N$) and give non-negligible contributions. The second reason becomes clear after the more attentive looking at the structure of the Eq. (1) of the old-fashioned perturbation theory. Since the intermediate state $|n\rangle$ can include or not the external particles N or N' (with the same momenta and spins), the sum (1) is splitted into corresponding four terms of the cluster expansion :

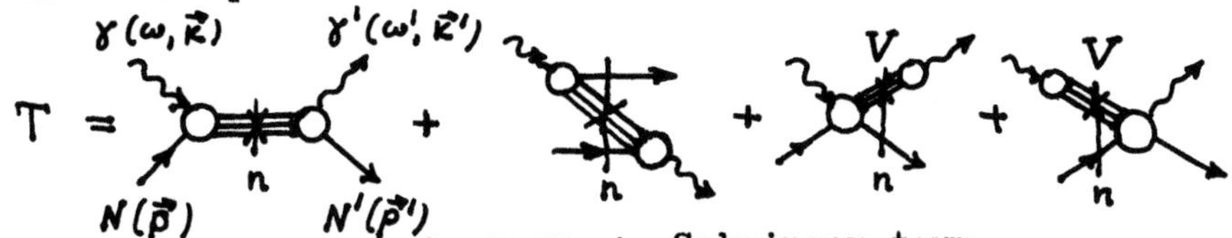

+ cross + Schwinger term . (2)

Here the first term is a usual resonance contribution T_{res} . The second one is the contribution of baryon-antibaryon pairs (Z-diagrams) T_Z ; in the Dirac theory namely this contribution gives the threshold (Thomson) value of the scattering amplitude $T(\omega=0)=-e^2/M$. The third and the fourth terms T_3 and T_4 are assosiated with the photoproduction of vector states V and the reverse process. (Note that because of the possibility to produce the on-shell state V by virtue of the $\gamma N \to V'N \to VN'$ transitions the terms T_3 and T_4 have non-physical singularities due to the propagator of the V', e.g. for the forward scattering, which are compensated mutially in the sum $T_V=T_3+T_4$.) Owing to the terms T_3 and T_4 the γN-scattering amplitude is enriched with such important t-channel exchanges as

+ cross .

As the result the finding of the scattering amplitude even at small energies $\omega \sim 10^2$ MeV needs not only knowledge of the nearest resonance contribution but also of the vector meson photoproduction amplitudes at higher energies $\omega \sim M_V \sim 10^3$ MeV, that implies, in accordance with the vector dominance, the knowing the Compton amplitude in the GeV-region. An importance of such ingredients of the γ-scattering is illustrated by what they provide more than half of the static electric polarizability of the nucleon /9/.

The complicated structure of the scattering amplitude (1)-(2) and the involving the high-lying intermediate states is related only in part with the use of the old-fashioned perturbation theory, in which the energy is not conserved and the intermediate particles are on-shell and the relevant transitions are produced by virtual photons. When one attempts to derive the scattering from photoabsorption the similar difficulties appear as well in covariant approaches such as the dispersion theory, in which the photons are real throughout and cannot transform into vector particles. The scattering amplitude formally obeys the dispersion relation (DR)

$$\mathrm{Re}\, T(s,t) = \frac{1}{\pi} \int_{-\infty}^{\infty} \frac{d\tilde{s}}{\tilde{s}-s}\, \mathrm{Im}\, T(\tilde{s},t) \qquad (3)$$

which revives the amplitude from its imaginary part, i.e. photoabsorption. However, because of the growing of T at $s \to \infty$ the integral (3) diverges and the DR must be recasted into a subtracted form, e.g.

$$\mathrm{Re}\, T(s,t) = T(s_0,t) + \frac{s-s_0}{\pi} \int_{-\infty}^{\infty} \frac{d\tilde{s}}{(\tilde{s}-s)(\tilde{s}-s_0)} \mathrm{Im}\, T(\tilde{s},t). \quad (4)$$

The subtraction function $F(t) = T(s_0,t)$ carries an information about high-energy behaviour of the amplitude T and thereby about high-lying state contributions as seen at low energies. The function F(t) can be found from the DR in t which, as a Regge-analysis shows, does not require any subtractions :

$$F(t) = \frac{1}{\pi} \int_{-\infty}^{\infty} \frac{d\tilde{t}}{\tilde{t}-t} \mathrm{Im}_t\, F(\tilde{t}) \; . \quad (5)$$

As the result the γN-scattering is represented by s- and u-channel contributions from photoabsorption and t-channel contributions from the photon scattering off virtual formations presented near the nucleon :

T = 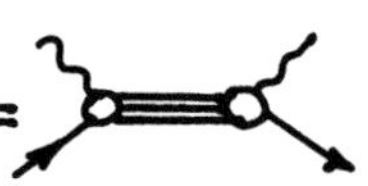 + cross +  . (6)

Because of the convergence of the integral (5) the main contribution to (5) at small t is given by the scattering off one or two pions, so that the above-mentioned formations are mainly the light pion periphery of the nucleon.

To evaluate the pridictive power of such approach it is essential that the one-pion exchange contribution is determined by the experimentally well-known (possibly, apart from relative sign) couplings of πNN and $\pi^0\gamma\gamma$ and therefore is calculated reliably, whereas the two-pion contribution determined by the $\gamma\gamma \to \pi\pi$ and $\pi\pi \to N\bar{N}$ amplitudes (the latter taken below physical threshold) /15/ is understood not so well. Data on the $\gamma\gamma \to \pi\pi$ obtained at e^+e^--colliders are not explained up to now within simple models usually applied at low energies and predicting ~ 2 times as small as experimental cross section /16/ ; possibly, these data provide an evidence for a large two-photon decay width of the effective σ-meson.

Technically some different way is possible to insert a subtraction into (3) /8,17/, which explicitly uses the value of the γN-scattering amplitude at high energies where a Regge-parametrization is applicable. Within such an approach the additional t-channel part of (6) is expressed through Regge couplings of mesons with $N\bar{N}$ and $\gamma\gamma$ . .

The summary is that the photon scattering on nucleons is produced by the excitation of resonances, what can be interpreted as an interaction with the nucleon quark core, and by the scattering on light pion periphery of the nucleon. Respectively, under studying the γN-scattering one can pursue an object of exploring both the resonances and the meson cloud.

3. Relativistic versus non-relativistic dispersion relations

An essential point in predictions of the γN-scattering amplitude is the using of the relativistic DR (3) since non-relativistic DRs for separate multipoles used in nuclear physics prove to be not quite adequate. The difference between these DRs can be easy explained for a spinless particle.

In this case there are two helicity scattering amplitudes $T_{1\pm 1}$ which are expanded in electric and magnetic multipoles as

$$T_{1\pm 1} = \sum_J (2J+1)\, d^J_{1\pm 1}(\theta)\, (EJ \pm MJ) \quad . \tag{7}$$

Let us write down fixed-t DRs for the invariant amplitudes

$$B_1 = t^{-1}\, T_{1\,-1} \quad , \qquad B_2 = (su-M^4)^{-1}\, T_{1\,1} \tag{8}$$

free from kinematical zeros (at $\theta = 0$ when $t=0$ and at $\theta = 180°$ when $s\cdot u = M^4$; $u \equiv 2M^2 - s - t$) and even under the crossing $s \leftrightarrow u$:

$$\mathrm{Re}\, B_i(s,u,t) = \frac{1}{\pi}\int_{M^2}^{\infty} \left(\frac{d\tilde{s}}{\tilde{s}-s} + \frac{d\tilde{s}}{\tilde{s}-u} \right) \mathrm{Im}\, B_i(\tilde{s},\tilde{u},t) \tag{9}$$

(we ignore a possible subtraction for the amplitude B_1). Integrating the amplitudes $T_{1\pm 1}$, defined by (8)-(9), with the weight $d^J_{1\pm 1}(\theta)d(\cos\theta)$, using (7) and taking into account that the l.h.s. and the integrand in the r.h.s. of (9) correspond to different scattering angles,

$$\sin^2\frac{\theta}{2} = \frac{-st}{(s-M^2)^2} \quad , \qquad \sin^2\frac{\tilde{\theta}}{2} = \frac{-\tilde{s}t}{(\tilde{s}-M^2)^2} \quad , \tag{10}$$

we arrive at DRs for separate multipoles. Up to $O(1/M)$ and dipole terms they look like

$$\mathrm{Re}\begin{Bmatrix} E1 \\ M1 \end{Bmatrix}(\omega) = \frac{1}{\pi}\int_0^{\infty} \frac{\omega^2}{\tilde{\omega}^2}\Bigg[\left(\frac{1}{\tilde{\omega}-\omega} + \frac{1}{\tilde{\omega}+\omega}\right)\cdot\Bigg(\mathrm{Im}\begin{Bmatrix} E1 \\ M1 \end{Bmatrix}(\tilde{\omega})$$
$$+ \frac{\tilde{\omega}-\omega}{M}\,\mathrm{Im}\begin{Bmatrix} E1-M1 \\ M1-E1 \end{Bmatrix}(\tilde{\omega})\Bigg) + \frac{\omega^2}{M(\tilde{\omega}+\omega)^2}\,\mathrm{Im}\begin{Bmatrix} E1-\frac{1}{2}M1 \\ M1-\frac{1}{2}E1 \end{Bmatrix}(\tilde{\omega})\Bigg]\, d\tilde{\omega} \tag{11}$$

$$+ \text{ higher multipoles } + O(M^{-2}) \; .$$

Here the lab. energy $\omega=(s-M^2)/2M$ is used instead of s. The factor $\omega^2/\tilde{\omega}^2$ is not essential and simply means that the DR (11) is of a subtracted form. More significant are corrections O(1/M) and higher multipoles which just differ the relativistic and non-relativistic DRs. Characteristic feature of the relativistic DR (11) is that the real part of the electric multipole E1 is contributed by a magnetic absorption. This is rather important for the γN-scattering because of the magnetic excitation of the Δ-isobar.

The origin of recoil mixing of different multipoles can be understood as following. To be expanded in multipoles the amplitude must be considered in the c.m. frame. But then only the states $|n\rangle$ of the direct term with $\vec{p}_n = \vec{p} + \vec{k} = 0$ in (1) will be at rest, whereas intermediate states in the cross term with $\vec{p}_n = \vec{p} - \vec{k}'$ will be moving. The mixing of the order of ω/M appears because the orbital momentum and therefore multipolarity is not invariant under boosts.

Note that if we consider the virtual photons with $\vec{k} = \vec{k}' = 0$, when there is no recoil and retardation and the scattering is of pure E1-type, the only ω-dependence in (1) will be in the energy denominators Then we immediately find the exact relativistic DR

$$\mathrm{Re}\left[E1(\omega)\right]_{\substack{\text{no recoil}\\ \text{no retard.}}} = \frac{1}{\pi}\int_0^\infty \left(\frac{1}{\tilde{\omega}-\omega} + \frac{1}{\tilde{\omega}+\omega}\right) \mathrm{Im}\left[E1(\tilde{\omega})\right]_{\substack{\text{no recoil}\\ \text{no retard.}}} d\tilde{\omega} \tag{12}$$

having the same form as nonrelativistic one.

Thus, the complicated structure of the Eq. (11) is related with taking into account the recoil and retardation. In practice, more simple formulae like (9) may be used instead of relations like (11) if we decline to compute preliminary all separate multipoles.

4. Relativistic dispersion theory of γN-scattering

Due to the nucleon spin there are six invariant amplitudes, A_i /8/, of the γN-scattering. Four of them satisfy the unsubtracted DR (9) and two others (A_1 and A_2), entering with the helicity-flip structures $\bar{u}'u\, F_{\mu\nu}F_{\mu\nu}$ and $i\bar{u}'\gamma_5 u\, F_{\mu\nu}F_{\rho\tau}\,\varepsilon_{\mu\nu\rho\tau}$ and containing contributions of scalar (σ) and pseudoscalar (π^0) exchanges, **need a subtraction.**

Imaginary parts of the amplitudes A_i are determined by additive contributions from different intermediate states. The simplest state consists of the single nucleon. Corresponding term in A_i is the so-called Born term, 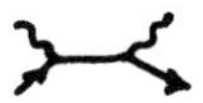 + cross, expressed through the nucleon mass M, electric charge e and anomalous magnetic moment $\varkappa$.

Non-trivial contribution to the imaginary parts arises from meson photoproduction. The finding Im A_i requires a knowledge of not simply the total photoabsorption cross section but also separate partial waves. For the πN-channel that provides no troubles since there are reliable partial wave analyses of the $\gamma N \to \pi N$ reaction both in the Δ-region most important for us and also up to ~ 2 GeV. However, the double pion photoproduction contributing at $\omega \gtrsim 320$ MeV can be taken into account only in framework of rather rough quasi-two-particle models like $\gamma N \to \pi\Delta \to \pi\pi N$ with one-pion-exchange. As the result the predictions of the DRs are well founded up to the Δ-resonance where the double pion contribution to the γN cross section is $\lesssim 10\%$; above the Δ-region it fastly grows and equals $\sim 30\%$ already at 400 MeV.

The most delicate point is the estimation of subtraction functions for the amplitudes A_1 and A_2. Cauchy formula for a closed path of large but finite size ω_M gives the modified DR /17/:

$$\mathrm{Re}\, A(\omega,t) = \frac{1}{\pi} \int_{-\omega_M}^{\omega_M} \frac{d\tilde{\omega}}{\tilde{\omega}-\omega}\, \mathrm{Im}\, A(\tilde{\omega},t) + \frac{1}{\pi}\, \mathrm{Im} \int_C \frac{d\tilde{\omega}}{\tilde{\omega}-\omega}\, A(\tilde{\omega},t) \qquad (13)$$

C = 

where the amplitude at complex energies $\tilde{\omega}$, $|\tilde{\omega}| = \omega_M$ can be evaluated within Regge model. Respectively, the amplitude A(ω,t) is splitted into the resonance contribution A^{res} and the asymptotic one, A^{asym}. The profit of such an approach is in that the photon helicity-flip vertexes for leading Reggeons (such as Pomeron) are small (that is referred to as an approximate s-channel helicity conservation) and hence need not in refined computations. That smallness is illustrated by the asymmetry measurement /19/ of the scattering of linearly-polarized photons at 3.5 GeV :

$$\Sigma = \frac{d\sigma_\perp - d\sigma_\parallel}{d\sigma_\perp + d\sigma_\parallel} = \frac{2\,\mathrm{Re}(T_{11}^* T_{1-1})}{|T_{11}|^2 + |T_{1-1}|^2} \approx 2\,\mathrm{Re}\, \frac{T_{1-1}}{T_{11}} \;. \qquad (14)$$

The result obtained,

$$\frac{\Sigma(t)}{t} = 0.08 \pm 0.18 \quad \mathrm{GeV}^{-2} \qquad (15)$$

at $-t \lesssim 0.5$ GeV^2, agrees with the assumption that the Pomeron does not change the photon helicity (natural scale of (15) is ~ 3 GeV^{-2}, cf.(17)).

Actually, the asymptotic contribution depends on ω very slowly at $\omega^2 \ll \omega_M^2$ and, in fact, is a function of the single variable t which, in turn, is completely similar to the subtraction function in (4). For the amplitude A_2 the contribution A_2^{asym} is mainly determined by π^o-exchange,

$$A_2^{asym} \approx A_2^{\pi^o} = \frac{g_{\pi NN} F_{\pi\gamma\gamma}}{t-m_\pi^2} , \tag{16}$$

due to the small pion mass. For the amplitude A_1 the asymptotic contribution is mainly governed by two-pion or σ exchange. Because of the large mass of the effective σ-meson the contribution A_1^{asym} can be parameterized at energies $\lesssim 300$ MeV as

$$A_1^{asym} = C \cdot \exp \frac{bt}{2} \tag{17}$$

with the slope parameter, $b \simeq 2m_\sigma^{-2} \approx 6\ \mathrm{GeV}^{-2}$ taken from t-dependence of the γp-scattering cross section in the few GeV-region. The constant

$$C \approx \frac{g_{\sigma NN}\, g_{\sigma\gamma\gamma}}{m_\sigma^2} , \qquad g_{\sigma\gamma\gamma} = -\left[\frac{64\pi}{m_\sigma^3}\Gamma_{\sigma\to\gamma\gamma}\right]^{1/2} \tag{18}$$

may be estimated within some models for the decay $\sigma\to\gamma\gamma$. However, relevant uncertainties are rather large and it is reasonably to consider the value C as a free parameter of the dispersion theory. At low energies $\omega \lesssim 100$ MeV both resonance and asymptotic contributions become to be almost energy-independent and unite to give the only parameter which determines the deviation of the amplitude $A_1 = A_1^{res} + A_1^{asym}$ from its Born value (i.e. one-nucleon contribution) and is interpreted as the difference of electric α and magnetic β nucleon polarizabilities:

$$A_1 = A_1^{Born} - 2\pi(\alpha-\beta) + O(\omega^2) . \tag{19}$$

The corresponding low-energy expansion of the scattering amplitude looks, in the lab. system, like

$$T = T^{Born} + \omega\omega'(\vec{\varepsilon}\vec{\varepsilon}\,')\,\alpha + (\vec{k}\times\vec{\varepsilon})(\vec{k}'\times\vec{\varepsilon}\,')\,\beta + \ldots \tag{20}$$

and the differential cross section is

$$\frac{d\sigma}{d\Omega} = \frac{d\sigma^{Born}}{d\Omega} - \frac{e^2}{M}\left(\frac{\omega'}{\omega}\right)^2 \omega\omega' \left[\alpha\,(1+z^2) + 2\beta z + O(\omega\omega')\right] \tag{21}$$

where $z=\cos\theta$ and

$$\frac{d\sigma^{Born}}{d\Omega} = \left(\frac{e^2}{M}\right)^2 \left(\frac{\omega'}{\omega}\right)^2 \Big\{ \frac{1+z^2}{2} + \frac{\omega\omega'}{2M^2} \Big[(1+2\varkappa)\,(1-z)^2 \tag{22}$$

$$+ \varkappa^2\left(\frac{9+z^2}{2} - 5z\right) + \varkappa^3\,(3-2z-z^2) + \varkappa^4\,\frac{3-z^2}{4}\Big]\Big\} .$$

The sum $\alpha+\beta$ is related with a helicity-non-flip invariant amplitude which does not need in subtractions and can be found from photoabsorption together with $\alpha+\beta$:

$$\alpha+\beta = \int_{\omega_0}^{\infty} \frac{d\omega}{2\pi^2\omega^2}\,\sigma_{tot}(\omega) . \tag{23}$$

Numerically /8/

$$\alpha+\beta = \begin{cases} 14.4 & \text{for proton} \\ 16 & \text{for neutron} \end{cases} \tag{24}$$

(uncertainties are ± 0.5; here and henceforth we use the units of 10^{-4} fm^3 for polarizabilities). As for the difference $\alpha-\beta$, the resonance energy region gives contributions strongly compensating and the main part of the $\alpha-\beta$ originates from asymptotic region.

This dispersion approach hopefully allows to describe γN-scattering at energies up to 300-400 MeV with the only free parameter C (or $\alpha-\beta$ +known constant) common for proton and neutron. Fitting /17/ of existing γp-scattering data below the pion threshold where the sensitivity of the theory predictions under variations of such ingredients as the slope parameter b is minimal, gives

$$\alpha_p-\beta_p \approx 14 \quad . \tag{25}$$

However, this result is not well-established since different measurements of the cross sections are not in quite agreement, especially for backward scattering. To improve the knowledge of the polarizabilities new measurements are desirable; possibly, the new data will appear from Mainz /20/ and Illinois /21/.

The dispersion theory successfully describes data above the pion threshold (Fig.1) though there are discrepancies near 200 MeV and at the Δ-resonance and above. Data indicate a narrow dip at $\theta^*=90°$ in the angular distribution at the Δ-resonance energy. Such dip cannot be understood within the discussed theory or other analogous approaches /15,22/; the M1-excitation of the delta gives an angular distribution $\sim 7+3\cos^2\theta^*$ and interference of E1 and M1 leads to a correction $\sim \cos\theta^*$. Accurate new measurements of the angular dependence would

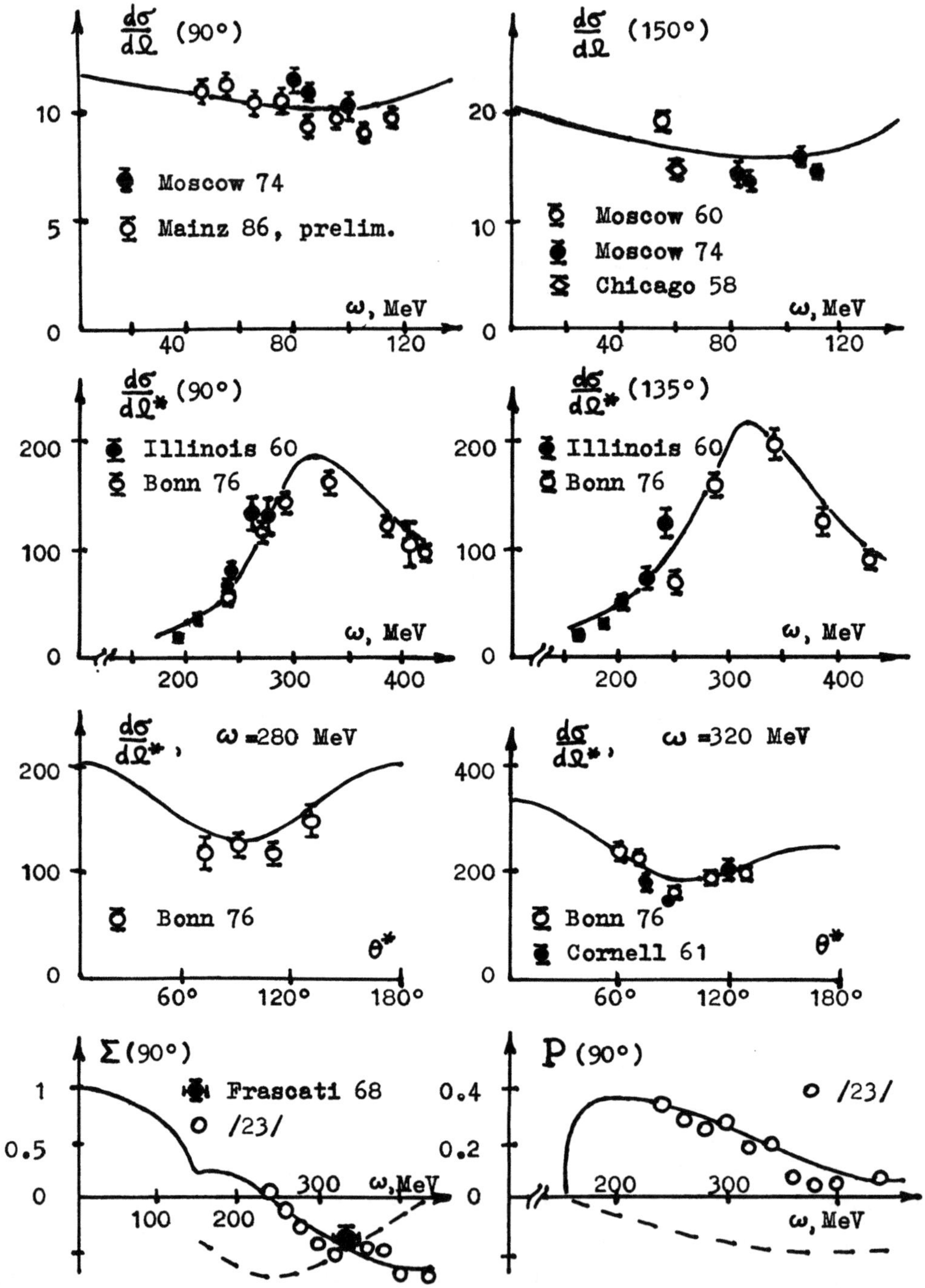

Fig.1 Differential cross section (nb/sr), asymmetry Σ and proton polarization for γp-scattering. Solid line: DRs with $\alpha_p - \beta_p = 14$ /17/, dashed lines: the resonance model /24/.

be very valuable as well as new data from the threshold up to 200-220 MeV where some problems with normalization of cross sections were in old experiments.

Measurements of the asymmetry Σ and proton polarization P are also very important for testing and improving the theory. The results of the dispersion computations for Σ and P are depicted on Fig.1 together with the predictions of a model-dependent partial-wave analysis /23/ and of the resonance model /24/, in which the amplitude T was represented as the sum of resonance contributions (calculated through radiative coupling constants of nucleon resonances) and of the Born term scaled with a phenomenological form factor. This model is able to describe differential cross section in the resonance region but its discrepancy with the DR predictions illustrates the sensitivity of Σ and P to various non-resonance contributions to the amplitude.

5. The nucleon polarizability

At low energies the two-photon nucleon structure is described by polarizabilities. Explicit formulae /8/ for them can be easy derived from the forward scattering amplitude for virtual transverse photons /25/. In this case the low-energy expansion (20) transforms into

$$T = T^{Born} + \omega^2 \alpha + \vec{k}^2 \beta + \dots \qquad (26)$$

where the Born (or, in other words, nucleon pole) term depends on the nucleon formfactors because $k^2 = \omega^2 - \vec{k}^2 \neq 0$:

$$T^{Born} = -\frac{e^2}{M}\left[\frac{G_E(k^2) - \frac{k^2}{4M^2} G_M(k^2)}{1 - \frac{k^2}{4M^2}}\right]^2 + \frac{e^2}{M}\,\frac{(k^2)^2\, G_M^2(k^2)}{4M^2\omega^2 - (k^2)^2}\,. \qquad (27)$$

If $\omega, \vec{k} \to 0$ then $G_E \to 1$, $G_M \to 1+\varkappa$ and the Born amplitude (27) provides the Thomson limit $-e^2/M$. The electric polarizability α is obtained from (26) as

$$\alpha = \frac{1}{2}\frac{\partial^2}{\partial\omega^2}\left[T(\omega,\vec{k}) - T^{Born}(\omega,\vec{k})\right]_{\vec{k}\to 0}\,. \qquad (28)$$

Using (1) and taking into account that the one-nucleon contribution in (1) vanishes at $\vec{k} \to 0$ and only the energy denominators are ω - dependent, one gets :

$$\alpha = \alpha_0 + \Delta\alpha \tag{29}$$

$$\alpha_0 = 2 \sum_n{}' (2\pi)^3 \delta^3(\vec{p}_n) \frac{|\langle n|J_z|N\rangle|^2}{(E_n - E_N)^3} = 2 \sum_n{}' \frac{|\langle n|D_z|N\rangle|^2}{E_n - E_N} \frac{1}{\langle N|N\rangle} \tag{30}$$

$$\Delta\alpha = \frac{e^2}{3M} \langle r_E^2 \rangle + \frac{e^2}{4M^3}(1+\varkappa^2) \quad . \tag{31}$$

Here the value α_0 has the same form as the usual expression from non-relativistic quantum mechnics. The relativistic correction $\Delta\alpha$ appears due to the subtraction of the Born term in (28) and is given by the charge radius $\langle r_E^2 \rangle$ and anomalous magnetic moment $\varkappa$; it is absent in non-relativistic theory since there the size effects are contained in the corresponding Born term of the total amplitude T and described by formfactors depending on $\vec{k}$, not on ω.

Analogous formula exists for β too /25/.

One can prove that within relativistic consideration the value α_0 contributes to any observable always together with the addend $\Delta\alpha$ as in (29), including the case of energy shift in external fields . In this sense the α is a genuine polarizability. The term $\Delta\alpha$ compensates those implicit pieces of α_0 which are hidden in the Born term. E.g. in the Dirac theory the electron can transit to the negative energy state under the dipole operator D_z . For this reason the value α_0 (30) for the electron is not zero but $-e^2/4M^3$ (negative value !). Nevertheless, the electric polarizability α equals 0 due to the compensation term $\Delta\alpha = e^2/4M^3$.

Being the derivative of the amplitude T the value α_0 has a cluster structure like (2):

$$\alpha_0 = \alpha_{res} + \alpha_z + \alpha_3 + \alpha_4 \quad . \tag{32}$$

The above mentioned polarizability $\alpha_0 = -e^2/4M^3$ is hidden in the Z-diagram part of (32). Within vector meson dominance model the contributions α_3 and α_4 from intermediate states V also have a piece $-e^2 \langle r_E^2 \rangle /3M$ which is exactly compensated by the corresponding counter-term in (31). This piece is governed by masses of the particles V, not the radius of VN-interaction. A deviation of the VN-vertexes from point-like will give additional contributions to α_3 and α_4 which are as hard to determine as structure effects in γN-interaction, i.e. polarizability. For nucleons these contributions seem to be even greater than α_{res} /9/.

The DRs enable us to link the polarizabilities with photoabsorption cross sections. In the spinless case the relevant formulae /8/ look like

$$\alpha = \int_{\omega_0}^{\omega_M} \frac{d\omega}{2\pi^2\omega^2}\Big[\Big(1+\frac{\omega}{M}\Big)\sigma_{E1} - \frac{\omega}{M}\sigma_{M1} + \Big(2+3\frac{\omega}{M}\Big)\sigma_{E2} - \Big(1+3\frac{\omega}{M}\Big)\sigma_{M2} + \dots\Big] + \alpha^{asym}, \tag{33}$$

$$\beta = \int_{\omega_0}^{\omega_M} \frac{d\omega}{2\pi^2\omega^2}\Big[\Big(1+\frac{\omega}{M}\Big)\sigma_{M1} - \frac{\omega}{M}\sigma_{E1} + \Big(2+3\frac{\omega}{M}\Big)\sigma_{M2} - \Big(1+3\frac{\omega}{M}\Big)\sigma_{E2} + \dots\Big] + \beta^{asym}.$$

Here the terms $\sim 1/M$ and higher multipoles are related, as in (11), with recoil and retardation. Such terms and asymptotic addend allow to escape the conclusion /26/ based on non-relativistic DRs that $\alpha < \beta$ for nucleons. In fact, the integral contributions similar to (33) are equal to /8/

$$\alpha_p^{int} \approx 8, \qquad \beta_p^{int} \approx 6,$$
$$\alpha_n^{int} \approx 10, \qquad \beta_n^{int} \approx 6 \tag{34}$$

at ω_M=1.5 GeV (uncertainties related with double pion photoproduction are $\sim 30\%$). Then the asymptotic contributions, mainly associated with the σ-exchange : $\alpha^{asym} \approx -\beta^{asym} \approx -C/4\pi$, must be

$$\alpha_p^{asym} \approx \alpha_n^{asym} \approx -\beta_p^{asym} \approx -\beta_n^{asym} \approx 6 \tag{35}$$

to reproduce the "experimental" value (25). Corresponding width of the decay $\sigma(700) \to \gamma\gamma$ will be ~ 2.5 keV or even several times as much due to off-shell effects ("Reggezation" /8,17/). Such width appears to be too large from the view point of quark models, where they expect the width to be ~ 1 keV, although it does not contradict a near-threshold enhancement of the $\gamma\gamma \to \pi\pi$ cross section /27/.

An explanation of the large value $\alpha_N \sim 14$ meets difficulties in non-relativistic quark models, in which the electric polarizability has the form (30) and involves only the direct (resonance) term among the whole set (31)-(32) :

$$\alpha = 2 \sum_n{}' \frac{|\langle n | D_z | N\rangle|^2}{E_n - E_N}. \tag{36}$$

For instance, the quark model with oscillator potential gives

$$\alpha = \frac{2}{3}\,\frac{e^2}{m_q \Omega^2} \approx 4.4 \qquad (37)$$

where m_q=340 MeV is quark mass and $\Omega \approx$ 500 MeV is the oscillator frequency (energy level splitting). Simultaneously the model leads to very small proton charge radius, $\langle r_E^2 \rangle = (m_q \Omega)^{-1}$=0.23 fm^2 (exp: 0.66 fm^2). Similar small values are resulted with another potentials as well as from MIT-bag-like models, if potential parameters are chosen to fit the observed spectrum of nucleon resonances. A plausible reason of this discrepancy consists in what both values α and $\langle r_E^2 \rangle$ are essentially generated by additional $q\bar{q}$-pairs (pion periphery of the nucleon) rather than valence quarks /9,28,29/. In particularly, the contribution of the diagrams

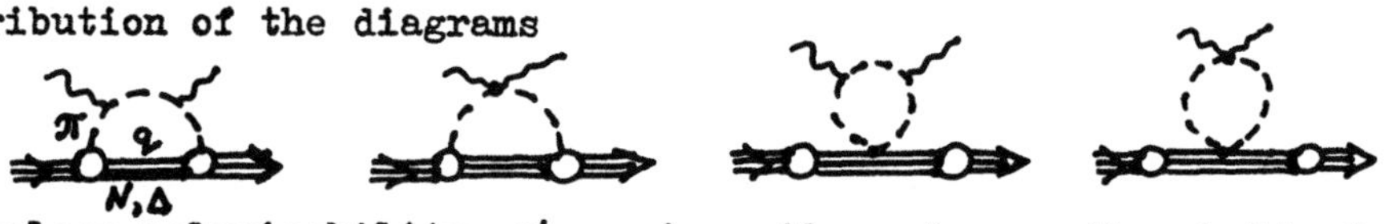

to the nucleon polarizability α_N is $\approx$ 10 and more than half of this value is given by the contact pion-quark interaction associated not with the resonance contribution in (2) but rather with vector intermediate states and Schwinger term.

There is an attractive possibility to test experimentally the significance of the pion periphery for producing the visible size and polarizability by studying the nucleon properties in nuclear medium, in which the pions suffer from rescattering and absorption. Qualitative estimate /10/ based on an analysis of quasideuteron contributions to the sum rule (23) shows that the polarizability of the bound nucleon might be 2-3 times enlarged. However, the measurements /3/ of the photon scattering on lead and carbon beyond the giant dipole resonance region does not support such a large effect and thereby indicates that the predicted increase of the polarizability is caused mainly by the dynamic nuclear polarizability rather than static polarizability of the bound nucleon. Evidently, more elaborated theoretical models are required here.

At the same time the more accurate determination of free proton polarizability and finding of the neutron one stays yet to be an actual aim. This problem can be solved by more precise measuring of the differential cross section of γp-scattering. An independent way to determine the proton polarizabilities consists of measurement of the asymmetry Σ at low energies where we have the following low-energy expansion /30,31/:

$$\Sigma = \left(\frac{d\sigma_\perp}{d\Omega} - \frac{d\sigma_\parallel}{d\Omega}\right)/2\frac{d\sigma}{d\Omega} = \Sigma^{Born} - \frac{1-z^2}{(1+z^2)^2}\cdot\frac{4Mz}{e^2}\omega\omega'\beta + \ldots,$$

$$\frac{d\sigma_\perp}{d\Omega} - \frac{d\sigma_\parallel}{d\Omega} = \left(\frac{e^2}{M}\right)^2(1-z^2)\left(\frac{\omega'}{\omega}\right)^2\left[1+\frac{\omega\omega'}{4M^3}(\varkappa^2+2\varkappa)^2 - 2\frac{M}{e^2}\omega\omega'\alpha + \ldots\right]. \quad (38)$$

The asymmetry depends on the magnetic polarizability β if $\theta \neq 0°$, $90°, 180°$ so that β_p can be found within uncertainties $\sim \pm(1-2)$ from measurements of Σ with an accuracy $\sim \pm 0.02$ (see Fig.2 where the predictions of DRs including $O(\omega^4)$ corrections are depicted; by using the DRs one can obtain the polarizability from higher-energy data, e.g. at 250 MeV).

Up to now a measuring of the neutron polarizability α_n, expected to be as large as α_p, is unresolved problem. Attempts to determine the α_n from precise measurements of the nA-scattering cross section extracting a tiny contribution of the polarization interaction $-\frac{1}{2}\alpha_n E^2$ with the nucleus Coulomb field on the background of nuclear forces had lead only to upper bounds $\alpha_n = 8 \pm 10$ /32/ and 12 ± 10 /33/. Therefore, it is worth to try to obtain the neutron polarizability from the γn-scattering cross section $\sigma_{\gamma n}$ which, in turn, can be studied in the reaction

$$\gamma + d \rightarrow \gamma' + n + p_s \quad (39)$$

in the region of quasi-free scattering off the neutron when the proton is a spectator. Since the neutron electric charge is zero and hence the interference term $\sim e^2\omega^2\alpha_n$ in the $\sigma_{\gamma n}$ is absent (cf.(21)), the polarizability appears in $\sigma_{\gamma n}$ only in the fourth order in ω. Similar terms of the same order originate with another mechanisms like π^0-exchange and that makes finding of the polarizability be impossible without any models. As such a model the DRs can be used and that enables us to evaluate reliably enough all unknown values except $\alpha_n - \beta_n$ and quantitatively interpret the cross section $\sigma_{\gamma n}$ in terms of the polarizability /34/.

Computations of the process (39) in DWIA+MEC with the DR's γN-amplitudes as input (Fig.3) indicate /35/ that the subprocess of γn-scattering dominates when the proton-spectator momentum is $\lesssim 50$ MeV/c and the recoil neutron one is $\gtrsim 100-150$ MeV/c (Fig.4). Besides, the MEC contribution is rather small ($\sim -5\%$ at 100 MeV) and the main correction at energies up to ~ 250 MeV results from a part of the FSI

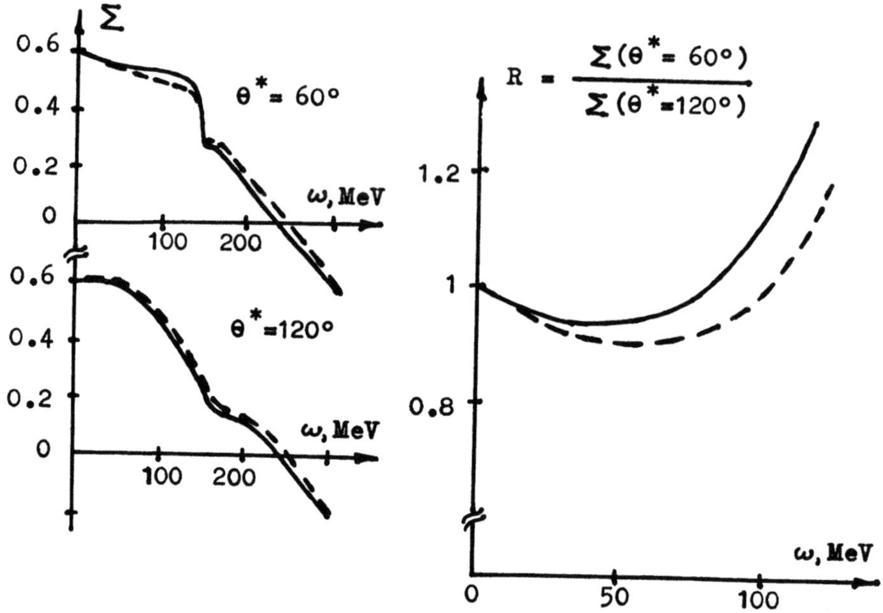

Fig.2 Asymmetry Σ of the $\vec{\gamma}$p-scattering at $\beta_p = 2$ (solid lines) and $\beta_p = -1$ (dashed lines).

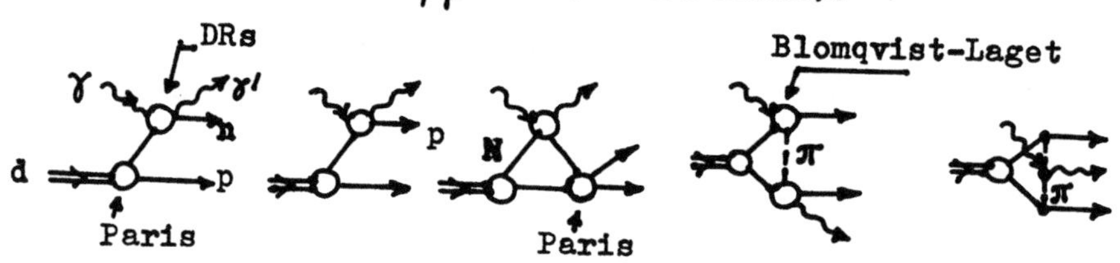

Fig.3 Diagrams of the d(γ, γ'n)p process.

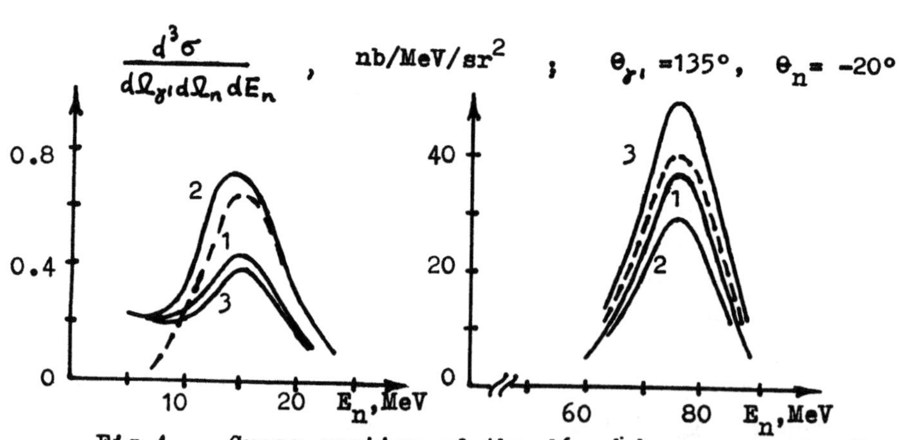

Fig.4 Cross section of the d(γ,γ'n)p process in the neutron quasi-free peak /35/. Dashed lines: scattering on the neutron only ($\alpha_n = 10$), solid lines 1,2,3: all diagrams (respectively $\alpha_n = 10$, 20 and 0).

related with the γp-scattering followed by the charge exchange pn $\rightarrow$ np.

Recent measurement by Göttingen-Mainz group of the reaction (39) with $\sim$20% accuracy at $\theta_{\gamma'}$=90° and 135° and $\omega\sim$110 MeV has resulted to $\alpha_n = 11\,^{+3}_{-11}$ /2/ close to theoretical expectations and improving the upper bounds for α_n. Moreover, because of high sensitivity of the γn-scattering cross section with respect to the relative sign of the coupling constants $g_{\pi NN}$ and $F_{\pi\gamma\gamma}$ determining the π^0-exchange diagram, an evidence was obtained in favour of the negative sign, in accordance with current algebra requirements. It is rather probable that a continuation of this experiment at $\sim$100 and $\sim$250 MeV will result at last to the first establishing the neutron polarizability.

6. Elastic photon scattering on nuclei

Photon scattering on nuclei provides wide possibilities to study both nucleon properties and non-nucleon degrees of freedom. To illustrate this point let us consider the forward γd-scattering /36-38/. The scattering amplitude can be found from total photoabsorption cross section by the DR

$$\mathrm{Re}\, f(\omega) = -\frac{Z^2 e^2}{AM} + \frac{\omega^2}{2\pi^2} \int_{\omega_0}^{\infty} \frac{\sigma(\tilde{\omega})\, d\tilde{\omega}}{\tilde{\omega}^2 - \omega^2} \,. \tag{40}$$

Inserting to (40) the total cross section or the photodisintegration cross section (i.e. without real pions) or theoretical cross section of the photodisintegration without virtual pions (no Siegert theorem) one after another, they can observe the following /37/. At energies above the giant dipole resonance the amplitude found without both real and virtual pions comes fastly to the constant $-Ze^2/M$ which is the coherent sum of the scattering amplitudes off constituent point-like nucleons (Fig.5). Virtual pions in the photodisintegration (or MEC in the γ-scattering) result in $\sim$20% amplifying of the scattering amplitude what looks like a scattering off additional pions inside the nucleus /10,39/. At last, real pion contribution connected mainly with inner structure of nucleons becomes the most important correction at $\omega \gtrsim 50$ MeV. It is extremely interesting that there is an energy region, $\omega \simeq 40$-100 MeV, in which nuclear excitation effects are already almost died out but nucleon structure begins just to be seen. In this region one can hope to study quantitatively non-nucleon (pion) component of nuclei.

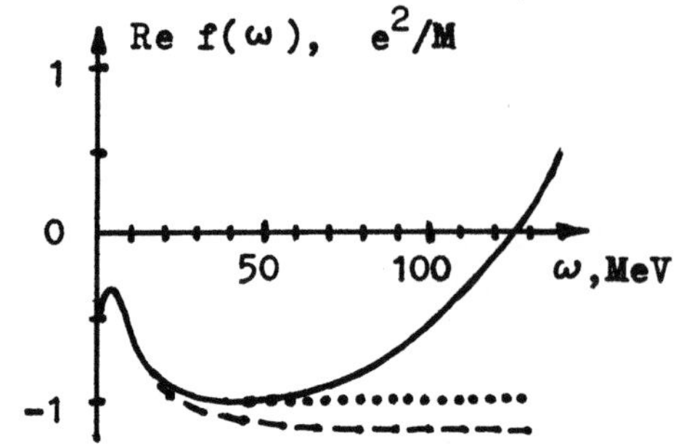

Fig.5 Forward γd-scattering amplitude.
Solid line: total.
Dashed line: no real pions.
Dotted line: no real and virtual pions.

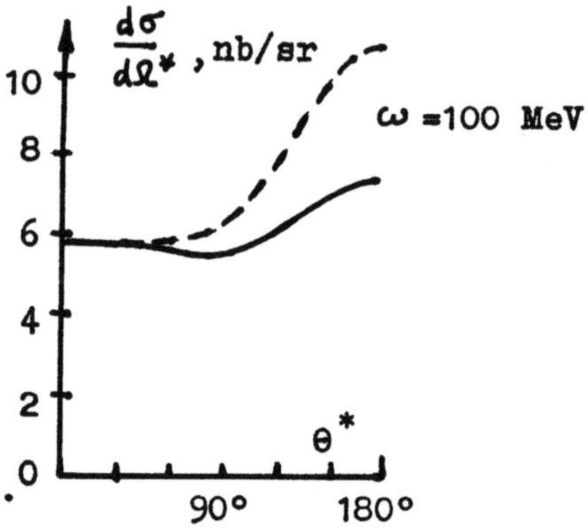

Fig.6 Cross section of γd-scattering in IA with $\alpha_N = \frac{1}{2}(\alpha_p + \alpha_n) = 15$ (solid line) and $\alpha_N = 10$ (dashed line).

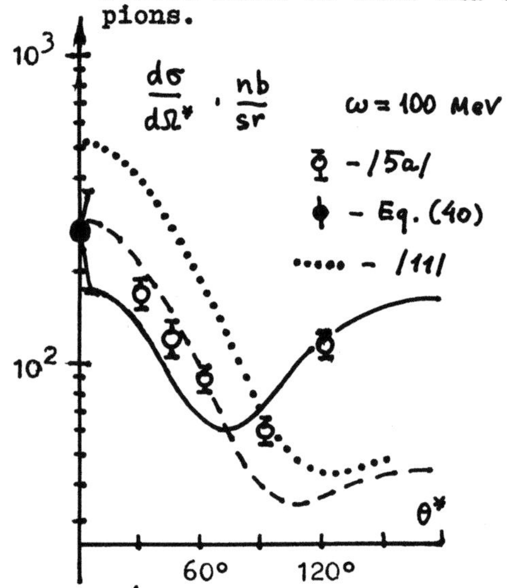

Fig.7 $\gamma\,^4$He-scattering cross section.
Dashed line: IA, solid line: IA+MEC.

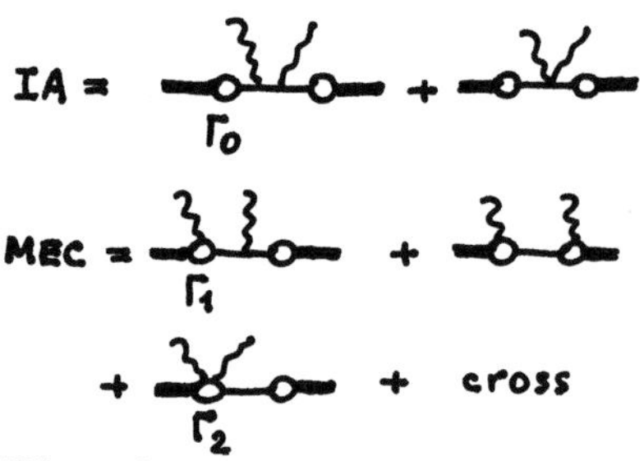

Fig. 8 Diagrams of γA-scattering.

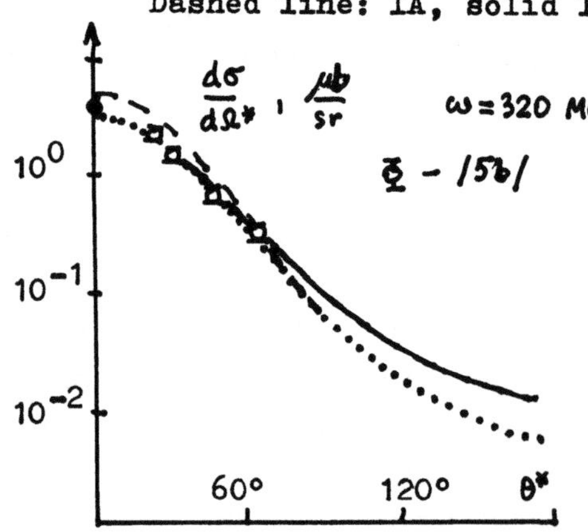

Fig.9 The same as on Fig.7.
Solid line: diagrams 10a and c.

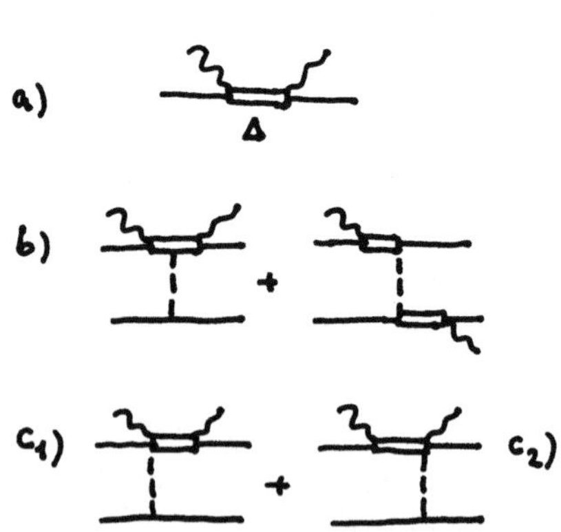

Fig.10 Mesonic corrections to resonance scattering

Photon-nucleus scattering at $\omega \gtrsim 100$ MeV is especially sensitive to the nucleon structure. The γd-scattering cross section $\sigma_{\gamma d}$ calculated in IA (without MEC) with DR's γN-amplitudes is shown in Fig.6. According to /38,40/ MEC contributions increase the cross section by ~ 40-50% at 100 MeV but the nucleon structure effects reduce $\sigma_{\gamma d}$ by factor of ~ 1/5 . The forward γN-scattering amplitude has a positive sign above the pion threshold. Since MEC contribution is negative yet, it diminishes $\sigma_{\gamma d}$ there.

Selection of elastic and inelastic γA-scattering is very difficult task /3,4,12/ which was succeeded to solve for energies below pion threshold. The only successful experiment carring out this selection above the pion threshold was done for ^{4}He at 180 and 320 MeV /5/ apart from very old and rough data on γd-scattering /41/. The experiment /5a/ revealed a strong inconsistency of the isobar-hole model /11/ predictions at 180 MeV both in absolute value and angular dependence (Fig.7). One of the drawbacks of the model /11/ responsible for the discrepancy at forward angles seems to be in treating with E1-scattering which was estimated through a non-relativistic DR like (12). Indeed, impulse approximation with more accurate γN-amplitudes found from relativistic DRs describes the data at $\theta^* \lesssim 60°$ quite well. Nevertheless, both approaches fail to explain the observed increase in the cross section at backward semisphere because of ignoring MEC.

Since direct allowing for MEC for such nucleus as ^{4}He is a complicated problem and because the energy 180 MeV lies far from the magnetic Δ-resonance and hence the main part of the MEC contribution is of E1-type, a simple phenomenological prescription can be used /13/. This receipt was applied successfully to deuteron photodisintegration /42/ and consists of minimal substitution

$$\vec{p} \to \vec{p} - e\,\vec{A}(\vec{r}) \tag{41}$$

in the effective vertex

$$\Gamma_0(p) = \left(-\frac{p^2}{2M} - \varepsilon\right)\phi(p) \tag{42}$$

describing a bound nucleon of the energy ε and wave function $\phi(p)$. Due to p-dependence of the vertex Γ_0 such substitution generates additional e.m. vertexes

$$\Gamma_1(\vec{p},\vec{k}_1) = -e\int_0^1 ds_1 \left(\vec{\varepsilon}_1 \frac{\partial}{\partial \vec{p}}\right) \Gamma_0(\vec{p} + s_1\vec{k}_1) \,, \tag{43}$$

$$\Gamma_2(\vec{p},\vec{k}_1,\vec{k}_2) = e^2 \int_0^1 ds_1 \int_0^1 ds_2 \left(\vec{\varepsilon}_1 \frac{\partial}{\partial \vec{p}}\right)\left(\vec{\varepsilon}_2 \frac{\partial}{\partial \vec{p}}\right) \Gamma_0(\vec{p}+s_1\vec{k}_1+s_2\vec{k}_2) \quad (44)$$

which contribute to the two-photon process of γA-scattering (Fig.8). Here $\vec{\varepsilon}_i$ and $\vec{k}_i$ are polarizations and momenta of photons. Physically such receipt is justified because the additional vertexes (43)-(44) restore the gauge invariance of IA for off-shell nucleons without changing the charge density. Then Siegert-like arguments guarantee that MEC contributions to electric multipoles in the long wavelength limit are coming into account. Alternative although essentially equivalent approach /38,40,43/ consists in special choosing of the e.m. potential gauge to suppress an interaction with the current density. Diagrams of Fig.8 computed with an oscillator wave function and high-energy approximation for the nucleon propagator (at $\varepsilon \ll \omega \ll M$) provide necessary increase in the ^{4}He-scattering at backward angles (Fig.7). At small angles the MEC corrections diminish the cross section as expected.

In the Δ-resonance region the main correction to IA is not electric but magnetic contribution caused in part by modifying of the Δ-isobar propagation in nuclear medium /11/. The scale of this effect is illustrated by that IA with DR-amplitudes gives the forward cross section ~ 1.5 times greater than predictions of the isobar-hole model and optical estimation

$$\frac{d\sigma^{opt}}{d\Omega^*}(0^\circ) = \left| \frac{\omega^*}{4\pi} \sigma_{tot} \right|^2 \simeq 3.6\ \mu b \quad (45)$$

(real part of the γA-amplitude is negligible at resonance). An accurate consideration of isobar properties in nuclei fulfilled in /11/ results in agreement with the data /5b/ obtained at $\theta^* \lesssim 60°$ (Fig.9). At higher angles one can expect to find another mechanism of modifying isobar properties related with emission or absorption at the $\gamma N\Delta$ -vertex of virtual pions (Fig.10c). Corresponding contact interaction appears due to momentum-dependence of $\pi N\Delta$ vertex (cf.(43)) and is mainly of E1-type in contrast to the rest resonance contributions (Fig.10a-b). Diagrams 10c$_1$ and 10c$_2$ compensate each other if both photons have the same helicity and therefore do not contribute at forward directions . But for backward directions they double approximately the cross section (see dashed line in Fig.9 which gives the result of calculations of diagrams 10a and 10c ; instead of taking into account diagrams 10b modifying the Δ-propagator the propagator parameters, E_Δ =300 MeV and Γ_Δ =140 MeV, were chosen in such a way to reproduce roughly a shift and broading of the resonance peak /11,44/.

We may summarize by saying that impulse approximation with γN-amplitudes from relativistic dispersion relations provide a reasonable ground for discussing γA-scattering. Mesonic effects here are usually of the order of 50-100% or even higher in some cases and that gives a favourable possibility for quantitative studying of them.

This possibility is fortified by general progress in construction of c.w. electron machines, high-intensity tagged photon beams and large cristall γ-detectors of high resolution. All that gives a hope that studying of γN and γA-scattering at intermediate energies will become far more systematic and realize its powers drowsed.

References

1 P.S.Baranov et al. Phys.Lett. 52B (1974) 122.
2 K.W.Rose et al. Phys.Lett. 234B (1990) 460; Nucl.Phys.A .
3 A.Baumann et al. Phys.Rev. C38 (1988) 1940;
K.P.Schelhaas et al. Nucl.Phys. A489 (1988) 189; A506 (1990) 307.
4 E.Hayward, B.Ziegler. Nucl.Phys. A414 (1984) 333.
5 E.J.Austin et al. a) Phys.Rev.Lett. 57 (1986) 972;
b) ibid. 61 (1988) 1922.
6 Yu.M.Antipov et al. Phys.Lett. 121B (1983) 445; Zeit.Phys. C26 (1984) 39 and 495.
7 T.A.Aibergenov et al. Czech.J.Phys. B36 (1986) 948.
8 V.A.Petrun'kin. Fiz.Elem.Ch.At.Yad. (Sov.J.Part.Nucl.) 12 (1981) p. 692
9 A.I.L'vov, V.A.Petrun'kin. Preprint FIAN No.258 (1988).
10 M.Ericson. Progr.Theor.Phys.Suppl. 91 (1987) 235.
11 J.H.Koch, E.J.Moniz, N.Ohtsuka. Ann.Phys. 154 (1984) 99.
12 J.Vesper, D.Drechsel, N.Ohtsuka. Nucl.Phys. A466 (1987) 652.
13 A.I.L'vov, V.A.Petrun'kin. VANT (Vopr.At.Nauki i Tekhn.), ser.YFI 8 /8/ (1989) 84.
14 L.S.Brown. Phys.Rev. 150 (1966) 1338.
15 D.M.Akhmedov, L.V.Filkov. Yad.Fiz. (Sov.J.Nucl.Phys.) 33 (1981) p. 1083.
16 D.Morgan, M.R.Pennington. Phys.Lett. 192B (1987) 207.
17 A.I.L'vov. Yad.Fiz. 34 (1981) 1075.
18 H.D.I.Abarbanel, M.L.Goldberger. Phys.Rev. 165 (1968) 1594.
19 G.Buschhorn et al. Phys.Lett. 37B (1971) 211.
20 J.Ahrens et al. in "Physics with MAMI A", Institut für Kernphysik, Mainz (1988) 1.
21 A.M.Nathan, private communication.
22 I.Guiasu, C.Pomponiu, E.E.Radescu. Ann.Phys. 114 (1978) 296.
23 W.Pfeil, H.Rollnik, S.Stankowski. Nucl.Phys. B73 (1974) 166.
24 T.Ishii et al. Nucl.Phys. B165 (1980) 189.
25 A.I.L'vov. Preprint FIAN No.344 (1987).
26 E.Hayward. Phys.Lett. 223B (1989) 133.
27 A.Courau et al. Nucl.Phys. B271 (1986) 1;
S.Minami. Progr.Theor.Phys. 81 (1989) 1064.
28 E.Oset, R.Tegen, W.Weise. Nucl.Phys. A426 (1984) 456.
29 R.Weiner, W.Weise. Phys.Lett. 159B (1985) 85.
30 A.I.L'vov, V.A.Petrun'kin. VANT, ser.OYF 2/35/ (1986) 56.

31 L.C.Maximon. Phys.Rev. C39 (1989) 347.
32 L.Koester, W.Waschkowski, J.Meier. Zeit.Phys. A329 (1988) 229.
33 J.Schniedmayer, H.Rauch, P.Riehs. Phys.Rev.Lett. 61 (1988) 1065.
34 A.I.L'vov. Yad.Fiz. 42 (1985) 919.
35 M.I.Levchook, A.I.L'vov, V.A.Petrun'kin. VANT, ser.OYF 2/38/-3/39/ (1987) 96.
36 J.Ahrens, L.S.Ferreira, W.Weise. Nucl.Phys. A485 (1988) 621.
37 N.V.Maksimenko, L.G.Teplyakov; V.A.Petrun'kin. VANT, ser.OYF 2/35/ (1986) 96.
38 M.Weyrauch, H.Arenhövel. Nucl.Phys. A408 (1983) 425.
39 P.Christillin, M.Rosa-Clot. Nuovo Cim. 28A (1975) 29.
40 M.Weyrauch. Phys.Rev. C41, No.2 (1990).
41 R.S.Jones et al. Phys.Rev. 128 (1962) 1357.
42 A.I.L'vov. VANT, ser.OYF 2/35/ (1986) 51,53.
43 J.L.Friar, S.Fallieros. Phys.Rev. C34 (1986) 2029.
44 P.Christillin. IPNO/TH 87-90 (1989); Phys.Reports .

Photon and Pion Scattering on the Basis of Quark Models

M.G. Huber, B.Ch. Metsch and H.R. Petry

Institut für Theoretische Kernphysik der Universität Bonn
Nußallee 14-16, D-5300 Bonn, Germany

Abstract: Within the framework of constituent quark models of simple and complex hadrons it is possible to investigate the interrelations of scattering observables and dynamical assumptions about the underlying quark structure of the participating hadrons. In the case of simple hadrons, here baryons, the inclusion of the coupling of bound q^3-states to meson-baryon decay channels, accounts for the full complexity of the fact that all excited baryonic states are in fact resonances. We present here such a description of the baryonic excitation spectrum within the non-relativistic quarkmodel with *all* terms of the Fermi-Breit interaction. In this case the very hadronic vertices that are responsible for the dressing of hadrons also provide a semi realistic description of various scattering observables for pion scattering and photo-production of pions. The treatment of more complex hadronic systems within the framework of a quark shell model of nuclei is then used to investigate the role of the quark Pauli principle in intermediate energy excitations of nuclei with pionic or photonic probes.

1. Introduction

Constituent quark models allow for an *effective* description of the internal dynamics of simple and complex hadrons in terms of those degrees of freedom (i.e. the valence quarks) that are relevant for the hadronic excitations upto excitation energies of roughly 1 GeV. This goes along with the assumption that gluon degrees of freedom can be eliminated in favor of interactions between valence quarks, which indeed seems to be appropriate, since upto now there is no evidence for the *explicit* excitation of gluonic degrees of freedom in the above mentioned energy domain. Such an approach is especially very useful in relating a variety of experimental phenomena, such as e.g. mass spectra and static properties like radii and electromagnetic moments to an underlying model-Hamiltonian, where specific dynamical assumptions concerning quark confinement, quark-quark interactions and hadronic coupling vertices enter.

In the present contribution the interrelation of specific assumptions concerning the underlying quark dynamics and in particular observables for pion and photon scattering on hadrons will be examplified by an investigation of pion-nucleon scattering and photoproduction of pions within a constituent quark model of baryons, with explicit coupling of baryonic resonances to meson-baryon channels as well as by a study of the role of the quark Pauli principle for pion and photon induced intermediate energy excitations of nuclei within the framework of a quark shell model.

Concerning the first topic the central issue of the present contribution is a discussion of the relevance of the coupling of (bound) states of three quarks (*"bare baryons"*, within the framework of the non-relativistic quark model which is briefly sketched in section 2) to various baryon-meson decay channels for the structure of the baryons themselves (*"dressing of baryons"*, outlined in section 3) and for the description of meson-baryon scattering and photoproduction of mesons, which is discussed in section 4. After a brief overview of a quark shell model of nuclei in section 5, where it is expected that excitations in specific channels should not occur because of the quark Pauli principle the detectability of such an effect is investigated by a case study of pion and photon scattering off ^{4}He in section 6. A summary and some conclusions constitute section 7.

2. Baryons as q^3-configurations

Within the framework of the constituent quark model [1-5] the internal Hamiltonian for a baryon as a system of three quarks is given by

$$\hat{H}_3 = M + T + V^{\text{conf.}} + V^{BF}, \tag{2.1}$$

where the mass and the kinetic energy of the relative motions are given by

$$M = \sum m_i \tag{2.2}$$

$$T = \sum p_i^2/(2m_i) - T_{cm}. \tag{2.3}$$

The dynamics is described by the central and non-central parts of the confinement potential

$$V^{\text{conf.}} = \sum_{i<j} \left[V_c^{\text{conf.}}(i,j) + V_{ls}^{\text{conf.}}(i,j) \right], \tag{2.4}$$

where

$$V_c^{\text{conf.}}(i,j) = a + b \cdot r_{ij} \tag{2.4a}$$

is a linear confinement potential for large distances and

$$V_{ls}^{conf.}(i,j) = -\frac{V_c^{conf.}(i,j)'}{r_{ij}} \left\{\frac{1}{2m_i^2}(\mathbf{r}_{ij}\times\mathbf{p}_i)\cdot\mathbf{S}_i - \frac{1}{2m_j^2}(\mathbf{r}_{ij}\times\mathbf{p}_j)\cdot\mathbf{S}_j\right\} \tag{2.4b}$$

is the corresponding Thomas precession term. The residual quark-quark interaction is assumed to be given by the Fermi-Breit interaction:

$$V^{BF}=\sum_{i<j}\left[V_c^{BF}(i,j)+ V_\sigma^{BF}(i,j) + V_t^{BF}(i,j) + V_D^{BF}(i,j) + V_{o-o}^{BF}(i,j) +V_{ls}^{BF}(i,j)\right], \tag{2.5}$$

with the Coulomb term

$$V_c^{BF}(i,j) = -\frac{2}{3}\,\alpha_s\,\frac{1}{r_{ij}} \tag{2.5a}$$

and all the spin dependent terms

$$V_\sigma^{BF}(i,j) = \frac{2}{3m_i m_j}\,\mathbf{S}_i\cdot\mathbf{S}_j\left[\Delta_{(r_{ij})}V_c^{BF}\right], \tag{2.5b}$$

$$V_t^{BF}(i,j) = -\frac{1}{3m_i m_j}\left[3(\mathbf{S}_i\cdot\hat{\mathbf{r}}_{ij})(\mathbf{S}_j\cdot\hat{\mathbf{r}}_{ij}) - \mathbf{S}_i\cdot\mathbf{S}_j\right]\left[D_{(r_{ij})}V_c^{BF}\right], \tag{2.5c}$$

$$V_D^{BF}(i,j) = \frac{1}{8}\left[\frac{1}{m_i^2} + \frac{1}{m_j^2}\right]\left[\Delta_{(r_{ij})}V_c^{BF}\right], \tag{2.5d}$$

$$V_{o-o}^{BF}(i,j)=-\frac{1}{2m_i m_j}\left\{\left\{\mathbf{p}_i\cdot\left[V_c^{BF}\mathbb{1} - \mathbf{r}_{ij}\,\frac{V_c^{BF\prime}}{r_{ij}}\,\mathbf{r}_{ij}\right]\cdot\mathbf{p}_j\right\}\right\}, \tag{2.5e}$$

$$V_{ls}^{BF}(i,j) = \frac{V_c^{BF\prime}}{2r_{ij}}\left[\frac{1}{m_i^2}(\mathbf{r}_{ij}\times\mathbf{p}_i)\cdot\mathbf{S}_i - \frac{1}{m_j^2}(\mathbf{r}_{ij}\times\mathbf{p}_j)\cdot\mathbf{S}_j - \frac{2}{m_i m_j}\left[(\mathbf{r}_{ij}\times\mathbf{p}_j)\cdot\mathbf{S}_i - (\mathbf{r}_{ij}\times\mathbf{p}_i)\cdot\mathbf{S}_j\right]\right]. \tag{2.5f}$$

Here we used the abreviations

$$D_\rho = \left[\frac{d^2}{d\rho^2} - \rho^{-1}\frac{d}{d\rho}\right],$$

$$V_c^{BF\prime}(\rho) = \frac{d}{d\rho}\,V_c^{BF}(\rho)$$

and

$$\left\{\left\{\mathbf{p}\cdot\mathbf{F}\cdot\mathbf{q}\right\}\right\} = \frac{1}{4}\left[\mathbf{p}\cdot\mathbf{q}:\mathbf{F} + \mathbf{p}\cdot\mathbf{F}\cdot\mathbf{q} + \mathbf{q}\cdot\mathbf{F}\cdot\mathbf{p} + \mathbf{F}:\mathbf{p}\cdot\mathbf{q}\right]. \tag{2.6}$$

The parameters a, b and α_s (the effective strong coupling constant) and the quark masses m_f are considered as free adjustable parameters. The eigenstates of the Hamiltonian (2.1) can be determined by diagonalization within a large harmonic oscillator basis, using the variational principle. In such a non-pertrubative treatment the Hamilton operator has to be regularized by an appropriate procedure. Here the Coulomb term is approximated by a sum of Gaussians in the interval $[r_{min}, r_{max}]$:

$$V_c^{BF}(i,j) = -\frac{2}{3}\alpha_s \frac{1}{r_{ij}} \cong \sum_{n=1}^{5} \alpha_n \, e^{-\gamma_n^2 r_{ij}^2}, \quad r_{ij} \in [r_{min}, r_{max}]$$

which leads to a potential with a finite depth of about 1 GeV at the origin.

3. Dressing and Decay of Baryons

Upto now the baryon has been treated as a bound (q^3)-configuration. We extend this concept by introducing explicitely the coupling to meson-baryon decay channels. The elimination of the meson fields then leads to an effective Hamilton operator for the baryon system which is manifestly energy dependent and is constructed such as to describe both stable baryons and baryon resonances on the same footing.

We consider the Hamiltonian

$$H = \hat{H}_3 + \hat{H}_2 + V_{BM} + V_{MB} \tag{3.1}$$

where $\hat{H}_3$ and $\hat{H}_2$ describe the internal dynamics of the bare baryon (see Eq.(2.1)) and the bare meson, respectively, with corresponding eigenstates $|\nu\rangle$ and $|\mu\rangle$:

$$\hat{H}_3 |\nu\rangle = \varepsilon_\nu |\nu\rangle, \tag{3.2a}$$

$$\hat{H}_2 |\mu;\mathbf{k}\rangle = \omega_\mu(\mathbf{k}) |\mu\rangle, \tag{3.2b}$$

where

$$\omega_\mu(\mathbf{k}) = (m_\mu^2 + \mathbf{k}^2)^{1/2}$$

and ν and μ denote a complete set of quantum numbers that uniquely specify a hadronic state and where $\mathbf{k}$ denotes the meson momentum and m_μ its mass. The operator $V_{BM} = V_{MB}^\dagger$ (to be specified below) describes the coupling of the (bare) baryon to a baryon-meson channel. The meson baryon scattering operator can be written as

$$T_{MM}(z) = V_{MB}\ \mathcal{G}_B(z)\ V_{BM} \tag{3.3}$$

where the resolvent projected on the baryon space is given by

$$\mathcal{G}_B(z) := \left[z - H_3(z)\right]^{-1}, \tag{3.4}$$

with the effective baryon Hamilton operator

$$H_3(z) := \hat{H}_3 + V_{BM} \left[z - H\right]^{-1} V_{MB}. \tag{3.5}$$

Upon inserting the eigenmodes of $H_3(z)$

$$H_3(z)\ |B_\nu(z)\rangle = \lambda_\nu(z)\ |B_\nu(z)\rangle \tag{3.6a}$$

$$\langle\tilde{B}_\nu(z)|\ H_3(z) = \lambda_\nu(z)\ \langle\tilde{B}_\nu(z)| \tag{3.6b}$$

one finds for the transition amplitude

$$\langle\mu,\nu;\mathbf{k}'|T_{MM}(z)|\mu,\nu;\mathbf{k}\rangle = \sum_{\tilde{\nu}} \langle\mu,\nu;\mathbf{k}'|V_{MB}|B_{\tilde{\nu}}(z)\rangle \frac{1}{z - \lambda_{\tilde{\nu}}(z)} \langle\tilde{B}_{\tilde{\nu}}(z)|V_{BM}|\mu,\nu;\mathbf{k}\rangle \tag{3.7}$$

In view of Eq.(3.7) we define the position E^R_ν of a resonance by the condition that

$$\mathcal{Re}\left[\lambda_\nu(E^R_\nu)\right] = E^R_\nu. \tag{3.8a}$$

Then the corresponding width is given by

$$\frac{1}{2}\Gamma^R_\nu = -\mathcal{Im}\left[\lambda_\nu(E^R_\nu)\right]. \tag{3.8b}$$

As a first approximation for the evaluation of the matrix elements of the effective Hamilton operator $H_3(z)$ of Eq.(3.5) the following assumptions are made: first of all we restrict the intermediate states to the (π,η)-(N,Δ) channels only. Furthermore the mesons are treated as elementary fields that couple pointlike to the quarks in the baryon via a non-relativistic reduction of the usual γ_5-coupling:

$$V_{BM} + V_{MB} = \sum_{\mu=\pi,\eta} \left[\mathcal{L}_{\mu qq} + \mathcal{L}^\dagger_{\mu qq}\right], \tag{3.9}$$

where $\mathcal{L}^\dagger_{\mu qq}$ is a sum of single quark operators

$$\mathcal{L}^\dagger_{\mu qq} = \sum_{i=1}^{3} \mathcal{L}^\dagger(i) = 3\ \mathcal{L}^\dagger(3)$$

with

$$\mathfrak{L}^{\dagger}(3) = - i \frac{g_{\mu qq}}{2m_q} [2\pi]^{-3/2} [2\omega_\mu(k_\mu)]^{-1/2} \vec{\sigma}_3 \cdot \mathbf{K}_\mu \mathfrak{F}_\mu(3)$$
$$\delta^3(\mathbf{p}_1-\mathbf{p}_1') \, \delta^3(\mathbf{p}_2-\mathbf{p}_2') \, \delta^3(\mathbf{p}_3 - \mathbf{p}_3' - \mathbf{k}_\mu) \qquad (3.10)$$

whre m_q is the constituent quark mass and $\mathfrak{F}_\mu$ is the flavour operator.

We investigated two versions of these vertex functions: in the so called static approximation we set $\mathbf{K}_\mu = \mathbf{k}_\mu$ (Version 1) or alternatively, taking into account recoil, we put $\mathbf{K}_\mu = \mathbf{k}_\mu-(m_\mu/m_q)\mathbf{p}_3'$ (Version 2).

We thus arrive at the following approximate expression for the matrix elements for the resolvent $(z\text{-}H)^{-1}$ in Eq.(3.5):

$$\langle\mu,\bar{\nu};\mathbf{k}\,|(z\text{-}H)^{-1}|\,\mu,\bar{\nu};\mathbf{k}\rangle = \left[z - \sqrt{(M_{\bar{\nu}}^2+\mathbf{k}^2)} - \omega_\mu(\mathbf{k}) \right]^{-1} \qquad (III.11)$$

where $M_{\bar{\nu}}$ is the observed mass of the baryon in the intermediate state $|\bar{\nu}\rangle$. Here it has been assumed, that the eigenvalues of H can be suitably approximated by substituting the physical masses. This we feel to be important to get the correct energy dependence of our effective Hamilton operator, especially in the vicinity of the various baryon-meson thresholds.

As discussed above for the spectrum of the bare Hamiltonian $\hat{H}_3$ the eigenstates of the effective Hamilton operator $H_3(z)$ are obtained by diagonalization in a finite basis including 6 major oscillator shells, where the resonance positions and widths are defined by Eqs.(3.8).

The resulting spectra for nucleon and delta resonances, which are quantatively almost identical for both versions of the vertex operator of Eq.(3.10) are compared to experimental data in Figs.1-4. As can be seen from these figures the coupling to the meson baryon channels leads to appreciable energy shifts that are comparable in magnitude to the splittings induced by the Fermi-Breit interaction. Moreover, it is found that only by taking into account this coupling *all* spin dependent terms of the Fermi-Breit interaction can be included. In addition also a satisfactory description of the partial decay widths of especially the lowest positive and negative parity resonances into the considered channels is obtained. Furthermore, the coupling to meson baryon channels produces a selective downward shift of the first excited nucleon resonance compared to the other excited positive parity states, thus yielding a satisfactory description of the splittings between these resonances, but the centroid of this group of states is calculated to high by roughly 200 MeV. Pilot studies indicate that this is probably not due to the restricted configuration space. More likely the approximations made for the calculation of the vertex matrix elements, in particular the single particle nature of the coupling given in Eqs.(3.10) are the origin of this defect. This is also reflected by the fact that the decay widths of higher lying resonances are in general calculated too small.

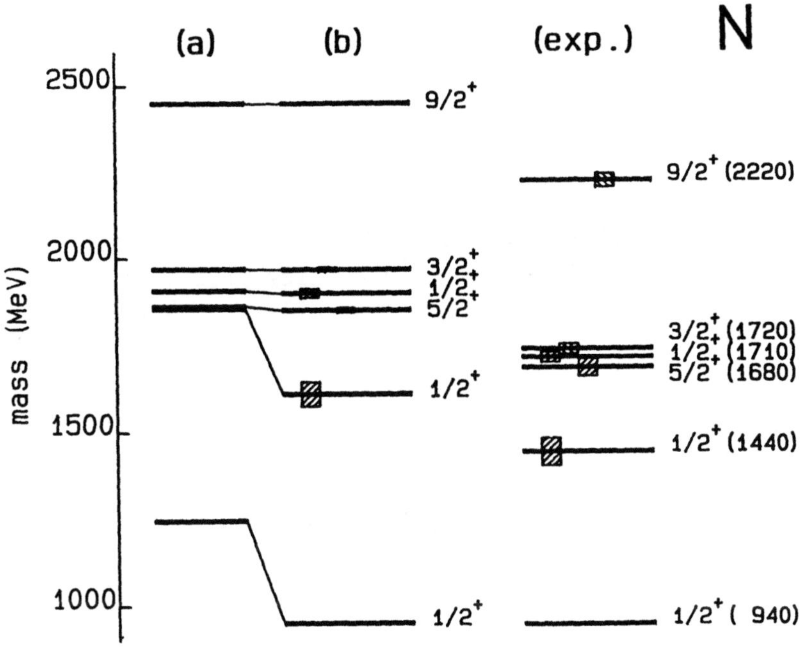

Fig.1. Comparison of calculated and experimental positions and widths of positive parity *nucleon resonances*. Column (a) shows the spectrum of the bare Hamiltonion of Eq.(2.1) and (b) the spectrum of the effective Hamiltonian of Eq.(3.5). Experimental data are taken from [6]. Hatched rectangles represent the partial width for π-decay.

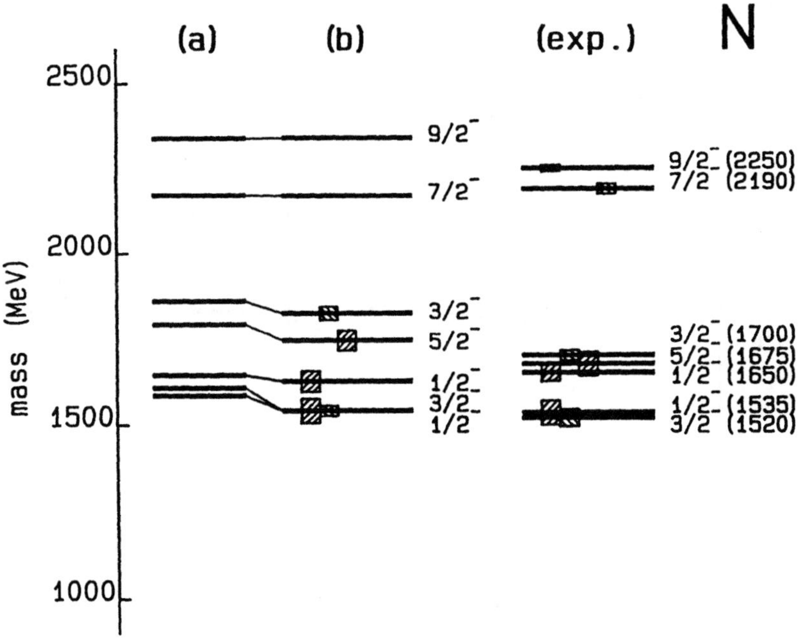

Fig.2. Spectrum of negative parity *N-resonances*, see also Fig.1.

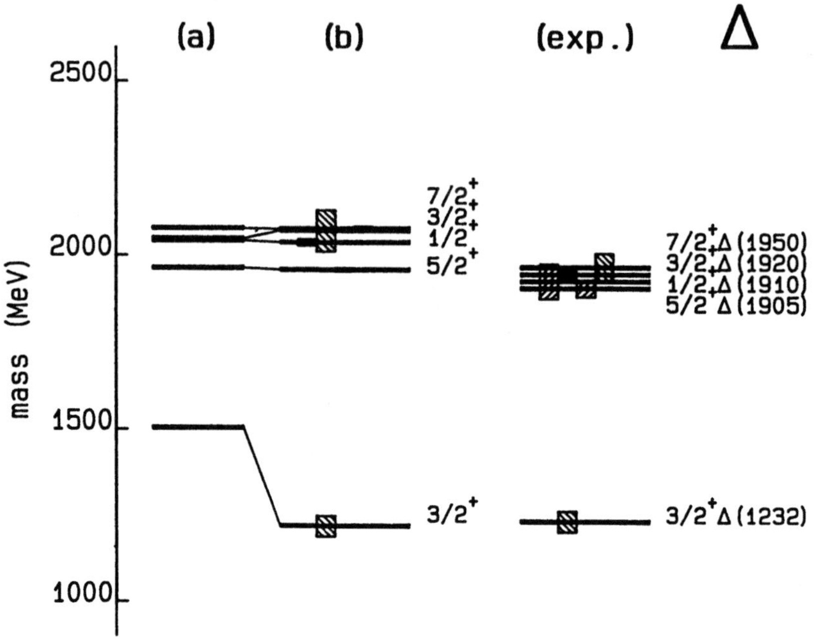

Fig.3. Spectrum of positive parity Δ-*resonances*, see also Fig.1.

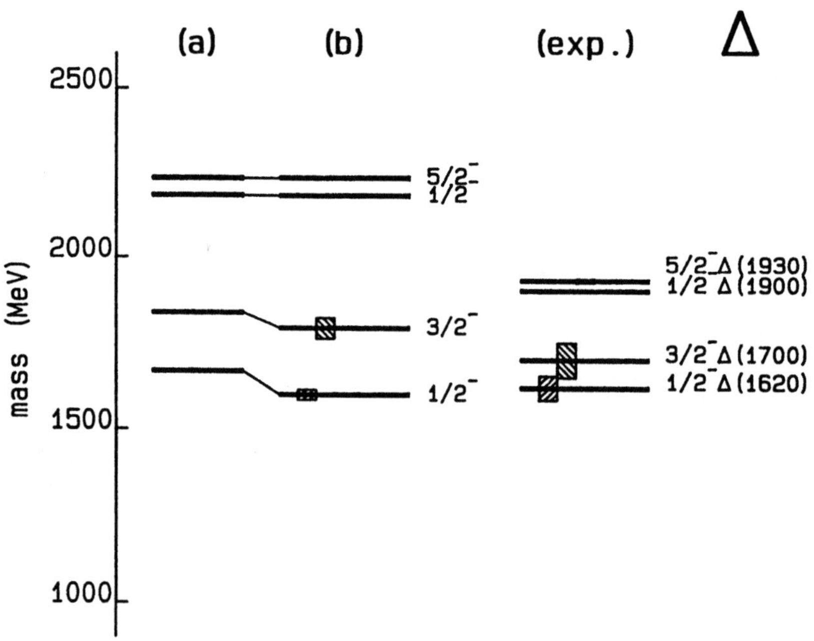

Fig.4. Spectrum of negative parity Δ-*resonances*, see also Fig.1.

4. Pion-Nucleon Scattering and Photoproduction of Mesons

As discussed above the scattering amplitude for resonant pion-nucleon scattering can be written as

$$<\pi,N;\mathbf{k}'|T_{MM}(E)|\pi,N;\mathbf{k}> =$$
$$\sum_{\bar{\nu}} <\pi,N;\mathbf{k}'|V_{MB}|B_{\bar{\nu}}(E)> \frac{1}{z-[M_{\bar{\nu}}(E)-i\Gamma_{\bar{\nu}}(E)/2]} <\tilde{B}_{\bar{\nu}}(E)|V_{BM}|\pi,N;\mathbf{k}> \quad (4.1)$$

It should be noted, that in the present treatment the vertex functions, apart from the usual momentum dependence of the form factors are manifestly energy dependent as are the positions and widths of the intermediate resonances.

The total elastic cross sections for π-p scattering is displayed in Fig.4.1. Although the description in the delta resonance region is quite satisfactory at higher energies all calculated cross sections are too low, in consonance with the remarks made above for the coupling strengths of the scattering channels to resonances at higher energies. Similar remarks apply to the analysis of the differential cross sections where especially the structures found at backward angles cannot be reproduced.

Within the same framework it is also possible to calculate the resonant contribution (see Fig. 4.2a) for the photoproduction of pions. However, especially for charged particle production the dynamics must be supplemented by non-resonant scattering effects, like the coupling of the photon to the pion cloud, the seagull photon-pion-nucleon coupling, motivated by gauge invariance, and the crossed diagram, see Fig. 4.2b-d. The relative importance of the various contributions to the differential cross section is displayed in Fig.4.3. Although for charged particle production the non-resonant contributions are by far dominant it should be stressed that they are calculated with the same vertices that also determine the resonant contribution and thus also constitutes a test of the wave functions calculated in the present model. As can be seen from Fig.4.4 a reasonable description of not only the differential cross section, but also of the polarization observables is obtained.

In Fig.4.5 the total cross sections for pion production of eta mesons and eta-photoproduction on the nucleon, calculated with both versions of the meson nucleon vertices of Eq.(3.10), are compared to the (scarce) experimental data. In contrast to all processes considered sofar a clear distinction between the two types of coupling appears. An analysis of these processes shows, that this is mainly due to the fact that only by the inclusion of the recoil term in the meson-baryon vertex, which enters both in the effective Hamiltoninan as well as in the scattering amplitude, the lowest $N(1/2^-)$-resonance is strongly coupled to the eta-nucleon channel in accordance with experiment [6]. In this manner a semi-realistic description of the experimental data is in fact obtained.

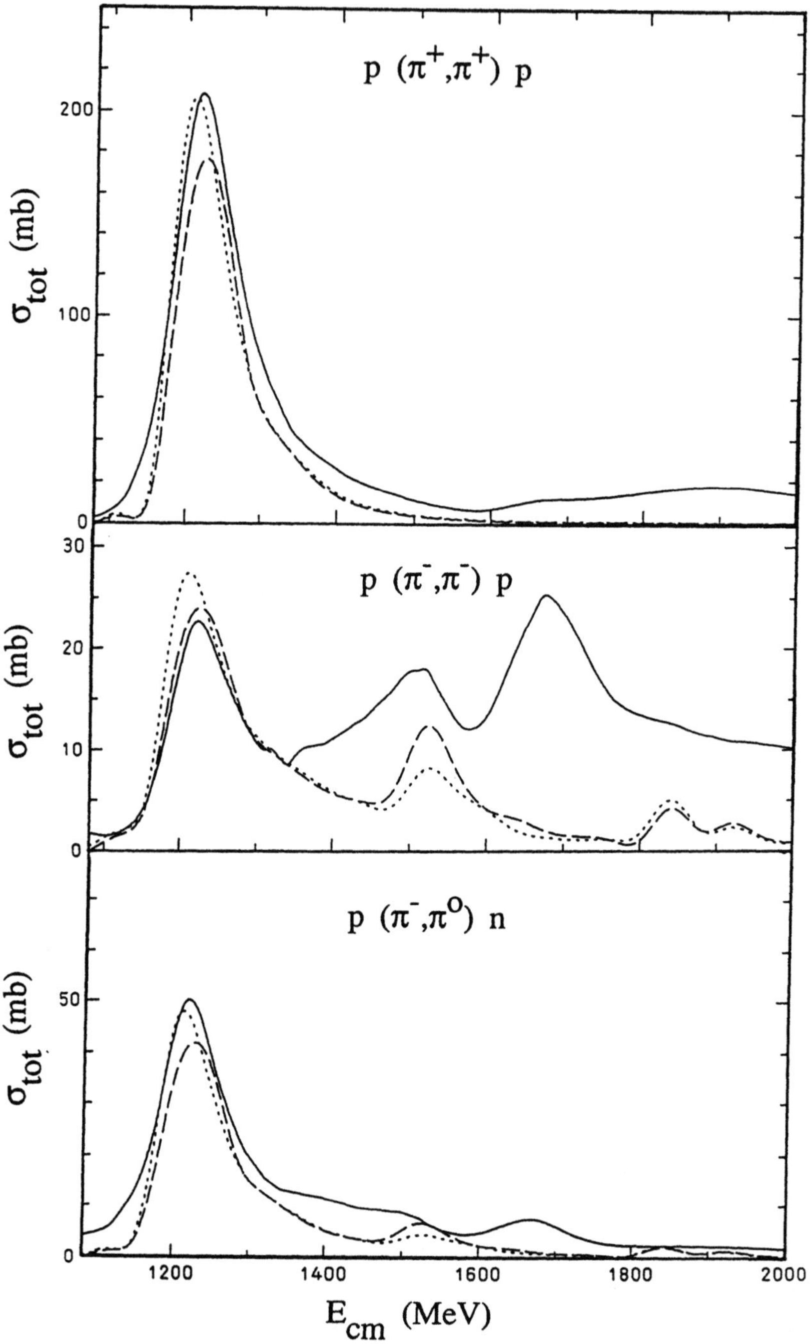

Fig.5. Comparison of the calculated and experimental total cross section for elastic π-N scatter ing. Experimental data are obtained from the Höhler-phase shift analysis and does not include Coulomb interactions [7]. Dashed and dotted lines correspond to version 1 and 2 of the meson baryon vertex (see Eq.(3.10)).

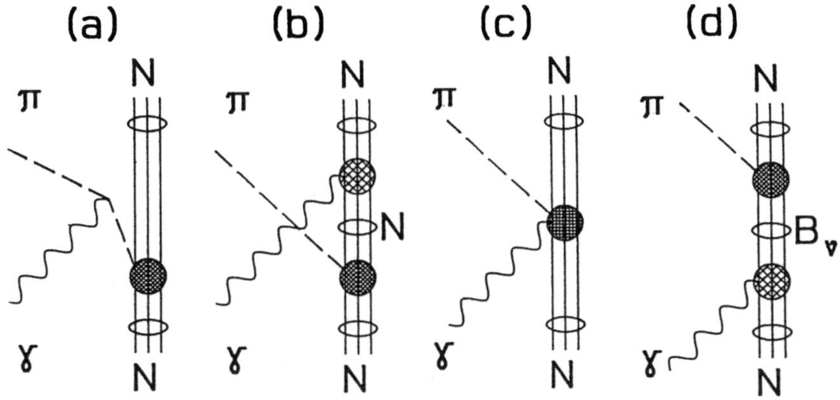

Fig.6. Contributions to the photo-induced pion production on the nucleon. (a) Pion knock-out, (b) crossed graph, (c) seagullgraph, (d) resonant contribution. (a) and (c) contribute to charged pion production only.

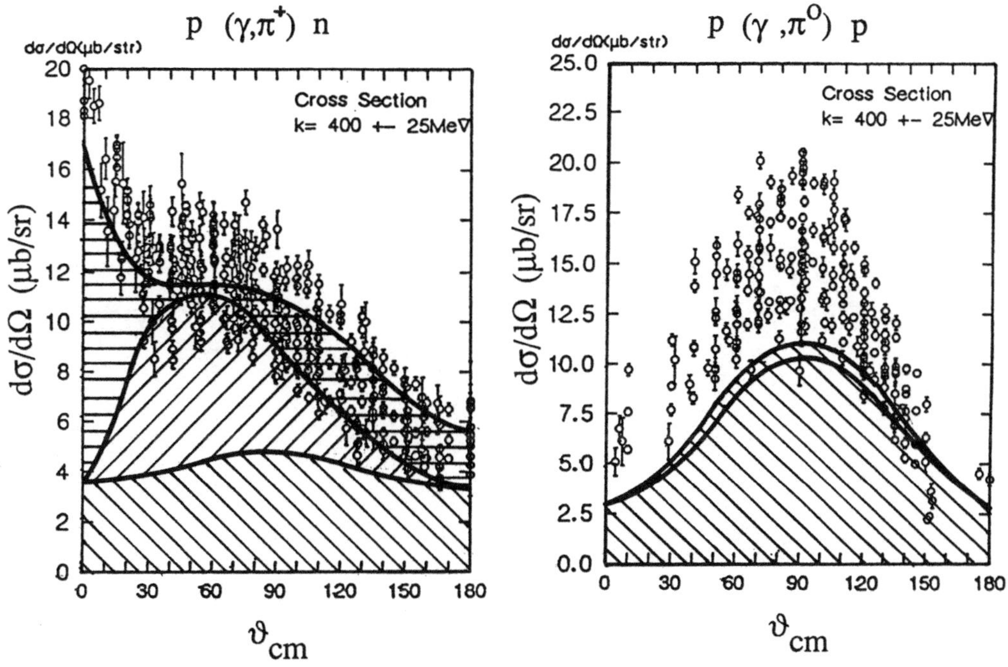

Fig.7. Comparison of calculated and experimental differential cross sections for the $\gamma + p \rightarrow \pi^{+} + n$ and the $\gamma + p \rightarrow \pi^{0} + p$ reaction. The backward hatched area represents the resont contribution, the forward hatched area also includes pion knock-out and the sum of all contributions is given by the horizontally hatched area. Experimental data from [8].

$n\,(\gamma,\pi^-)\,p$

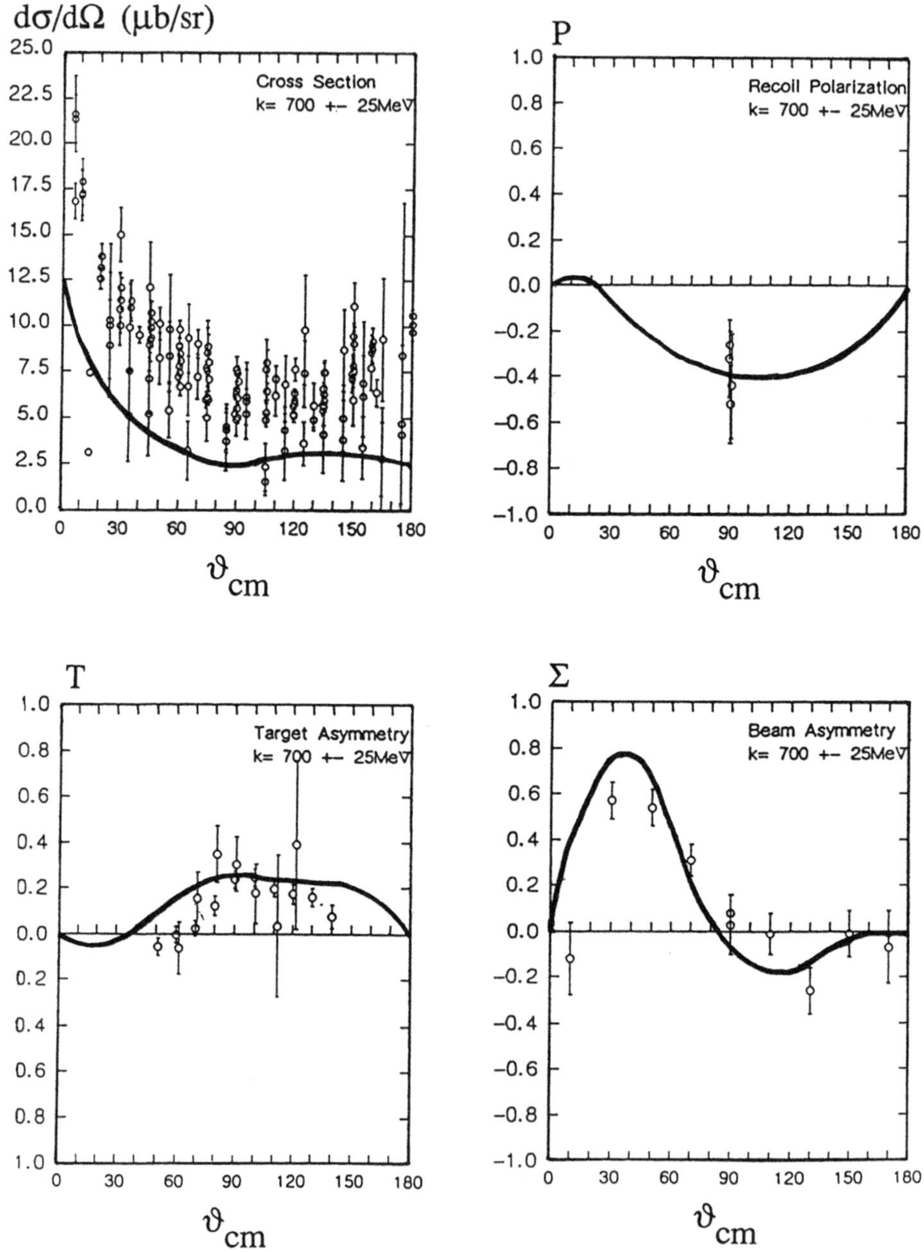

Fig.8. Comparison of calculated and experimental differential cross section and polarization observables for the $\gamma + n \rightarrow \pi^- + p$ reaction. Experimental data from [8].

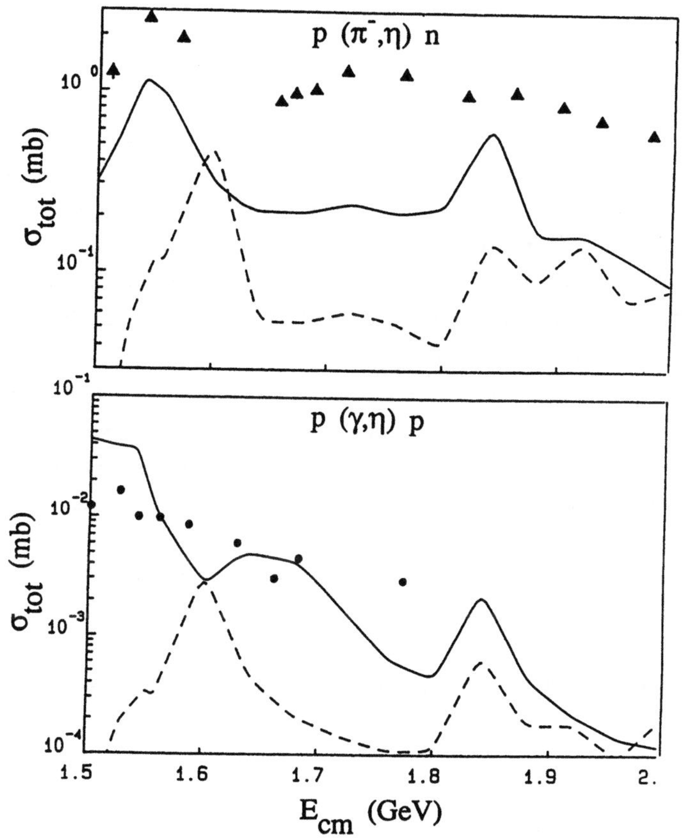

Fig.9. Comparison of the calculated and experimental total cross section for the $p(\pi^-,\eta^0)n$ reaction (top); Experimental data are obtained from [9] and for the $p(\gamma,\eta^0)p$ reaction (bottom). Experimental data are from [7]. Solid and dashed lines correspond to version 1 and 2 of the meson baryon vertex (see Eq.(3.10).

It should be stressed, that the considerations given above are not meant to state that the present calculation favours one specific ansatz for the meson baryon vertex, but rather form an example how certain issues can be discussed within a coherent treatment of both hadronic reactions and the underlying hadronic structure within a constituent quark model that is supplemented by the coupling to meson baryon channels.

5. A Quark Shell Model of Nuclei

The conventional description of complex hadrons like nuclei generally assumes that the valence quarks are locally confined into baryons which interact by relatively weak forces, commonly given by meson exchange (*multihadron picture*). In spite of the fact that in view of baryonic and nuclear sizes in principle some overlap of internal quark wave functions is to be expected a succesful description of the low energy nuclear excitations in terms of nucleons and nucleonic interactions seems to be established. However, this not neccessarily has to apply to intermediate energy excitations of nuclei, where within such a multihadron picture nucleons can be polarized by excitation into baryonic resonances and the corresponding baryon-baryon interactions can lead to more tighter bound configurations were the overlap of internal quark wave functions might be stronger. It therefore seems worthwhile to investigate an alternative description of complex hadrons where the local confinement of quarks is given up on favour of a global confinement in a bag of nuclear size and where the hadronic correlations are produced by a residual quark-quark interaction (*multiquark picture*) [10]. In order to study the consequences of quark wave function overlap we will briefly discuss a quark shell model of nuclei.

We start from the many-quark Hamilton operator for a nucleus of mass number A:

$$H = H_0 + V \tag{5.1}$$

where H_0 is a sum of single particle operators

$$H_0 = \sum_{i}^{3A} h_0(i) \tag{5.2}$$

with the eigenstates ψ_α with energies ε_α of h_0 being the solutions of the well-known MIT-Dirac equation, within a bag of nuclear radius. Here α is a collection of quantumnumbers that uniquely specifies a single-quark state. The configuration space consists of all positive energy single particle states with $\varepsilon_\alpha < \varepsilon_{max}$. All colourfree multiquark states can then be written as a linear combination of the vectors

$$\prod_{r=1}^{A} \mathcal{B}^\dagger(\alpha_r,\beta_r,\gamma_r) \mid > \tag{5.3}$$

where we have defined the (colourfree) quasi baryon creation operators

$$\mathcal{B}^\dagger(\alpha_r,\beta_r,\gamma_r) = \varepsilon^{ijk}\, a^\dagger_{i\alpha_r} a^\dagger_{j\alpha_r} a^\dagger_{k\alpha_r} \tag{5.4}$$

with $a^\dagger_{i\alpha}$ denoting the creation operator for a quark of colour i in state α.

The residual quark-quark interaction is taken to be a pairing force:

$$V = - g_P \mathcal{A}_i^\dagger \mathcal{A}_i, \tag{5.5}$$

with

$$\mathcal{A}_i^\dagger = \varepsilon_{ijk} \left[a_{j\alpha}^\dagger \times a_{k\alpha}^\dagger \right]^{(J=T=0)} \tag{5.6}$$

so the force works in the spin-singlett, isospin singlett channel only.

By introducing the mean single particle energy $\bar{\varepsilon}$ in the truncated model space $\bar{\varepsilon}$ the corresponding Hamilton operator $\bar{H}$ can be diagonalized exactly by group theoretical methods and the term $\Delta H = H-\bar{H}$ is then treated in perturbation theory. The resulting spectrum is given schematically in Fig.10.

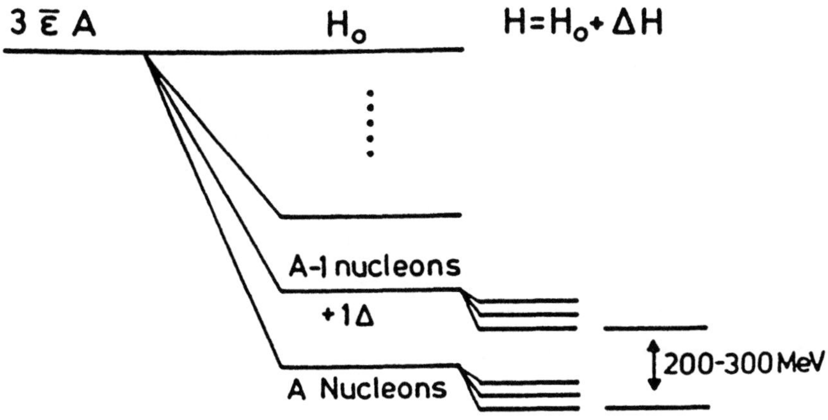

FIG.10. Schematic picture of the energy splitting produced by the pair force $\bar{H}$ and the perturbation by the deviation ΔH from the mean single particle energy $\bar{\varepsilon}$.

The wavefunctions of the lowest band can be written as

$$\prod_{r=1}^{A} \mathcal{N}^\dagger(\alpha_r) \mid > \tag{5.7}$$

where $\mathcal{N}^\dagger(\alpha_r) = a^\dagger_{i\alpha_r} \mathcal{A}_i^\dagger$ is a quasi nucleon creation operator. In the present model the ground state wave functions are thus interpreted as the states of A (partially deconfined) quasi nucleons in the form of A diquarks and A valence quarks. The excited bands correspond to states where succesivly diquark pairs are broken, which is interpreted as the excitation of more and more baryonic resonances. For instance the lowest state in the first excited band is given by

$$\Delta^\dagger_{s_{1/}} \cdot \prod_{r=1}^{A-1} \mathcal{N}^\dagger(\alpha_r) \mid >$$

with

$$\Delta^\dagger_{s_{1/2}} = \varepsilon^{ijk} \left[a^\dagger_{is_{1/2}} \times a^\dagger_{js_{1/2}} \times a^\dagger_{ks_{1/2}}\right]^{(J=T=3/2)}$$

and thus is interpreted as the lowest nuclear Δ-excitation.The process of Δ-formation from the nuclear ground state can therefore be described by the breakup of a diquark and the release of two additional valence quarks in the lowest $s_{1/2}$-orbital. The allowed quantum numbers for these nuclear Δ-excitations are restricted by the total antisymmetry requirement on the multiquark state. In this way one finds e.g. for the lowest (ΔT=1) Δ-excitation in ^{4}He, restricting to ΔL=0 excitations, only a $J^\pi=1^+$ state, in contrast to the multibaryon picture, where for ΔL=0 one would expect states with $J^\pi = 1^+$ and $J^\pi = 2^+$, corresponding to the antiparallel and parallel coupling of the Δ and the nucleon-hole, respectively. Since in any two-body reaction with pions and ^{4}He only unnatural parity states can be excited, the appropriate reaction for to investigating the existence of a low lying $J^\pi = 2^+$ Δ-excitation in ^{4}He, is Compton-scattering. To this end we will in the next section study the realibility of a description of scattering processes with light particles within the present nuclear quark shell model by a calculation of the ^{4}He(π,π)^{4}He and ^{4}He(γ,π)^{4}He reactions and will then analyse the (scarce) data for ^{4}He(γ,γ)^{4}He.

6. Two Body Reactions in a Quark Shell Model

As in Section 3 we will assume a local pseudoscalar coupling of pions to quarks

$$\mathcal{L}_{\pi qq}(i) = g_{\pi qq}\, \bar{\psi}_q(i)\, \gamma_5\, \vec{\tau}\, \psi_q(i)\, \vec{\phi}_\pi \tag{6.1}$$

with an effective coupling constant $g_{\pi qq}$, which is determined by a best-fit to π-N scattering. As for the photon quark coupling we used

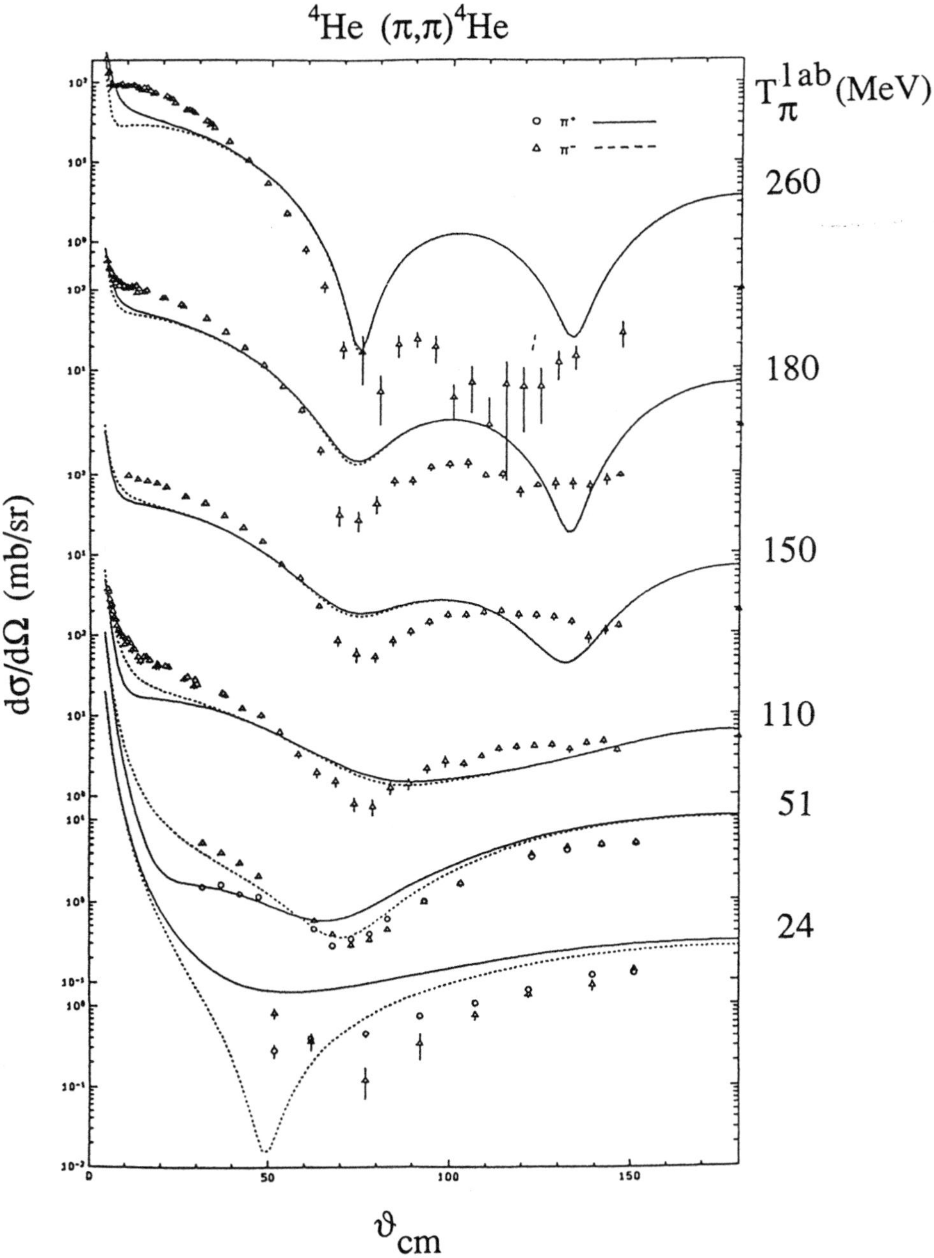

Fig. 11. Comparison of the calculated and experimental differential cross sections for elastic π-^{4}He scattering. Experimental data are from [12,13].

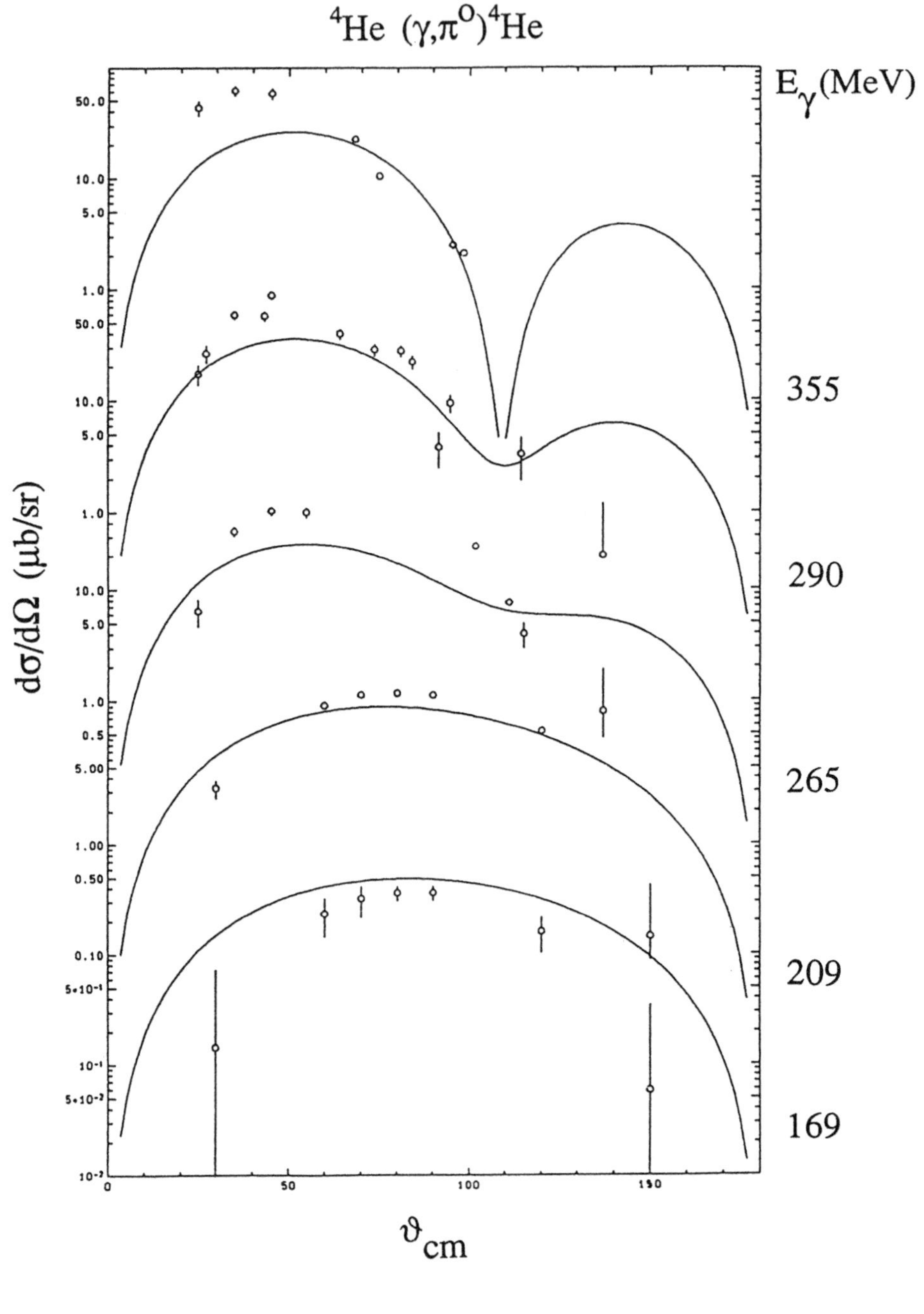

Fig. 12. Comparison of the calculated and experimental differential cross sections for photoproduction of pions on ^{4}He. Experimental data are from [14-16].

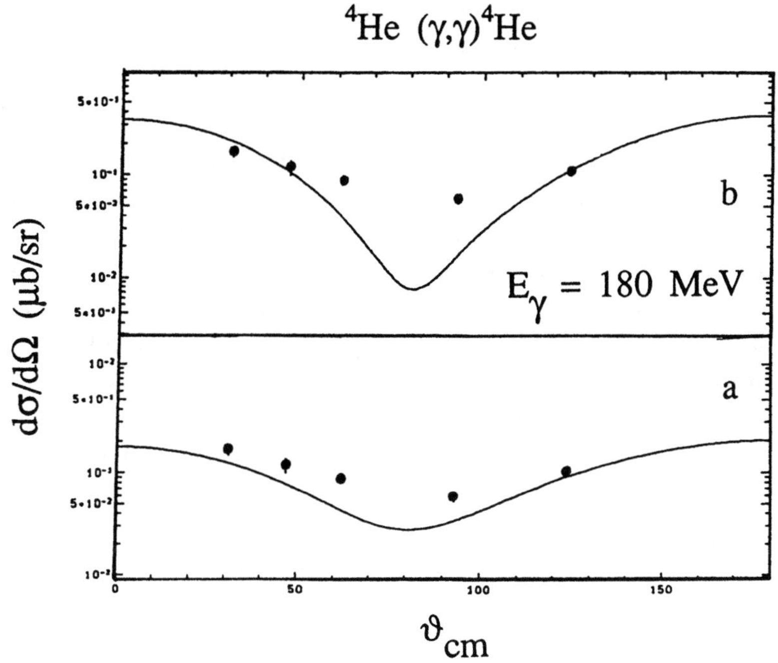

Fig. 13. (a) Comparison of the calculated and experimental differential cross section for Compton scattering off ^{4}He at E_γ=180 MeV. Experimental data are from [17]. (b) As (a) but after introduction of an artificial 2^+ nuclear Δ-resonance (see also main text).

$$\mathcal{L}_{\gamma qq}(i) = e\ \bar{\psi}_q(i)\gamma_\mu\ Q\ \psi_q(i)\ A^\mu + \frac{e}{4m_q}\ \bar{\psi}_q(i)\ \sigma_{\mu\nu}\ (a\mathbb{1}+b\tau_3)\ \psi_q(i)\ (\partial^\mu A^\nu-\partial^\nu A^\mu) \tag{6.2}$$

where Q is the charge operator and where the second term, motivated by vector meson dominance, amounts up to introducing an effective, anomalous magnetic moment for the quarks, which is fixed to the known electromagnetic formfactors of the nucleon. The amplitudes are then calculated with the quark model wavefunctions for the ground state and the Δ-excitation in ^{4}He. In order to account for the inelasticity in two body reactions on nuclei in the Δ-region we adopt $\Gamma_l^{tot}(E) = 3\ \Gamma_l^{\pi}(E)$ for each partial wave, since experimentally $\sigma^{tot}(E) = 3\ \sigma^{el}(E)$.

The resulting cross sections for the ^{4}He(π,π)^{4}He and ^{4}He(γ,π)^{4}He reactions are shown in Figs. 11 and 12, respectively. These figures show, that, apart from very low energies, where the assumed Δ-dominance is not expected to be relevant and for high energies, where due to technical reasons it was not possible to take into account sufficiently high partial waves, an in general good description for the intermediate energies is obtained.

Within the same framework also the ^{4}He-Compton scattering can be computed. The differential cross section for E_γ=180 MeV, where in view of what was said in the preceeding paragraph our assumptions are expected to hold, is displayed in Fig. 13a. From this figure it can be seen that the absence of a low lying a $(J^\pi,T)=(2^+,1)$ Δ-excitation within the present quark shell model does not impede a satisfactory description of the available experimental data. Indeed, if we introduce artificially a resonance in the 2^+-partial wave and assume it to be degenerate with the 1^+-resonance the description actually deteriorates (see Fig.13b). Of course, the issue can be settled only if a complete partial wave analysis of experimental Compton scattering is available. Also it should be remarked that in a multibaryon picture, even although a low-lying 2^+-resonance is in principle allowed the exciation of this partial wave can of course be dynamically suppressed. The main purpose of the present discussion is, however, to show that assumptions about the underlying quark dynamics of a hadronic system (here mainly the global confinement and the special quark-quark interaction) do again lead to clearcut observable effects, which should encourage a systematic study of such experimentally difficult reactions like intermediate energy nuclear Compton scattering.

7. Summary and Conclusion

In the present treatment we have pointed out that constituent quark models of simple and complex hadrons form a suitable framework for relating assumptions about the underlying quark dynamics (concerning e.g. local versus global confinement, the residual quark-quark interactions and hadronic creation and annihilation vertices) to scattering observables for exclusive two-body reactions involving photons and light mesons.

In the first part we have sketched an attempt to describe consistently pion scattering and the photoproduction of pions on the nucleon with a potential quark model of baryons that dynamically takes into account the coupling of baryon resonances to baryon meson decay channels. This was found to be important for a unified description of the baryon spectrum and the reaction amplitudes. We found that the coupling to these decay channels induces shifts for the resonance positions that are comparable to the splittings due to the residual quark-quark interaction and in addition provides a semi realistic description of hadronic decay widths for the lower baryon resonances. An analysis of scattering at higher energies indicates that it neccessary to improve the present treatment (where it was assumed that meson couple pointlike to quarks) by a more microscopic treatment of hadronic vertices that accounts for the internal structure of the mesons and better describes the important many body aspects of quark-pair creation and annihilation processes.

In the second part we have pointed out that it is possible to obtain a satisfactory description of photo- and pion induced reaction on nuclei in the region of the Δ-excitation, within the framework of a quark shell model of nuclei with global confinement and dynamical correlations between quarks. It was shown that within such a framework the partial deconfinement of quarks constrains the allowed quantum numbers of low lying intermediate Δ-nuclear resonances. The best way to study relevance of this effect experimentally is to aim at a sufficiently systematic partial wave analysis of nuclear Compton scattering in this energy region.

Acknowledgements

The authors like to thank for the help of W. Blask, W. ten Brink, H. Weber and R. Scholl in performing calculations and preparing the figures. This work has been supported by the Bundesministerium für Forschung und Technologie and the Deutsche Forschungs Gemeinschaft.

References

[1] A. DeRujula, H. Georgi and S.L. Glashow, Phys. Rev. **D12** (1975) 147
[2] D. Gromes in:*"The quark structure of matter"*, *Eds.* N. Isgur, G. Karl and P.J. O'Donnell, World Scientific, Singapore (1985) 1-58
[3] N. Isgur, G. Karl, Phys. Rev. **D18** (1978) 487, **D19** (1979) 2653, **D20** (1979) 1191
[4] W. Blask, M.G. Huber and B.C. Metsch, Zeit.f.Phys. **A326** (1987) 413
[5] K.S. Sharma, W.H. Blask, B. Metsch and M.G. Huber, Phys. Rev. Lett. **62**, (1989) 2913
[6] Particle Data Group, Phys. Lett. **204B** (1988) 1
[7] G.Höhler, Landoldt-Börnstein, Volume **9b**, Part 2, *Ed. H.Schopper*, Springer, Berlin, Heidelberg, New-York (1983) 1
[8] K. Ukai and T. Nakamura, INS-TEC-22, Institute for Nuclear Study, University of Tokyo, (1985)
[9] R.M. Brown *et. al.*, Nucl. Phys. **B153** (1979) 89
[10] H. Hofestädt, S. Merk and H.R. Petry, Zeit.f.Phys. **A326** (1987) 391
[11] M.E. Nordberg and K.F. Kinsey, Phys. Lett. **20** (1966) 693
[11] F. Binon *et. al.*, Nucl. Phys. **A298** (1978) 499
[13] A.S. Belusov, Sov. Journ. Nucl. Phys. **4** (1967) 78
[14] P.S. Anan'in, Yad. Fiz. **41** (1985) 1393
[15] D.R. Tieger *et. al.*, Phys. Rev. Lett. **53** (1984) 755
[16] E.J. Austin *et.al.*, Phys. Rev. Lett. **57** (1986) 972

PION PHOTOPRODUCTION AND COMPTON SCATTERING IN QUARK MODELS

M. Weyrauch*
Institut für Kernphysik, Johannes Gutenberg-Universität Mainz
D-6500 Mainz, FRG

Abstract

I review quark model studies for pion photoproduction and Compton scattering off nucleons. The important role of unitarity and electromagnetic gauge invariance is stressed.

I. Introduction

The study of electromagnetic processes serves as a useful test of our understanding of nuclear and subnuclear dynamics. Here I want to investigate selected electromagnetic processes with nucleons emphasizing the question: "What do we learn about quarks and their interactions from different models and approaches?" Some particular consideration will be given to the cloudy bag model (CBM) and calculations I have performed for pion photoproduction and Compton scattering in the Δ resonance region [1].

Until recently electromagnetic processes with nucleons (apart from the e.m. form factors) have not attracted the same attention as strong processes (i.e. pion scattering in particular) in investigations of fundamental quark dynamics. It seems therefore appropriate to collect issues of fundamental interest in this field and to point out possible future developments. Certainly my selection of issues and references is by no means complete. In particular, I will not discuss e.m. form factors and moments.

Different quark models have been proposed for the low-energy dynamics of quarks in an attempt to better understand Quantum Chromodynamics (QCD) in the nonperturbative regime. The oldest is the simple constituent quark model as proposed originally by Gell-Mann and Zweig [2] and studied extensively by Isgur and Karl [3]. This model has been remarkably successful phenomenologically especially in explaining particle spectra. However, it does not include coupling to mesonic channels, which - when consistently included - tend to renormalize the spectrum. Some progress in this direction has been recently achieved by Huber and collaborators (see the contribution by Metsch in these proceedings).

The cloudy bag model [4], which has been developed as an extension of the MIT bag model [5], takes coupling to mesonic channels into account. But here the "bare" spectrum is usually fitted in order to reproduce certain meson scattering observables. A really successful "marriage" of the particle spectrum with meson scattering observables has still

---

*) supported by the Deutsche Forschungsgemeinschaft (SFB 201)

to be achieved. Attempts in this direction are chiral soliton models, which have been extensively discussed recently [6].

Clearly all these models are phenomenological in nature, and their connection to QCD is more or less obscure. Essential parameters and ingredients (like the confinement size) are hotly debated and no general agreement has yet been reached. A careful study of electromagnetic processes may be an important contribution in this debate, in order to justify or reject different model assumptions. In the absence of a "good" theory consideration of fundamental symmetries plays an important role in order to constrain the necessary phenomenology. In particular, careful consideration has to be given to unitarity, gauge invariance of the e.m. interaction, and (approximate) chiral symmetry. In some early attempts to calculate observables of e.m. processes the importance of these fundamental symmetries has been overlooked.

My considerations here will encompass the energy region from below π production threshold up to the Δ resonance. Therefore, on- and off-shell pion scattering plays an important role. In fact, unitarity considerations dictate that π scattering is considered *simultaneously* with photoproduction and Compton scattering. Implementation of e.m. gauge invariance in phenomenological quark models is still in its beginning. This has been pointed out by Drechsel and Giannini [7] for the E2/M1 ratio of Δ e.m. decay. Furthermore, without correct implementation of e.m. gauge invariance the low-energy theorem for Compton scattering will be violated.

In the following I will study the role of symmetries in section II, pion photoproduction in section III and Compton scattering in section IV.

II. Symmetries

In this section I want to briefly discuss the role of (a) unitarity and (b) electromagnetic gauge invariance in models for $\gamma\pi$ and $\gamma\gamma$ processes with the nucleon.

(a) Unitarity

Unitarity relates pion scattering and the $\gamma\pi$ and $\gamma\gamma$ processes with each other. Therefore, e.m. and strong interactions can actually not be described independently. Explicitly, below 2π threshold the matrix of possible processes

$$T = \begin{pmatrix} t_{\pi\pi} & t_{\pi\gamma} \\ t_{\gamma\pi} & t_{\gamma\gamma} \end{pmatrix} \tag{1}$$

has to fulfill the unitarity condition

$$T + T^{+} = TT^{+}. \tag{2}$$

As a consequence $t_{\pi\pi}$ and $t_{\pi\gamma}$ are related by Watson's theorem (if one also requires time-reversal invariance), and $t_{\pi\gamma}$ and $t_{\gamma\gamma}$ by the optical theorem. A consistent approach, therefore, has to treat pion scattering, photoproduction, and photon scattering simultaneously. A perturbative approach like the Blomqvist-Laget [8] phenomenological model for pion photoproduction violates unitarity and should therefore be considered cautiously.

A phenomenological treatment of some photoproduction partial waves in a consistent unitary approach has been given in ref. [9]. In ref. [1] I follow a technically very similar approach and include photon scattering. Furthermore, I connect with the cloudy bag model (CBM) of Thomas, Théberge and Miller [4]. The basis of these calculations is a (coupled) Lippmann-Schwinger equation for the T-matrix defined in eq. (1),

$$T = V + V\, G_0 T, \tag{3}$$

with a model for the Born terms or driving potentials V and the free propagator G_0. While ref. [9] takes a purely phenomenological ansatz for V, in the CBM approach one calculates V from the CBM Lagrangian. Quark dynamics essentially enters via pion-baryon form factors (see the following section).

The coupled equations (3) can be easily decoupled (up to second order in the e.m. interaction) in terms of the off-shell pion scattering T-matrix $t_{\pi\pi}$

$$t_{\pi\pi} = V_{\pi\pi} + V_{\pi\pi}\, G_0\, t_{\pi\pi} \tag{4}$$
$$t_{\pi\gamma} = V_{\pi\gamma} + t_{\pi\pi}\, G_0\, V_{\pi\gamma} \tag{5}$$
$$t_{\gamma\gamma} = V_{\gamma\gamma} + V_{\gamma\pi}\, G_0\, V_{\pi\gamma} + V_{\gamma\pi}\, G_0\, t_{\pi\pi}\, G_0\, V_{\pi\gamma}. \tag{6}$$

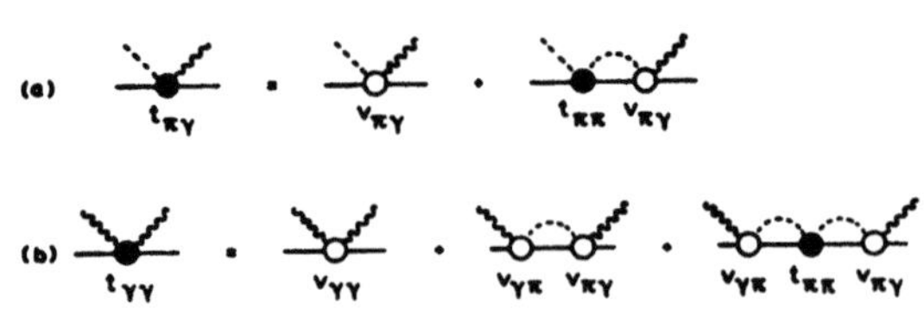

Fig. 1: The photoproduction amplitude Eq. (5) and the Compton scattering amplitude Eq. (6). The full circles represent amplitudes and the open circles driving potentials. Photons are represented by wiggly lines, pions by dashed lines, nucleons by full lines and Δ's by open boxes throughout all figures.

For G_0 all calculations [1,4,9] take an effective πN free propagator. Details are found in these references. A pictorial representation of eqs. (5) and (6) is given in Fig. 1.

The decisive advantage of the approach outlined above is that unitarity is automatically ensured. This is a consequence of the fact that the Lippmann-Schwinger equation satisfies two-particle unitarity. Watson's theorem and the optical theorem can be easily shown to hold explicitly (see ref. 1).

(b) Gauge invariance

While the Lippmann-Schwinger approach outlined above is conceptually very simple, easy to calculate with, and automatically unitary, electromagnetic gauge invariance is not easily implemented, and it is (usually) violated in explicit calculations. In fact, gauge invariance relates electromagnetic and strong interactions (described by H^S) as can be seen most easily from the charge conservation requirement

$$\vec{\nabla} \cdot \vec{j} = -i[H^s, \rho]. \tag{7}$$

That means that

$$V_{\pi\gamma} = \langle \pi N | \vec{\varepsilon}_\lambda \cdot \vec{j} | \gamma N \rangle \tag{8}$$

cannot be defined completely independently of H^s. Here $\vec{j}$ and ρ are the current and charge densities and λ the photon polarization. Of course (7) does not define $V_{\pi\gamma}$ uniquely, but is a necessary consistency condition. The essential problem is that the Born terms V are highly non-local due to two phenomenological ingredients: (i) the Baryon propagators and (ii) the form factors at the strong vertices. Even if the form factors are calculated in a quark model, they are essentially of phenomenological origin since the confining mechanism of a particular model is phenomenological.

This is an unfortunate situation and has not been addressed much in the literature. The only attempt dealing with this problem systematically in a phenomenological context (i.e. without considering quark dynamics) is a calculation of πN Bremsstrahlung by Heller, Kumano, Martinez and Moniz [10]. They introduce additional vertex currents to remedy the violation of gauge invariance introduced through the above mentioned nonlocalities. A similar approach should be attempted for photoproduction and Compton scattering.

Here I would like to propose a less ambitious approach, namely to use Siegert's theorem, which has been so successful and important in nuclear physics [11]. I will briefly outline the essentials. For a particular electric multipole L the current matrix element is related to the charge multipole via Siegert's theorem

$$V^L_{\pi\gamma} = \langle \pi N | \frac{1}{k} [H^s, C^L_\lambda] | \gamma N \rangle + b^L_{\pi\gamma}$$
$$= \frac{1}{k} \langle \pi N | H^s | N \rangle \, c^L_{\pi\gamma} + b^L_{\pi\gamma} \tag{9}$$

where the charge multipole is given by

$$C^L_\lambda = i^{L+1} \sqrt{\frac{L+1}{L}} \int d^3x \, \rho(\vec{x}) \, j_L(kx) \; Y_{L\lambda}(\hat{x}), \tag{10}$$

and $b^L_{\pi\gamma}$ is a current multipole (see ref. [11]). At low energies $b^L_{\pi\gamma}$ is of order $(kx)^{L+2}$ while c^L is of order $(kx)^L$ (see ref. [11]).

Now I rewrite eq. (5)

$$t^L_{\pi\gamma}(E) = \left[1 + t_{\pi\pi}(E) \; G_0(E)\right] \; V^L_{\pi\gamma}$$

and use

$$\left[1 + t_{\pi\pi}(E) \; G_0(E)\right] \; H^s = E\left[1 + t_{\pi\pi}(E) \; G_0(E)\right] \tag{11}$$

to show that

$$t^L_{\pi\gamma} = \left(1 + t_{\pi\pi} \, G_0\right)\left(c^L_{\pi\gamma} + b^L_{\pi\gamma}\right) . \tag{12}$$

To calculate $c^L_{\pi\gamma}$ one only needs to know the charge density, which is presumably better understood than the current density (Siegert's hypothesis). In practice, one just takes the "point" charge densities for pions and quarks. For $b^L_{\pi\gamma}$ one must take a simple model for those terms which cannot be incorporated by Siegert's theorem. An analogous approach has been very successful for e.g. deuteron photodisintegration below π threshold. Details will be considered in a forthcoming publication [12], where also Compton scattering is considered. For Compton scattering in a nonrelativistic phenomenological approach one also needs to consider gauge conditions for the two-photon amplitudes $V_{\gamma\gamma}$ in order to ensure overall gauge invariance. I will not consider this technical point here, but will just state that the low-energy theorem [13]

$$t_{\gamma\gamma} = \frac{e}{m_N} \, \vec{\varepsilon}^{\,\prime}_{\lambda} \cdot \vec{\varepsilon}_{\lambda} \tag{13}$$

is fulfilled in the approach outlined above.

III. Pion photoproduction

A comprehensive model covering both the resonance region and the threshold region is still missing. Here I collect results of calculations in different models and discuss resonance and threshold dynamics separately.

(a) Δ resonance region

There is a number of phenomenological calculations of the $\gamma\pi$ process (e.g. refs. [8,9]). Quark model calculations are contained in refs. 1, 14-16. Many calculations only address the E2/M1 ratio of the radiative decay of the Δ [7,17-19]. The latter issue is directly related to π photoproduction or its inverse. As a matter of fact, the E2/M1 ratio is a somewhat artificial number, since it can only be inferred indirectly by a model-dependent analysis of pion photoproduction data. Nevertheless many calculations only address the E2/M1 ratio for radiative decay; therefore it needs to be included here.

A calculation in the CBM which takes care of unitary symmetry is presented in ref. 1. It is an attempt to relate $\gamma\pi$ and $\gamma\gamma$ processes with π scattering. Starting point is eqs. (4) - (6). The Born terms, which are calculated from the CBM are graphically represented in Fig. 2.

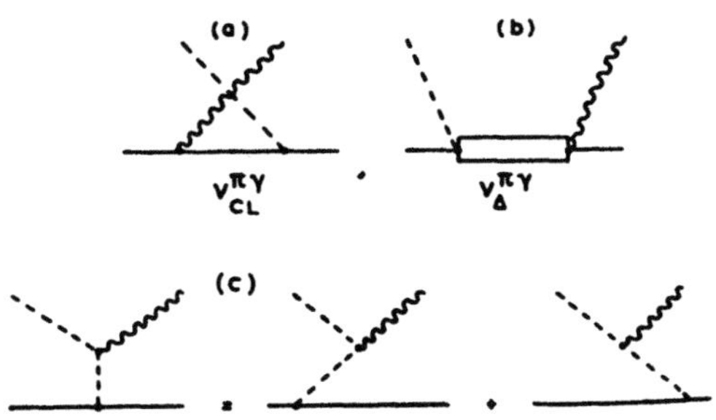

Fig. 2: Born terms for pion photoproduction in the 33 channel: (a) crossed Born-term, (b) "bare" Δ, (c) pion-pole term.

The quark model enters via the πN form factors. Explicit expressions are given in ref. [1]. The model parameters, bag radius R and "bare" Δ mass m_Δ are fitted in such a way that a reasonable description of the δ_{33} phase shift is obtained (R = 0.97, m_Δ = 1242 MeV). With an isovector magnetic moment x = 2.33 nuclear magnetons one obtains the M_{1+} ($T = \frac{3}{2}$) multipole as displayed in Fig. 3. In ref. 14 and 15 also other partial waves are considered, but neither unitarity nor gauge invariance requirements are addressed. The results therefore have to be treated cautiously. What concerns gauge invariance the same reservation also applies to the calculations in ref. 9.

Some quark model calculations only consider the E2/M1 ratio in radiative Δ decay [7,17-19]. The most important result of these investigations is to show how important gauge invariance considerations are [7]. Completely different results for E2/M1 are obtained, if one calculates E2 via the charge or current multipole. This finding has triggered the gauge invariance considerations presented in section II. Furthermore, the numbers obtained for E2/M1 vary wildly depending on specific dynamical assumptions. For example, in the consituent quark model with colour hyperfine and/or one pion exchange interaction one obtains values of -0.4 % to -0.8 % [17-19], while a recent phenomenological calculation [9] obtained +3.7 %. While in the calculations of refs. [17-19] the colour hyperfine interaction plays an essential role it is pion dynamics only, which determines the result of ref. [9]. Unfortunately, experimental numbers also vary greatly due to difficulties to extract the E2/M1 ratio from pion photoproduction data [20].

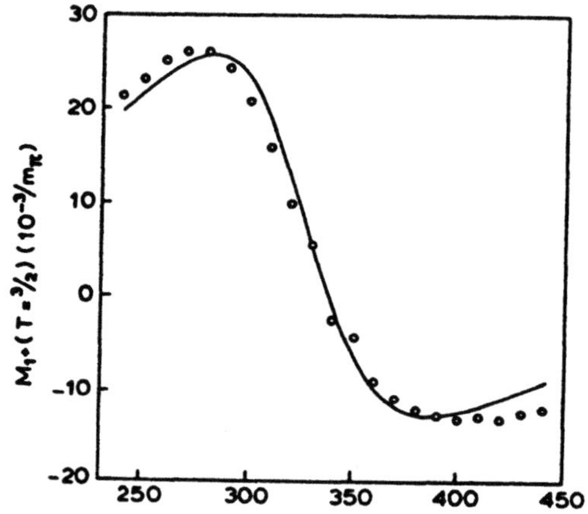

Fig. 3: The real part of the M_{1+} $\left(T = \frac{3}{2}\right)$ multipole (full line) compared to the data of ref. 15 (open circles).

(b) Threshold region

Very recently the threshold region of pion photoproduction has attracted a lot of attention. Experimental results obtained at Mainz and Saclay [21] seem to contradict the low-energy theorem [22]. This issue is outside the scope of the present paper and has been excellently reviewed recently by Drechsel [23]. The most exciting idea discussed in this reference is the possibility to eventually determine the current quark masses by threshold photoproduction [24].

IV. Compton scattering

Also in this section I want to separately study the Δ resonance region and the low-energy region.

(a) Resonance region

In the Lippmann-Schwinger approach described briefly in section II the nucleon Compton amplitude is straightforwardly obtained. Since in this approach π scattering, photoproduction and Compton scattering are described simultaneously, all parameters of the model (see section III) are determined by π scattering and photoproduction. Compton scattering is then a decisive prediction of the model without any new parameters (for details see ref. 1). The f^{1+}_{MM} multipole (which corresponds to the 33 partial wave in pion scattering) is shown in Fig. 4.

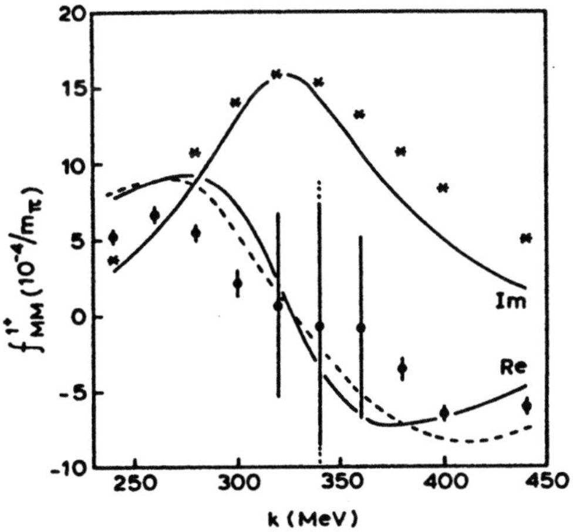

Fig. 4: The f^{1+}_{MM} multipole of the Compton scattering amplitude: the results of this calculation (full line) compared to the data as obtained by a multipole fit to the scattering cross sections (ref. [25]). Stars represent the imaginary part of the data, full circles represent the real part. The dashed line is the result of the dispersion-theoretical calculation reported in ref. [25].

Here the quality of the available data is much poorer than in the photoproduction case. Nevertheless, some general remarks can be made at this stage. The data given in Fig. 4 are taken from a fit of the multipoles to the available scattering data. As is explained in detail in that reference, the exact values for the f^{1+}_{MM} multipole strongly depend on the conditions under which the fit is performed, so that one may expect that the experimental situation will change when more and better scattering data become available. We also show in Fig. 4 results of a very elaborate dispersion theoretical calculation [25] of the real part of the f^{1+}_{MM} multipole. As is obvious, the result of the CBM is very similar to the one obtained in the dispersion calculation. That means that, if we take the experimental data at face value, there is a 30 - 40 % discrepancy between both calculations and the data. This is a serious problem which needs further attention. Note that there is no room to change the parameters in the CBM calculation in any significant way, since all parameters have been fixed to describe pion scattering and photoproduction. This situation certainly seems to be a challenge for both theory and experiment.

(b) Low-energy region

Low-energy Compton scattering is characterized by the (static) electric polarizability α and the magnetic susceptibility β. These structure constants enter to second order in the photon energy in an expansion of the Compton amplitude. The zeroth and first orders are model independently determined by the charge, mass and magnetic moment of the nucleon (low-energy theorem [13]). Typical experimental values for α and β are given in table 1 along with the results of several theoretical calculations. According to standard theory [26] one writes

$$\begin{aligned} \alpha &= \alpha_E + \alpha_R \\ \beta &= \beta_P + \beta_D, \end{aligned} \tag{14}$$

where α_E and β_P are related to nucleon excitations via

$$\alpha_E = 2 \sum_n \frac{|\langle 0|D_z|n\rangle|^2}{E_0 - E_n} \tag{15}$$

$$\beta_P = 2 \sum_n \frac{|\langle 0|M_z|n\rangle|^2}{E_0 - E_n} \tag{16}$$

with D_z and M_z the electric and magnetic dipole operators, respectively. α_R and β_D contain recoil and two-photon amplitude corrections and are therefore of order $1/m_N$. The separation in eq. (14) is purely formal and the different terms cannot be measured separately. Since the two terms are not separately gauge invariant one should not give too much physical significance to the different terms.

Table 1: Nucleon electric polarizability and magnetic susceptibility (in units 10^{-4} fm^3)

	exp Ref. [27]	CQM Ref. [28]	CBM Ref. [29]	CBM Ref. [1]
α	11.6 ± 2.5	13.6	7-9	
α_E	-	11.6		
α_R	-	2.0		
β	2.9 ± 2.4	1.5	<2.0	
β_D	-	7.5		9.2
β_P	-	-6.0		

CQM: constituent quark model, CBM: chiral bag model

On first sight it surprises that α is large as compared to β, since the first excited state of the nucleon is reached in a magnetic transition. The susceptibility β is small (or even negative) since β_P and β_D are roughly equal in magnitude, but of opposite sign. This is borne out nicely in calculations in the constituent quark model (column 2 in Table 1). The important point is that the nucleon mass and first excitation energy is of the same order of magnitude. Therefore, β_P and β_D are of similar magnitude. This is quite unlike nuclei, where mass and excitation energies are separated by 2 orders of magnitude at least! Therefore, recoil and two-photon amplitude terms which can be neglected in nuclear calculations become important for the nucleon.

A brief comment on Table 1: While it is quark dynamics only which determines the values for α and β in the constituent quark model, the result for α of ref. [29] is dominated by the pion cloud.

A final remark concerns a sum rule for $\alpha+\beta$. From the dispersion relation of Gell-Mann, Goldberger and Thirring [30] one can easily derive the sum rule

$$\alpha + \beta = \frac{1}{2\pi^2} \int_0^\infty \frac{\sigma^{tot}(k)}{k^2}\, dk.$$

If recoil and two-photon terms are negligible - as is the case for nuclei - then one approximately finds

$$\alpha \approx \frac{1}{2\pi^2} \int_0^\infty \frac{\sigma^{tot}(E1,k)}{k^2}\, dk \tag{17}$$

$$\beta \approx \frac{1}{2\pi^2} \int_0^\infty \frac{\sigma^{tot}(M1,k)}{k^2} dk \ . \qquad (18)$$

But eqs. (17) and (18) are *wrong* for the nucleon, since recoil and two-photon terms cannot be neglected. Unfortunately, uncritical use of (17) and (18) has led to confusing statements about α and β in ref. [31]. Indeed, if one calculates α and β via (17) and (18) from pion absorption data, one obtains results which are completely at variance with the results reported in Table 1. A detailed discussion will be published in ref. [12].

V. Concluding remarks

In this paper I have emphasized the important relations between $\pi\pi$, $\gamma\pi$ and $\gamma\gamma$ processes. It is crucial to treat these processes simultaneously with proper regard to unitarity and e.m. gauge invariance. At present only incomplete results are available and a comprehensive calculation is still missing. In particular, we need a calculation which applies to both the resonance and low-energy regions. This may well shed new light on some open questions touched upon in these pages.

REFERENCES

1) M. Weyrauch, Phys. Rev. D35 (1987) 1574
2) M. Gell-Mann, Phys. Lett. 8 (1964) 214
G. Zweig, Cern Report No. 8182/TH 401 (unpublished)
3) N. Isgur and G. Karl, Phys. Rev. D18 (1978) 4187, D19 (1979) 2653, D20 (1979) 1191
4) S. Thèberge, A.W. Thomas and G.A. Miller, Phys. Rev. D22 (1980) 2838; 23 (1981) 2106(E); A.W. Thomas, S. Thèberge and G.A. Miller, ibid 24 (1981) 216
5) A. Chodos, R.L. Jaffe, C.B. Thorne and V.F. Weisskopf, Phys. Rev. D9 (1974) 3471
6) L. Vepstas and A.D. Jackson, Justifying the Chiral Bag, Phys. Rep. (to be published)
7) D. Drechsel and M.M. Giannini, Phys. Lett. 134B (1984) 329
8) J.M. Laget, Physics Reports 69 (1981) 1 and references therein
9) H. Tanabe and K. Ohta, Phys. Rev. C31 (1985) 1876
10) L. Heller, S. Kumano, J.C. Martinez and E.J. Moniz, Phys. Rev. C35 (1987) 718
11) A.J.F. Siegert, Phys. Rev. 52 (1937) 787
H. Arenhövel and M. Weyrauch, Nucl. Phys. A457 (1986) 573
M. Weyrauch, Phys. Rev. C41 (1990) 880
12) M. Weyrauch, to be published
13) F.E. Low, Phys. Rev. 96 (1954) 1428
14) M. Araki and A.M. Kamal, Phys. Rev. D29 (1983) 1345
15) G. Kälbermann and J.M. Eisenberg, Phys. Rev. D28 (1983) 71

16) M. Araki and I.R. Afnan, Phys. Rev. C36 (1987) 250

17) S. Gershtein and G.V. Dzhikiya, Sov. J. Nucl. Phys. 34 (1981) 870

18) N. Isgur, G. Karl and R. Konink, Phys. Rev. D25 (1982) 2394

19) M. Weyrauch and H.J. Weber, Phys. Lett. B171 (1986) 13

20) R. Davidson, N.C. Mukhopadhyay and R. Wittmann, Phys. Lett. 142B (1984) 336

21) E. Mazzucato et al., Phys. Rev. Lett. 57 (1986) 3144
R. Beck, Ph.D. Thesis, Institut für Kernphysik, Mainz, 1989

22) F.E. Low, Phys. Rev. 110 (1958) 974

23) D. Drechsel, Proc. Yamada Conference XXIII on Nuclear Weak Processes and Nuclear Structure, Osaka 1989, eds. M. Morita et al., p. 384 (World Scientific, Singapore 1989)

24) L.M. Nath and S.K. Singh, Phys. Rev. C39 (1989) 1207

25) W. Pfeil, H. Rollnik and S. Stankowski, Nucl. Phys. B73 (1974) 166
S. Stankowski, diploma-thesis PIB 2-137, University of Bonn

26) J.L. Friar, Ann. Phys. 95 (1975) 170

27) V.A. Petrun'kin, Sov. J. Part. Nucl. 12 (1981) 278

28) D. Drechsel and A. Russo, Phys. Lett. 137B (1984) 294
G. Dattoli, G. Matone and D. Prosperi, Lett. Nuovo Cimento 19 (1977) 601

29) R. Weiner and W. Weise, Phys. Lett. 159B (1985) 85

30) M. Gell-Mann, M.L. Goldberger and W.F. Thirring, Phys. Rev. 95 (1954) 1612

31) E. Hayward, Phys. Lett. B223 (1989) 133

COMPTON SCATTERING OFF NUCLEONS NEAR THRESHOLD

B. König, D. Drechsel and L. Tiator
Institut für Kernphysik, Johannes Gutenberg-Universität Mainz
D-6500 Mainz, FRG

Abstract

The energy dependence of the magnetic polarizability β has been evaluated by use of the delta resonance in the s- and u-channel of the Compton scattering amplitude. It was the idea to extract the influence of higher order terms (higher than ω^2, $\omega = |k|$ being the photon energy) in the analysis of the polarizability. The static paramagnetic polarizability $\beta_{Para}(0)$ has been substituted by an energy dependent $\beta^{\Delta}_{Para}(k)$ and the result is compared with experimental data.

The elastic scattering of low-energy photons from nucleons is a process that has been studied through the years and developed from the photon scattering off atoms, from which the name Compton scattering stems. This elementary process is profoundly understood since the electromagnetic interaction of the photon is well known. The nucleon, being the simplest baryon, allows a lowest order check for subhadronic degrees of freedom.

When considering elastic scattering of photons by nucleons, one observes a difference from the Compton effect of electrons. This difference has its roots in the anomalous magnetic moment of the proton and neutron (1.79 and -1.91, respectively) and in the fact that there occur two structure functions $\alpha(E)$ and $\beta(E)$, the electric polarizability and the magnetic susceptibility. It is generally assumed that for photon energies below the (3,3)-resonance only the static parts α_0 and β_0 contribute. The idea behind the concept of electromagnetic polarizabilities can be easily understood by a comparison with classical physics. A dipole in an electric field is stretched according to the equation

$$\vec{D} = \alpha \vec{E}_0 = 2q\vec{r}, \tag{1}$$

where the constant of proportionality α is given by

$$\alpha = 2\frac{q^2}{K}$$

with the spring constant K.

Multiplying this with r^2 in numerator and denominator leads to the form (dipole moment)^2/energy which is similar to the quantum mechanical expression

$$\alpha = 2 \sum_{N \neq 0} \frac{|\langle N|D_z|0\rangle|^2}{E_N - E_0} \qquad (2)$$

with D_z the z-component of the electric dipole operator. It is illustrative that at low frequencies the photon contains a near static electromagnetic field which induces electric and magnetic dipole moments proportional to α and β.

The motivations for this calculation were the experiments on Compton scattering off the proton (e.g. at Mainz, [Ahr 87]). This experiment was performed at energies up to pion threshold, where the dynamic dependence of the polarizability on the photon frequency ($\omega = |\vec{k}|$) cannot be neglected anymore. With other words, corrections of orders higher than ω^2 will have to be considered in the low-energy expansion of the cross section.

The explicit energy dependence of the cross section which enters via the delta resonance contribution is used to determine the structure of the paramagnetic polarizability $\beta^{\Delta}_{Para}(k)$. It has been assumed that in the analysis of the above-mentioned experiment an implicit energy dependence is hidden, which might be taken care of by evaluating $\beta^{\Delta}_{Para}(k)$.

To extract the polarizabilities, the elastic Compton amplitude is expanded in powers of the incoming photon energy. An expression herefore is given by Friar [Friar 75]

$$T_c = \left\{\frac{-e^2}{m_p}\left(1 - \frac{\langle r^2\rangle}{6}(\vec{k}^2 + \vec{k}'^2)\right) + \alpha\,\omega\omega'\right\} \vec{\varepsilon}\cdot\vec{\varepsilon}\,' + (\vec{\varepsilon}\times\vec{k})\cdot(\vec{\varepsilon}\,'\times\vec{k}')\,\beta + \ldots \qquad (3)$$

with m_p the proton mass and $e^2 = \alpha = \frac{1}{137}$. It describes Compton scattering for a spinless particle by a photon of energy $\omega = k = |\vec{k}|$, $\vec{k}$ being the photon momentum and polarization $\vec{\varepsilon}$ in the lab system.

The Thomson- and Rayleigh-terms were kept (constant and quadratic in k) and in the latter there occur 3 structure dependent functions: $\langle r^2\rangle$, the rms radius of the charge density, α, β the electric polarizability and the magnetic susceptibility. The electric polarizability enters the scattering amplitude as the sum of the two contributions

$$\bar{\alpha} = \alpha_E + \Delta\alpha = \frac{e^2\langle r^2\rangle}{3m} + 2e^2 \sum_{N \neq 0} \frac{|\langle N|D_z|0\rangle|}{E_N - E_0} \,. \qquad (4)$$

The magnetic susceptibility β is treated as a constant, made up of a paramagnetic and a diamagnetic part,

$$\beta = \beta_{Para} + \beta_{Dia}$$

$$= 2e^2 \sum_{N \neq 0} \frac{|\langle N|\mu_z|0\rangle|}{E_N - E_0} - \left\{ \frac{1}{6} \langle \sum_i \frac{e_i^2 r_i^2}{m_i} \rangle + \frac{e^2}{2m} \langle \vec{D}^2 \rangle \right\} \tag{5}$$

Equations 4 and 5 show that $\overline{\alpha}$ is the sum of two positive contributions and thus always positive itself, while β contains a positive (β_{Para}) and a negative (β_{Dia}) part. This makes clear that sign and size of the magnetic susceptibility allow conclusions on the relative strength of dia- and paramagnetism in the observed particle.

A relativistic approach has been used to determine the differential cross section for Compton scattering on nucleons. This relativistic treatment ensures that particle as well as antiparticle states were included in the calculation. Fig. 1 shows the Feynman diagrams for the channels considered here. The amplitude is made up of the Born terms (s- and u-channel), the anomalie term (t-channel) and the delta resonance in s- and u-channel.

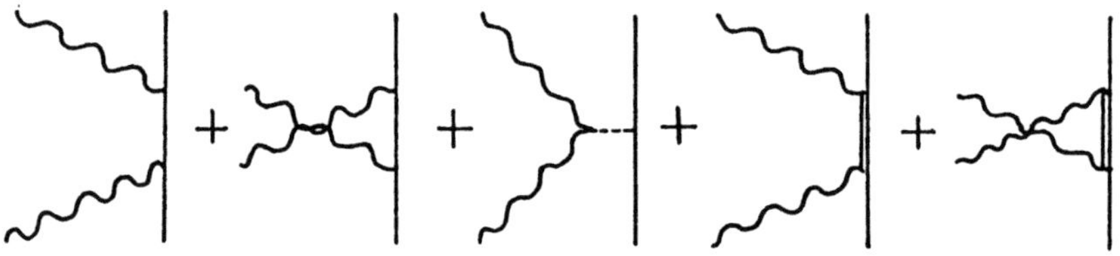

Fig. 1: Compton scattering for the respective channels.

The Born terms are evaluated according to Bjorken and Drell, whose conventions are adopted in the calculations. The invariant matrix element that includes the anomalous magnetic moment of the nucleon leads to the Powell [Pow 49] cross section. This cross section has been divided into spin flip and non-spin flip parts by Gomez and Walecka [GoWa 56], who made use of the relations between these amplitudes:

A_1 (up→up) = A_2^* (down→down),

A_3 (down→up) = A_4^* (up→down).

In the low-energy limit the result corresponds to the Thomson cross section with the characteristic dipole radiation, as is required by the Low Energy Theorems:

$$\left(\frac{d\sigma}{d\Omega_{lab}}\right)_{Th} = \frac{\alpha^2}{2m^2} (1 + \cos^2\vartheta_{lab}) \tag{6}$$

When omitting the anomalous magnetic moment of the nucleon, one obtains the famous Klein-Niskina formula [KlNi 29].

At low energies, the s- and u-channel can account for the experimental data. However, with an increase in energy these terms cannot come up with the experiment any more and polarizabilities have to be taken into account. In the late 50's Capps, Klein, Balding [Capp 57, Klei 55, Bald 59] and other showed that in the expansion of Compton scattering amplitude for spin-$\frac{1}{2}$ particles the terms quadratic in the photon energy ω are determined by the mass and magnetic moment of the particle and two structure functions: the polarizabilities α and β.

The triangle graph represents the exchange of a neutral pion which can be calculated either by use of fourth-order perturbation theory or with an effective lagrangian. Fig. 2 shows both variants of this process.

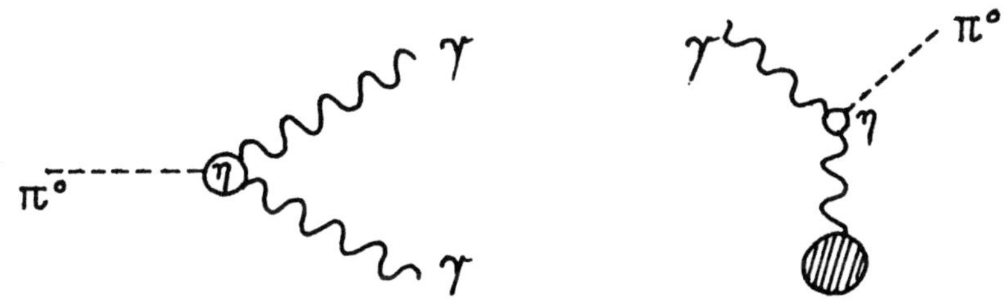

Fig. 2: Neutral pion decay and the inverse decay, the Primakoff effect.

In the first method, it is assumed that a structureless particle (proton or quark) propagates along the triangle. The quark description turned out to be a clear proof for the colour degeneracy, namely the concept that quarks have to be characterized by an additional quantum number "colour".

The assumption for the use of an effective lagrangian is that the $\gamma\pi^0$-vertex can be expressed in the same way as the π^0-decay or the inverse process, the Primakoff-effect. This leads to a "point interaction" that has been described by Belletini [Bell 65].

Both methods - fourth-order and effective lagrangian - lead to the same result. In particular the fourth-order perturbation theory uniquely determines the phase of this contribution relative to the s- and u-channel amplitudes. At this point, it should be mentioned that the two vertices - (N,π^0) and $(\pi^0,\gamma\gamma)$ - represent two different processes and should therefore be handled with care. While the $N\pi^0$ vertex is characterized by the pion-nucleon decay constant $g_{\pi NN} = \sqrt{14.5 \times 4\pi}$, the decay of the π^0 into two gammas is represented by $f_\pi \sim 93$ MeV. Therefore both constants should appear in the calculation (and not, as would happen in perturbation-theoretical approaches, only $g^2_{\pi NN}$). The Goldberger-Treiman relation can be used to convert either one in the respective other

$$g_A(0) = \frac{G_A}{G_V}\, g_{\pi NN}\, \frac{f_\pi}{m} \; . \tag{7}$$

The result obtained via perturbation theory is in agreement with the phase obtained by Guiasu [Gui 77] and corresponds to a negative product of f_π and $g_{\pi NN}$. It is also the one obtained with the $N\bar{N}$ model calculation of $\pi^0 \rightarrow \gamma\gamma$ of Goldberger and Treiman and with the Primakoff effect.

The delta resonance serves as an example for the excited states of the nucleon. The magnetic dipole operator $\hat{\mu}$ causes an M1-transition into the excited state $\Delta^{3,3}$ with a width of 109 MeV, mass of 1232 MeV, and spin and isospin $\frac{3}{2}$. Up until now, the magnetic polarizability has been treated as a constant $\beta = \beta(0)$. The idea of this calculation was to include the full energy dependence $\beta(k)$ by adding the resonant channels.

For the description of the resonance with spin and isospin $\frac{3}{2}$ in the Dirac algebra for spin-$\frac{1}{2}$ particles, Rarita-Schwinger spinors were used, whith explicit forms given by Arenhövel [Aren 75] and [WeAr 78]. Since the propagator described by Berends and Fronsdal [BeFr 57] contains some ambiguities and is uniquely determined only in the resonance ($P^2 = M_\Delta^2$), it has been expressed here by the projector

$$P^{\mu\nu} = \sum_\Lambda u^\mu(\vec{P},\Lambda)\, \bar{u}^\nu(\vec{P},\Lambda). \tag{8}$$

The Rarita-Schwinger propagator for the intermediate excited delta resonance reads

$$\Delta^{\mu\nu} = \frac{-2M_\Delta}{Q^2 - M_\Delta^2 + iM_\Delta\Gamma}\, P^{\mu\nu} \tag{9}$$

with $\Gamma = 109\, \frac{|q|^2}{|q_\Delta|^2}\, \frac{M_\Delta}{Q}\, \frac{1+(R|q_\Delta|)^2}{1+(R|q|)^2}$,

$R = 5.52 \times 10^{-3}\ \mathrm{MeV}^{-1}$ and the pion momentum in the isobar rest frame, q^*, which is $q_\Delta^* = 1.144\ \mathrm{fm}^{-1}$ at resonance. Note that the delta width vanishes for photon energies below pion threshold.

For the electromagnetic coupling of the nucleon to a spin-$\frac{3}{2}$ baryon there exist two separately gauge invariant possibilities: the electric quadrupole (E2) and the magnetic dipole (M1) coupling. It has been shown [Schi 67], however, that the electric coupling constant is only about 1/20 of the magnetic and that in addition a factor $\left(\frac{k}{m_N}\right)$ in the coupling operator reduces the electric contribution. The vertex is thus described by

$$\Gamma^{\mu\nu}_{N\Delta\gamma} = -ieG_1C_\gamma\, \frac{\gamma^\mu k^\nu - g^{\mu\nu}\not{k}}{m_N + M_\Delta}\, \gamma_5 \tag{10}$$

and the coupling constant G_1 is determined by the total photoabsorption cross section as $G_1 = g_1 \frac{M_\Delta + m_N}{m_\pi}$ with $g_1 = 0.31$. C_γ contains the isospin structure of the vertex which is $\sqrt{2/3}$ in this case.

With all these ingredients, the matrix element can be readily evaluated and the result in the s-channel reads

$$M_{fi}^{\Delta} = \frac{2}{3}\,\frac{e^2 G_1^2}{(M_\Delta + m_N)^2}\,\frac{2M_\Delta}{Q_s^2 - M^2 + iM_\Delta\Gamma}\left\{\frac{2}{3}\,(\vec{\varepsilon}\times\vec{K})\cdot(\vec{\varepsilon}\,'\times\vec{K}') + \frac{i}{3}\,\vec{\sigma}\cdot(\vec{\varepsilon}\times\vec{K})\times(\vec{\varepsilon}\,'\times\vec{K}')\right\} \tag{11}$$

For the u-channel, only the energy of the intermediate state $Q_s^2 = E(E+2\omega) \rightarrow Q_u^2 = E(E-2\omega')$ and the sign of the spin dependent term has to be changed.

A comparison of the amplitude made up of these contributions (represented in Fig. 1) with equation (3) allows one to extract the paramagnetic polarizability that arises from the delta resonance

$$\begin{aligned} M_{fi}^{\Delta} &= -\beta_M(k)\,(\vec{\varepsilon}\times\vec{K})\cdot(\vec{\varepsilon}\,'\times\vec{K}') \\ &= \frac{-2}{3}\,\frac{e^2 G_1^2}{(M_\Delta + m_p)^2}\,\frac{2}{3}\,(\vec{\varepsilon}\times\vec{K})\cdot(\vec{\varepsilon}\,'\times\vec{K}')\left(\frac{2M_\Delta}{Q_s^2 - M_\Delta^2 + iM_\Delta\Gamma} + \frac{2M_\Delta}{Q_u^2 - M_\Delta^2 + iM_\Delta\Gamma}\right). \end{aligned} \tag{12}$$

The static limit, i.e. the constant contribution usually considered, is given by

$$\begin{aligned} \beta_{Para}^{\Delta}(k\to 0) &= \lim_{k\to 0}\frac{1}{4\pi}\,\frac{4}{9}\,\frac{e^2 G_1^2}{(M_\Delta + m_p)^2}\left(\frac{2M_\Delta}{Q_s^2 - M_\Delta^2 + iM_\Delta\Gamma} + \frac{2M_\Delta}{Q_u^2 - M_\Delta^2 + iM_\Delta\Gamma}\right) \\ &= \frac{4}{9\pi}\,\frac{e^2 G_1^2}{(M_\Delta + m_p)^2}\,\frac{M_\Delta}{M_\Delta^2 - m_p^2} \sim 10 \times 10^{-4}\ \mathrm{fm}^3 \end{aligned} \tag{13}$$

in agreement with the value required by the dispersion relations for the partial cross section of pion photoproduction.

$$\beta(M1) = \frac{1}{2\pi^2}\int_0^\infty \frac{d\omega}{\omega^2}\,\sigma(M1,\omega) \sim 10.3 \times 10^{-4}\ \mathrm{fm}^3. \tag{14}$$

In order to use the experimental values for the electric and magnetic polarizabilities

$$\begin{aligned} \overline{\alpha}^{exp} &\sim (14.4 \pm 0.3) \times 10^{-4}\ \mathrm{fm}^3 \\ \beta^{exp} &\sim (0.0 \pm 0.2) \times 10^{-4}\ \mathrm{fm}^3, \end{aligned} \tag{15}$$

one has to subtract the static part $\beta_{Para}^{\Delta}(k\to 0)$ from the energy dependent $\beta_{Para}^{\Delta}(k)$. The amplitude therefore includes

$$\begin{aligned} T_c = {} & T_{s,u} + T_t + \left(\beta_{para}^{\Delta}(k) - \beta_{para}^{\Delta}(0)\right)\,(\hat{\varepsilon}\times\vec{k})\cdot(\hat{\varepsilon}\,'\times\vec{k}') \\ & + \alpha_{exp}\,\omega\omega'\,\hat{\varepsilon}\cdot\hat{\varepsilon}\,' + \beta_{exp}\,(\hat{\varepsilon}\times\vec{k})\cdot(\hat{\varepsilon}\,'\times\vec{k}'). \end{aligned} \tag{16}$$

The differential cross section that results from this amplitude can be expressed as

$$\frac{d\sigma}{d\Omega_{lab}} = \frac{1}{(4\pi)^2} \left(\frac{k'}{k}\right)^2 \frac{1}{2} \sum_{spin} \frac{1}{2} \sum_{pol} |T_c|^2.$$

The above value for β^{exp} contains the paramagnetic as well as the diamagnetic contribution. It corresponds to a value of $\beta^{Dia} \sim -10 \times 10^{-4}$ fm^3 which cancels with the paramagnetic part. This experimental value $\beta^{exp} \sim 0$ is appropriate, since with $\left(\beta^{\Delta}_{Para}(k) - \beta^{\Delta}_{Para}(k \to 0)\right)$ only the difference between static and energy dependent contribution has been accounted for. The "actual" paramagnetic part is furthermore represented by the experimental value β_{exp}.

The numerical results are given in figs. 3-7. In fig. 3 the cross sections are calculated for 3 different combinations of $\bar{\alpha}$ and β so that the sum $\bar{\alpha}+\beta = 14.4$ agrees with the photoabsorption sum rule

$$\bar{\alpha} + \beta = \frac{1}{2\pi^2} \int_0^\infty \frac{d\omega}{\omega^2} \sigma_{abs}(\omega) = (14.4 \pm 0.7) \cdot 10^{-4} \text{ fm}^3. \tag{17}$$

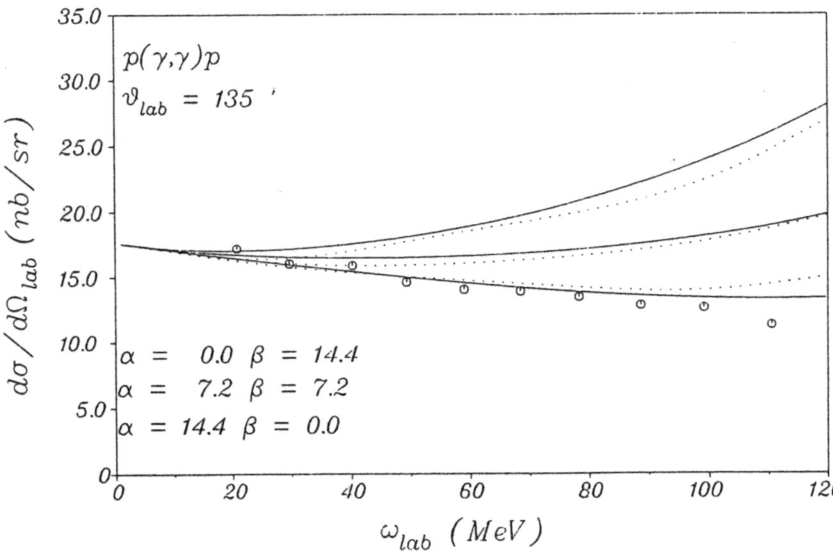

Fig. 3: Differential cross section for Compton scattering off the proton at $\vartheta_{lab} = 135^{\circ}$ as a function of photon lab energy. The upper, middle and lower curves are evaluated with $\bar{\alpha}$ = 0, 7.2, 14.4 and β = 14.4, 7.2, 0, respectively in order to be in agreement with $\bar{\alpha} + \beta = 14.4$ from the photoabsorption sum rule. The full lines show our results while the dashed lines show the calculations of L'vov [Ahr 90].

A good agreement with the experimental data is obtained for $\bar{\alpha}$ = 14.4 and β = 0 which is therefore used in the following calculations.

The angular distribution of the differential cross section is plotted in fig. 4 for the amplitude with constant polarizabilities and compared to the one that contains the energy dependent β(k).

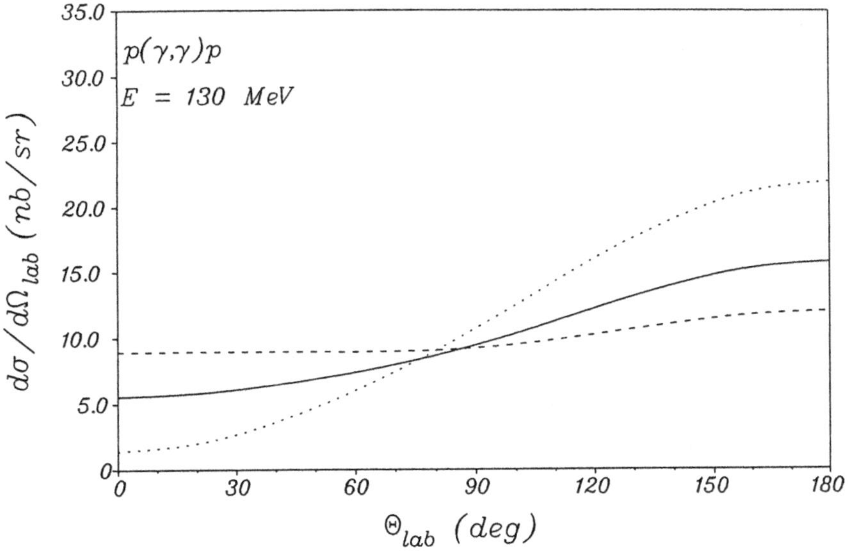

Fig. 4: Differential cross section for a lab energy 130 MeV. Dashed line: s-u-t channel with $\bar{\alpha}$ = 14.4, β = 0, full line: s-u-t plus Δ with α = 14.4, β = 0, dotted line: s-u-t plus Δ without additional polarizabilities.

It is apparent that the curves intercept at ϑ_{lab} = 86° which corresponds to ϑ_{cm} = 90°, and so the experiment will be least sensitive to this phenomenon of energy dependence for lab angles of 90°. This behaviour can be easily understood since the differential cross section around ϑ_{lab} = 90° is independent of β.

In figs. 5 and 6 the energy distributions are shown. For ϑ_{lab} = 90° there is hardly a noticeable difference between the two approaches in accordance with fig. 4. In fig. 6 for ϑ_{lab} = 135° the difference between the two approaches becomes apparent for energies above 100 MeV. The full circles are data from Mainz [Ahr 87] the from Baranov [Bar 74].

In the analysis of these theoretical descriptions for the respective cross sections, the standard deviations are evaluated with

$$\chi^2 = \frac{\left(\left(\frac{dN}{dk'}\right)_{Theo} - \left(\frac{dN}{dk'}\right)_{exp}\right)^2}{\left(\Delta\left(\frac{dN}{dk'}\right)_{exp}\right)^2}$$

and $\frac{dN}{dk'}$ is the number of photons measured per energy interval. Fig. 7 shows the contour of χ^2 to this calculation with experimental values ranging to 75 MeV, 100 MeV and 125 MeV respectively. With decreasing number of data points, the statistic deviations become larger and so do the contours. On the other hand, the calculation up to 75 MeV does not depend as much on the higher order terms ($>O(\omega^2)$). The description with the expansion of the differential cross section is thus more valid for the evaluation up to 75 MeV than up to 100 MeV. When varying the energy range of the analysis, the contours of χ^2 do not move. This shows that the calculation with the paramagnetic $\beta^{\Delta}_{para}(k)$ can properly describe the energy dependence of the polarizability.

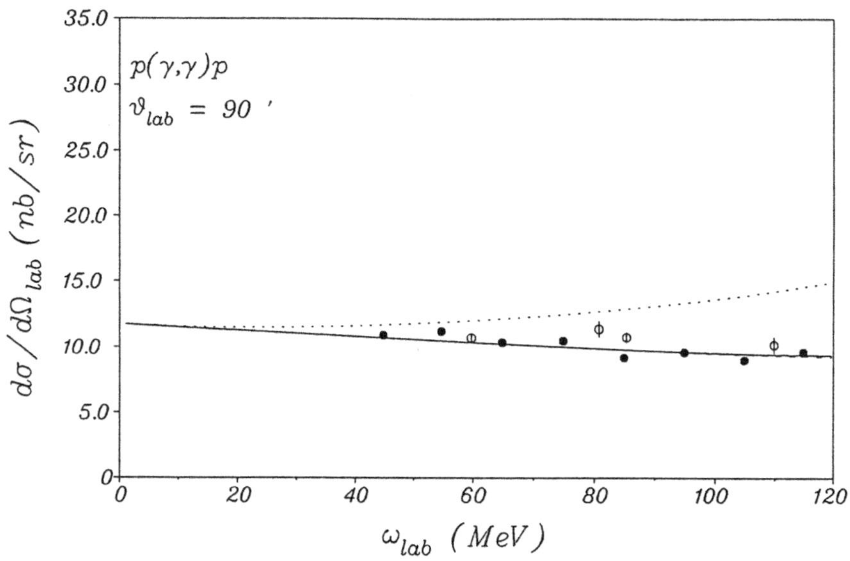

Fig. 5: Energy distribution of differential cross section in lab system for $\vartheta_{lab} = 90^\circ$. The full line shows the cross section of the s-, u- and t-channel amplitude with Δ and $\overline{\alpha} = 14.4$, $\beta = 0$. The dashed line represents s-, u- and t-channel with $\overline{\alpha} = 14.4$, $\beta = 0$. and the dotted s-, u-, t-channel plus Δ without additional polarizabilities.

In conclusion, it should be realized that the energy dependence contained in the delta resonance represents an important point in the analysis of nucleonic polarizabilities. The theoretical cross section describes the existing data and indicates that the presented calculation corresponds with the energy dependence inherent in the structure function β(k). At this point it should also be noted that no attempt was made to develop a full Compton scattering amplitude which would contain all processes that may occur in γN reactions. Much rather it is stated that in the analysis of the experiments at Mainz, performed at energies up to pion photoproduction threshold, the dynamic dependence of the polarizability on the photon frequency cannot be neglected. With other words, higher order corrections ($>\omega^2$) have been included in order to treat the phenomenon of nucleonic polarizability in a more physical way. However, it was not the aim to give a complete description of the Compton scattering

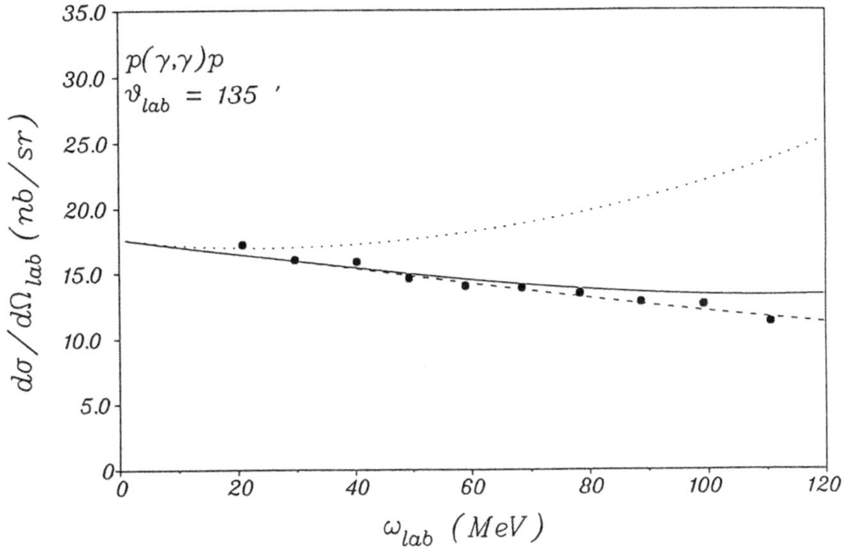

Fig. 6: The same as in fig. 5 for $\vartheta_{lab} = 135^o$.

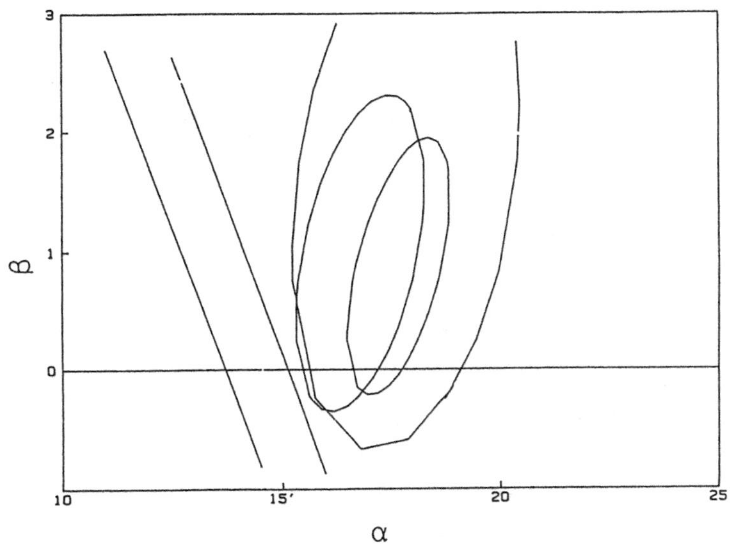

Fig. 7: Contours of χ^2 for the calculation with the energy dependent $\beta^{\Delta}_{para}(k)$ for various energy ranges (up to 75, 100, 125 MeV starting from the center curve).

amplitude. The intention was to include the energy dependence of β^{Δ}_{para} in the analysis of polarizabilities by an explicit description of the delta resonance. The most realistic assumption on nucleon polarizabilities is based on the dispersion relations for the Compton scattering amplitude

$$\bar{\alpha} + \beta = \frac{1}{2\pi} \int_{\omega_0}^{\infty} \frac{\sigma_{tot}(\omega)}{\omega^2}\, d\omega. \qquad (18)$$

About 70 % of the contribution is contained in the energy intervall between pion (photoproduction) threshold and delta resonance (1232 MeV)

$$\bar{\alpha} = \frac{1}{2\pi} \int_{\omega_0}^{\infty} \frac{d\omega}{\omega^2} \left[\left(1 + \frac{\omega}{m}\right)\sigma_{E1} - \frac{\omega}{m}\,\sigma_{M1} + \left(2 + \frac{3\omega}{m}\right)\sigma_{E2} - \left(1 + \frac{3\omega}{m}\right)\sigma_{M2} + \ldots\right] \qquad (19)$$

$$\beta = \frac{1}{2\pi} \int_{\omega_0}^{\infty} \frac{d\omega}{\omega^2} \left[\left(1 + \frac{\omega}{m}\right)\sigma_{M1} - \frac{\omega}{m}\,\sigma_{E1} + \left(2 + \frac{3\omega}{m}\right)\sigma_{M2} - \left(1 + \frac{3\omega}{m}\right)\sigma_{E2} + \ldots\right]. \qquad (20)$$

For the nucleon these formulae are further complicated by interferences of electric with magnetic transitions. Equations (19) and (20) show that for a better description of Compton scattering the energy dependences of various other multipoles will have to be added. In further analyses it will be necessary to include the energy dependence of the E1-contribution which influences the electric polarizability as was done by Nozawa [Noz 89] for γ,π in the framework of a dynamical model. These rescattering effects in pion production at the nucleon can further improve the agreement between theory and experiment not only with respect to resonant multipoles but also for the angular distributions and total cross section of photoproduction.

One of us (B. König) is very grateful for the continuous interest and help during this work by J. Ahrens.

REFERENCES

[Ahr 87] J. Ahrens et al., Physics with MAMI A, p. 1, Mainz 1988

[Ahr 90] J. Ahrens, private communication

[Aren 75] H. Arenhövel, Nucl. Phys. A247 (1975) 473

[Bald 59] A.M. Baldin, Nucl. Phys. 18 (1960) 310

[Bar 74] Baranov et al., Phys. Lett. 52B (1974) 122

[BeFr 57] R.E. Behrends, C. Fronsdal, Phys. Rev. 106 (1957) 345

[Bell 65] Belletini, Nuovo Cimento 11A (1965) 1139

[Capp 57] R. Capps, Phys. Rev. 106 (1957) 1037

[Fri 75] J.L. Friar, Annals of Physics 95 (1975) 170

[GoWa 56] R. Gomez, D. Walecka, Phys. Rev. 104 (1956) 1479

[Gui 77] I. Guiasu, C. Pompiniu, E.E. Radescu, Annals of Physics 117 (1977) 296

[Klei 55] A. Klein, Phys. Rev. 99 (1955) 998

[KlNi 29] O. Klein, Y. Nishima, Z. Physik 52 (1929) 853

[Noz 89] S. Nozawa, B. Blankleider, T.-S.H. Lee, Argonne preprint, PHY-6089-TH-89

[Pow 49] J.L. Powell, Phys. Rev. 75 (1949) 32

[Schi 67] D. Schiff, J. Tran Trahn Van, Nuovo Cimento 48A (1967) 1273

[WeAr 78] H.J. Weber, H. Arenhövel, Phys. Rep. C (1978) 277

ETA PHOTOPRODUCTION FROM NUCLEI*

C. Bennhold
Institut für Kernphysik, Johannes Gutenberg-Universität Mainz
D-6500 Mainz, FRG

Abstract

Low-energy eta photoproduction from complex nuclei is investigated in a distorted wave impulse approximation (DWIA) framework carried out in momentum spaces. The elementary operator is fixed by the hadronic reactions $\pi N \to \pi N$, $\pi N \to \pi\pi N$ and $\pi^- p \to \eta n$ as well as the process $\gamma N \to \pi N$ to constrain the electromagnetic vertex. The resonances $S_{11}(1535)$, $P_{11}(1440)$ and $D_{13}(1520)$ are used to generate s-, p- and d-wave interactions. This coupled channels approach fulfils unitarity and allows the prediction of $\gamma p \to \eta p$ cross sections. The η final state interaction has been included via a simple optical potential using the ηN t-matrix as an input. We find that coherent eta production is dominated by the $D_{13}(1520)$ isobar while spin-flip transitions to excited nuclear states are sensitive to the $S_{11}(1535)$ resonance. The influence of the $P_{11}(1440)$ appears to be marginal. Predictions are given for coherent production on ^4He, ^{12}C and ^{40}Ca, as well as incoherent production on ^6Li, ^{10}B, ^{12}C and ^{14}N.

The advent of high duty cycle electron accelerators at Mainz, Bates, NIKHEF and CEBAF will allow the study of reactions involving nucleon knockout and nuclear meson production in coincidence experiments which have not been performed up to now. One developing field of interest accessible through the higher energy range of the new accelerators is the photoproduction of eta mesons[1-3]. While some differential cross section data for the elementary process on protons have been measured 20 years ago[4,5] η-photoproduction from nuclear targets has not been studied experimentally. Especially with the upcoming start of the Mainz Microtron with its energy up to 850 MeV interest in this subject has been growing[6]. It is therefore timely to examine the dynamics of eta photoproduction from nucleons and nuclear targets. In this study we focus on the low-energy regime since, as can be seen in Fig. 1, the measured differential cross section for $\gamma p \to \eta p$ at higher energies is smaller by about a factor of five compared to the threshold region leading to very small

---

* Supported by the Deutsche Forschungsgemeinschaft (SFB 201)

counting rates when considering nuclear targets. Eta photoproduction threshold on the proton occurs at E_γ = 709.3 MeV (corresponding to W = 1487 MeV), but drops to E_γ = 592 MeV on a ^{12}C target (see Fig. 2 for a comparison of the kinematical situation).

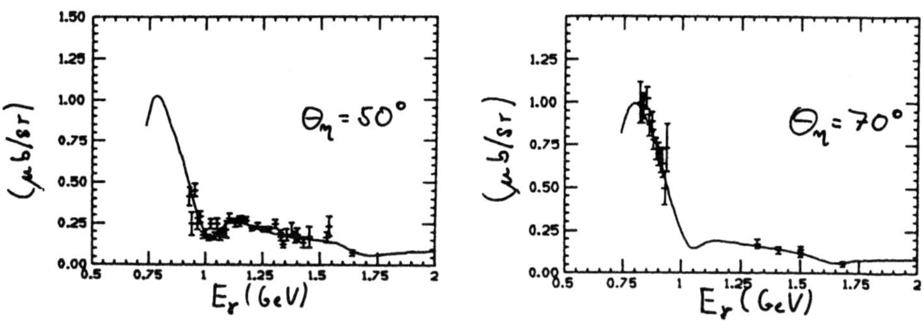

Fig. 1: Differential cross section for $\gamma p \rightarrow \eta p$ as a function of the photon lab energy. The figure is copied from ref. 1.

An impressive amount of information about the dynamical properties of the Δ(1232) resonance inside nuclei has been obtained through the study of (π,π) and (γ,π) reactions. Above the Δ resonance region, however, the overlapping of higher resonances prevents studying one specific resonance by pion production experiments. Just as the dominance of the Δ isobar in pion-nuclear reactions allowed the extraction of quantitative information on the Δ-nuclear potential, we would hope to learn about similar medium modifications of the S_{11} and other N^* resonances inside nuclei via photonuclear η production. Inelastic transitions such as $\gamma + A(T_i = 0, J_i = 0) \rightarrow \eta + A(T_f, J_f)$, where the etas are produced either by isoscalar ($T_f = 0$) or isovector ($T_f = 1$) photons, also offer the possibility of independently studying the spin-flip (S = 1) and non-spin-flip (S = 0) part of the production operator. The S_{11} resonance, for example, is excited by real photons through the E_{0+} multipole which contains only a spin-flip operator $\vec{\sigma} \cdot \vec{\varepsilon}$. Separating the S = 0 from the S = 1 part of the amplitude is not possible for the elementary process.

In this study we present calculations for coherent and incoherent η photoproduction from complex nuclei in a distorted wave impulse approximation framework. In this picture (see Fig. 3) of a single-particle model of the nucleus, the incident photon penetrates the nucleus and couples electromagnetically to an individual nucleon, causing it to emit η-mesons. The produced etas experience some rescattering before leaving the nucleus and reaching the detector. In order to describe η photoproduction off single nucleons we employ an operator developed in a coupled channels approach[2], in which the reaction (γ,η) is related to (γ,π) as well as the hadronic reactions (π,π), (π,η) and (η,η). This framework allowed the extraction of an $\eta N \rightarrow \eta N$ t-matrix[7] which is inserted into a simple optical potential in order to properly include the η final state interaction. While for coherent processes all nucleons participate

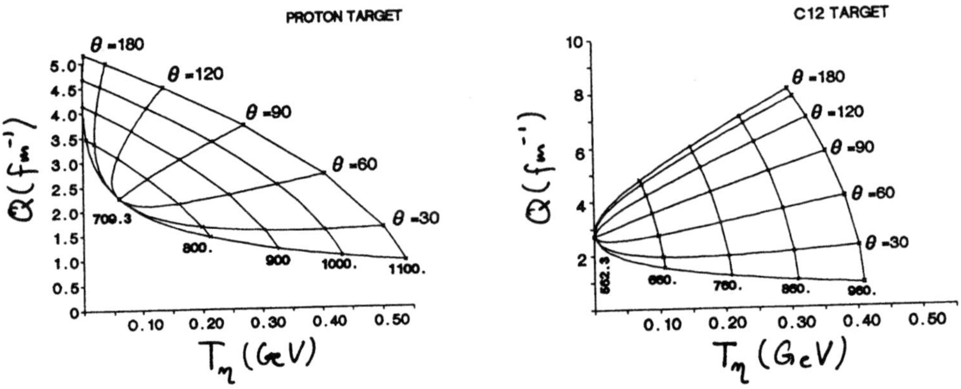

Fig. 2: The kinematical situation for the process $\gamma N \to \eta N$ and $^{12}C(\gamma,\eta)^{12}C$.

in the production process, we require the knowledge of nuclear many-body wave functions for incoherent reactions to describe the overall transition between specific initial and final nuclear states. This has been the subject of electron scattering for many years; we will limit ourselves to several low-lying well-studied excited states whose structure has in most cases been confirmed by pion photoproduction and other reactions.

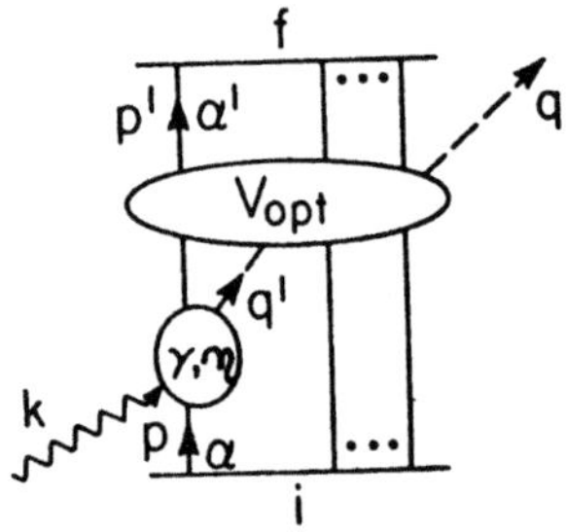

Fig. 3: Diagrammatic illustration of the DWIA process in momentum space. The momenta $\vec{p}$ and $\vec{q}'$ are integration variables, while $\vec{p}' = \vec{k} + \vec{p} - \vec{q}'$ is fixed by momentum conservation.

The differential cross section for eta photoproduction from complex nuclei $\gamma + A \rightarrow \eta + A^*$ in the center of momentum (cm) frame is given by[8,9]

$$\frac{d\sigma}{d\Omega_\eta^{c.m.}} = \frac{1}{16\pi^2}\frac{q^{cm}}{k^{cm}}\frac{m_i m_f}{W^2}\frac{F_{cm}}{2(2J_i+1)}\sum_{M_i,M_f,\lambda}|\langle J_f M_f, T_f N_f; \vec{q}|T|J_i M_i, T_i N_i; \vec{\varepsilon},\vec{k}\rangle|^2, \quad (1)$$

where we average over the initial spin projection M_i as well as the photon polarization λ and sum over the final spin projection M_f. The three-momenta of the photon and asymptotic eta are denoted by $\vec{k}$ and $\vec{q}$; the total spins and isospins are J_i, J_f and T_i, T_f, respectively, along with their projections M_i, M_f and N_i, N_f. The usual correction factor compensating for the lack of translational invariance of the shell model is given by $F_{cm} = \exp\left[b^2(\vec{k}-\vec{q})^2/2A\right]$, b being the harmonic oscillator parameter for the nucleus under study and A the nuclear mass number. The masses of the initial- and final-state nuclei are m_i and m_f, and W is the total energy in the cm system. The kinematics is depicted in Fig. 4. Assuming the elementary operator to be a one-body amplitude in the nucleon space we can write the nuclear matrix element

$$\langle J_f M_f, T_f N_f; \vec{q}|T|J_i M_i, T_i N_i; \vec{\varepsilon},\vec{k}\rangle = \sum_{\alpha\alpha'}\langle J_f M_f, T_f N_f|C^\dagger_{\alpha'}C_\alpha|J_i M_i, T_i N_i\rangle \langle\alpha';\vec{q}|t|\alpha;\vec{\varepsilon},\vec{k}\rangle \quad (2)$$

where in principle the sum extends over a complete set of single particle states α and α', and $C^\dagger_{\alpha'}$ and C_α are appropriate particle and hole creation and annihilation operators. In case of the coherent (γ,η) process we have $\alpha = \alpha'$ and sum over all occupied states since all nucleons participate in the production process.

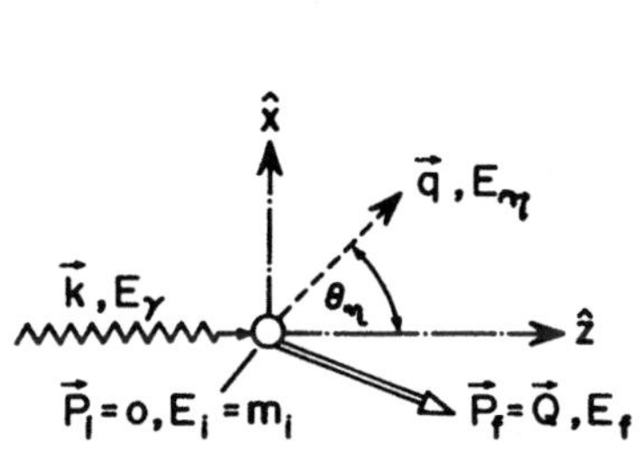

Fig. 4: The kinematics in the laboratory frame.

The required nuclear structure information is specified in the form of the double reduced one-body density matrix

$$\psi_{J;T}(a'a) = \hat{J}^{-1}\,\hat{T}^{-1}\,\langle J_f;T_f\,\|\,[C^{\dagger}_{a'}\times\tilde{C}_a]_{J;T}\,\|\,J_i\,;T_i\rangle \tag{3}$$

where $\hat{x} = \sqrt{2x+1}$ and J and T are the total spin and isospin transferred to the nucleus in the transition which, for the p-shell, can be J = 0, 1, 2, 3 and T = 0, 1.

All the dynamics of the photoproduction process is contained in the single-particle matrix element $\langle\alpha';\vec{q}\,|\,t\,|\,\alpha;\vec{\varepsilon},\vec{k}\rangle$ of Eq. (2) which in momentum space takes the form

$$\langle\alpha';\vec{q}\,|\,t\,|\,\alpha;\vec{\varepsilon},\vec{k}\rangle = \int d^3p\; d^3q'\;\psi^*_{\alpha'}(\vec{p}\,')\;\Phi^{*(-)}_{\eta}(\vec{q},\vec{q}\,')\;t_{\eta\gamma}(\vec{q}\,',\vec{p},\vec{p}\,',\vec{k})\;\psi_\alpha(\vec{p}) \tag{4}$$

where ψ_α is the single particle bound state wave function in momentum space and $\Phi^{(-)}_{\eta}(\vec{q},\vec{q}\,')$ denotes the wave function of the outgoing eta of asymptotic three-momentum $\vec{q}$. The four-momenta of the photon, η-meson, initial and final nucleon are denoted by $p_\gamma = (k,\vec{k})$, $q' = (E_{q'},\vec{q})$, $p = (E,\vec{p})$ and $p' = (E',\vec{p}\,')$ with $\vec{p}\,' = \vec{p} + \vec{k} - q'$, respectively. The photoproduction operator $t_{\eta\gamma}$ is given by

$$t_{\eta\gamma} = \sqrt{3}\; t^{T=0} - \frac{\hat{\tau}_0}{\sqrt{3}}\; t^{T=1} \tag{5}$$

where

$$t^T = L^T + i\vec{\sigma}\cdot\vec{K}^T = \sum_{S,m_s} i^S(-1)^{m_s}\,\sigma^S_{-m_s}\,(K^{S,T}_{m_s}) \tag{6}$$

with $\sigma^0 = 1$ and $K^{0,T} = L^T$. The spin-flip (S = 1) part which contains the $\vec{\sigma}\cdot\vec{\varepsilon}$ (Kroll-Ruderman) term plays a dominant role unless it is suppressed by nuclear matrix elements or selection rules.

In Fig. 5 we show the (γ,η) isovector and isoscalar as well as spin-flip and non-spin-flip total cross section separately. The spin-1 transition is totally dominated by the E_{0+} multipole which is given by the $S_{11}(1535)$ resonance. Only magnetic multipoles are allowed for the spin-0 transition operator which is dominated by the M_{2-} multipole of the $D_{13}(1520)$ resonance. Note that the T=0, S=0 cross section is more than a 500 times smaller than the T=1, S=1 cross section.

The operators L^T and $\vec{K}^T$ of Eq. (6) are given by

$$L^T = \frac{G_1}{2kM}\,\vec{q}\,'\cdot(\vec{\varepsilon}\times\vec{k}) \tag{7a}$$

$$\vec{K}^T = -\vec{\varepsilon}\left[G^T_3 - G^T_1 + \frac{G^T_1}{2kM}\left(\vec{k}\cdot\vec{p} + \vec{k}\cdot\vec{p}\,'\right)\right] + \vec{k}\left[\frac{G^T_1}{2kM}\left(\vec{\varepsilon}\cdot\vec{p}\,' + \vec{\varepsilon}\cdot\vec{p}\right) - G^T_4\right]$$
$$- \vec{p}\left(\frac{G^T_2 - kG^T_4}{2M}\right) + \vec{p}\,'\left(\frac{G^T_2 + kG^T_4}{2M}\right) \tag{7b}$$

with $G^T_1 - G^T_4$ given in terms of the A^T_i:

$$G^T_1 = k\,A^T_1 \qquad\qquad G^T_2 = 2\,A^T_2\,(\vec{\varepsilon}\cdot\vec{p}\,'\;k\cdot p - \vec{\varepsilon}\cdot\vec{p}\;k\cdot p')$$
$$G^T_3 = A^T_3\,k\cdot p + A^T_4\,k\cdot p' \qquad\qquad G^T_4 = A^T_3\,\vec{\varepsilon}\cdot\vec{p} + A^T_4\,\vec{\varepsilon}\cdot\vec{p}\,' \tag{8}$$

where in the lab frame we have made the usual gauge choice of $\varepsilon_0 = 0$, $\vec{\varepsilon}\cdot\vec{k} = 0$.

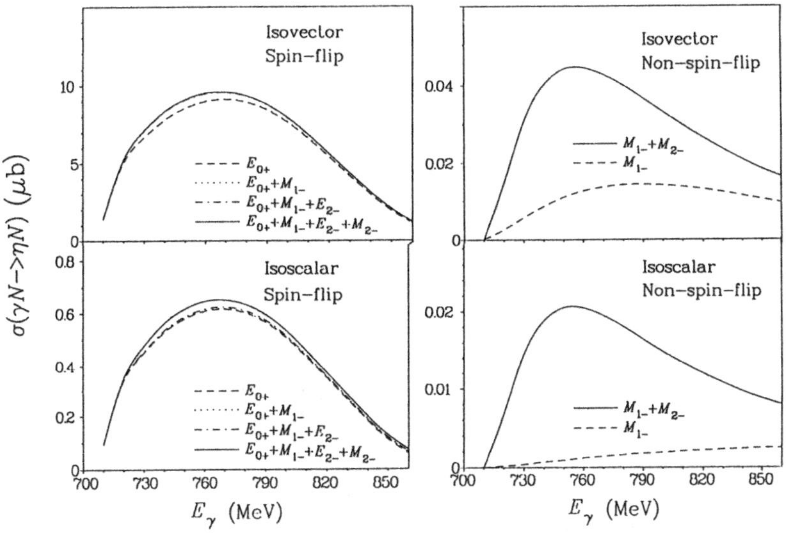

Fig. 5: **Total cross sections for the process $\gamma N \rightarrow \eta N$ showing the T=0 and T=1 as well as the S=0 and S=1 terms separately.**

We define the amplitudes A_i^T by

$$A_1^T = \frac{4\pi}{k^*\sqrt{2W}} \left(F_1^T \frac{W-M}{e^+} - F_2^T \frac{W+M}{e^-} \right) \tag{9a}$$

$$A_2^T = \frac{4\pi}{q^* k^* \sqrt{2W}} \left(\frac{F_4^T}{e^-} - \frac{F_3^T}{e^+} \right) \tag{9b}$$

$$A_3^T = \frac{4\pi\sqrt{2W}}{k^* W} \left(- \frac{F_1^T}{e^+} - \frac{F_2^T}{e^-} - \frac{(E'^* k^* + k^* q'^* \cos\Theta^*)}{q^*} \left(\frac{F_3^T}{e^+ (W-M)} + \frac{F_4^T}{e^- (W+M)} \right) \right) \tag{9c}$$

$$A_4^T = \frac{4\pi}{q'^* k^* \sqrt{2W}} \left(F_3^T \frac{W+M}{e^+} + F_4^T \frac{W-M}{e^-} \right) \tag{9d}$$

with $e^+ = \sqrt{E'^* + M}$ and $e^- = \sqrt{E'^* - M}$ where the asterisk marks the kinematical variables in the ηN c.m. system. Finally, the F_i are given in terms of the (γ,η) elementary multipoles E_{0+}, M_{1-}, E_{2-} and M_{2-} which correspond to the $S_{11}(1535)$, $P_{11}(1440)$ and $D_{13}(1520)$, respectively.

$$F_1 = E_{0+} + 3\, M_{2-} + E_{2-} \tag{10a}$$

$$F_2 = M_{1-} + 6 \cos\Theta^* \, M_{2-} \tag{10b}$$

$$F_3 = 0 \tag{10c}$$

$$F_4 = -3\,(M_{2-} + E_{2-}). \tag{10d}$$

The momenta $\vec{k}^*$ and $\vec{q}'^*$ in the ηN c.m. system are obtained by a Lorentz transformation from an arbitrary reference frame defined by $\vec{p}$ and $\vec{q}'$; thus for each value of the integration variables $\vec{p}$ and $\vec{q}'$ the amplitudes F_i^T are evaluated at the according momenta $\vec{k}^*$ and $\vec{q}'^*$. The explicit form of the multipoles is given in ref. 10.

To obtain a cross section we carry out the spin projection summations

$$\sum_{M_i, M_f, \lambda} |\langle J_f M_f, T_f N_f; \vec{q} | T | J_i M_i, T_i N_i; \vec{\varepsilon}, \vec{k} \rangle|^2 = \sum_{\substack{J,M \\ \lambda, T}} |F_M^{J,T}|^2, \tag{11}$$

where

$$F_M^{J,T} = \begin{pmatrix} T_f & T & T_i \\ -N_i & 0 & N_i \end{pmatrix} \frac{2\sqrt{3}\,(-1)^T}{\hat{T}} \sum_{a'a} \psi_{J;T}(a',a) \sum_{LS} (-i)^S (-1)^{\ell'+1}\, \hat{j}'\, \hat{j}\, \hat{L}\, \hat{S}$$

$$\begin{Bmatrix} \ell' & \frac{1}{2} & j' \\ \ell & \frac{1}{2} & j \\ L & S & J \end{Bmatrix} I_M^{(a',a)LSJ,T} \tag{12}$$

and the integrals are given by

$$I_M^{(a',a)LSJ,T} = \int d^3p\, d^3q'\, \rho_{a',a}(p',p)\, \Phi_\eta^{(+)}(\vec{q}',\vec{q})\, \left[[Y^{\ell'}(\hat{p}') \times Y^{\ell}(\hat{p})]^L \times K^{S,T} \right]_M^J. \tag{13}$$

The momentum overlap distribution ρ is given in general by

$$\rho_{a',a}(p',p) = \Phi^*_{n'\ell'j'}(p')\, \Phi_{n\ell j}(p). \tag{14}$$

Since in coherent (γ,η) processes all nucleons participate in the production process the reaction can proceed only through the S=0, T=0 part of the operator. For closed shell nuclei the nuclear many-body matrix elements of Eq. (2) therefore simplify

$$\langle J_i T_i; \vec{q} | T^{S=0,T=0} | J_i = 0\ T = 0; \vec{k}, \vec{\varepsilon} \rangle$$

$$= \sum_{i=1}^{A} \int d^3p\, d^3q'\, \psi_i^*(\vec{p}')\, \Phi_\eta^{(-)*}(\vec{q},\vec{q}')\, t^{T=0,S=0}(\vec{q}',\vec{k},\vec{p}',\vec{p})\, \psi_i(\vec{p}). \tag{15}$$

Rather than performing a full integration over the Fermi momentum $\vec{p}$ of the struck nucleon we invoke the "factorization approximation" of ref. 11 to arrive at

$$\langle J_i T_i; \vec{q} | T^{S=0,T=0} | J_i = 0\ T = 0; \vec{k}, \vec{\varepsilon} \rangle = A \int d^3q'\, \Phi_\eta^{(+)}(\vec{q},\vec{q}')\, F_0(\vec{q}'-\vec{k})\, t^{T=0,S=0}(\vec{q}',\vec{k},\langle\vec{p}\rangle) \tag{16}$$

in which the amplitude is evaluated at an averaged Fermi momentum

$$\langle \vec{p} \rangle = - \frac{(A-1)}{2A} (\vec{k} - \vec{q}\,') \tag{17}$$

This approximation is helpful in eliminating uncertainties contained in the wave functions, since in Eq. (16) the integral over the single particle wave functions is expressed in terms of the form factor $F_0(Q')$ were $\vec{Q}' = \vec{k} - \vec{q}\,'$. The nuclear form factor $F_0(Q')$ is the Fourier transform of the nuclear density $\rho(r)$ normalized to one.

As in ref. 11, we employ a phenomenological fit to nuclear electromagnetic form factors based on the symmetrized Fermi density which allows a unified description of charge form factors over a wide range of momentum transfers for light and heavy nuclei. In this model $F_0(Q')$ is given by

$$F_0(Q) = - \frac{3\pi b\left[\cos(Qc) - \pi b \sin(Qc) \operatorname{cth}(\pi b Q)/c\right]}{Qc^2 \operatorname{sh}(\pi b Q)\left[1 + \pi^2 b^2/c^2\right]} \exp\left(R^2 Q^2/6A\right) \tag{18}$$

where the parameters b, c and R are listed in Table 1 for ^{4}He, ^{12}C and ^{40}Ca.

Table 1

Symmetrized Fermi density parameters for ^{4}He, ^{12}C and ^{40}Ca (from ref. 11)

	b	c	R
^{4}He	0.406	1.231	1.806
^{12}C	0.478	2.220	2.462
^{40}Ca	0.537	3.573	3.287

In Fig. 6 we present differential cross sections for coherent (γ,η) production off ^{4}He, ^{12}C and ^{40}Ca calculated in the plane wave impulse approximation (PWIA). The curves have the typical shape of a non spin-flip transition being kinematically suppressed at forward angles. Using ^{4}He as a target yields the highest cross sections, up to 40 nb/sr in the lab system. For heavier nuclei there is a competition between the factor A of Eq. (16) enhancing the cross section and the nuclear form factor $F_0(Q)$ which falls off more rapidly for heavier nuclei. This leads to smaller cross sections for ^{12}C and ^{40}Ca, barely exceeding 10 nb/sr. Note that the angular distribution for ^{12}C has its minimum at smaller η angles than the one for ^{40}Ca. This is due to the fact that at this range of momentum transfer (Q = 300 - 500 MeV/c) the form factor exhibits the first minimum for ^{12}C but the second maximum for ^{40}Ca.

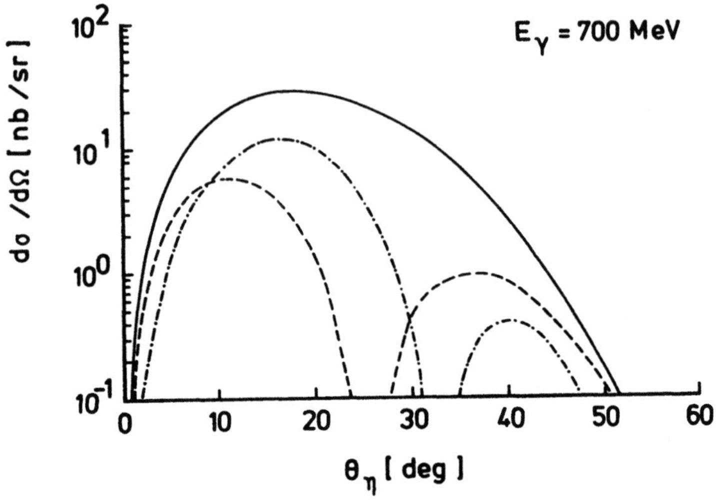

Fig. 6: Angular distribution at E_γ = 700 MeV for coherent η photoproduction on $^4He(^{12}C,^{40}Ca)$ shown by the solid (dashed, dash-dotted) curve in the PWIA.

Fig. 7 presents a number of angular distributions on ^4He for different photon lab energies. Since the momentum transfer increases faster for larger η angles at higher photon energies (Fig. 2) the respective cross sections fall off more rapidly. Optimal counting rates can be achieved around E_γ = 650 MeV.

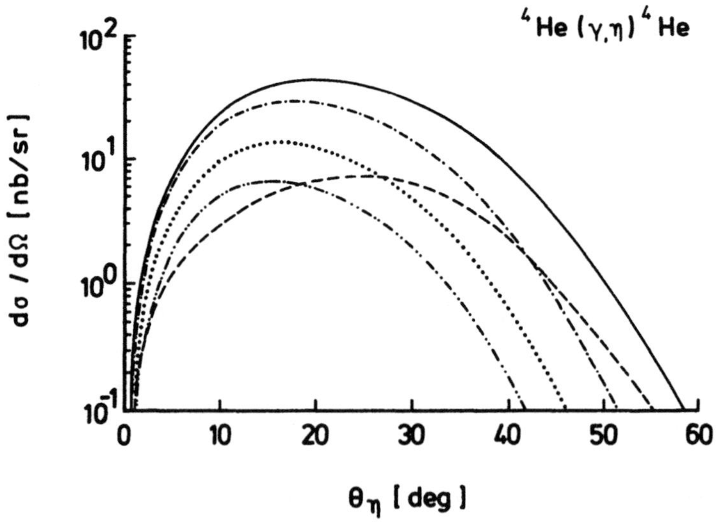

Fig. 7: Angular distributions for $^4He(\gamma,\eta)^4He$ at E_γ = 600 MeV (---), 650 MeV (——), 700 MeV (-·-·-), 750 MeV (······) and 800 MeV (-··-··-).

Fig. 8 shows coherent cross sections as a function of the photon lab energy comparing DWIA and PWIA calculations. As mentioned before, the threshold for (γ,η) production on nuclei is lowered below 600 MeV due to recoil. The curves clearly exhibit a resonant structure around E_γ = 650 MeV indicating that resonances which are excited in the elementary process just above threshold have shifted downwards due to Fermi motion. Above E_γ = 800 MeV higher resonances should be included in order to reliably predict nuclear cross sections.

The effect of η distortion is greatest around E_γ = 650 MeV and lowers the cross sections by up to 15 %. This effect becomes more important for heavier nuclei like ^{12}C and ^{40}Ca, lowering counting rates up to a factor of two in the latter case. Fig. 8 also shows the energy dependence of the η optical potential; at E_γ = 750 MeV DWIA computations are lower by only 5 %.

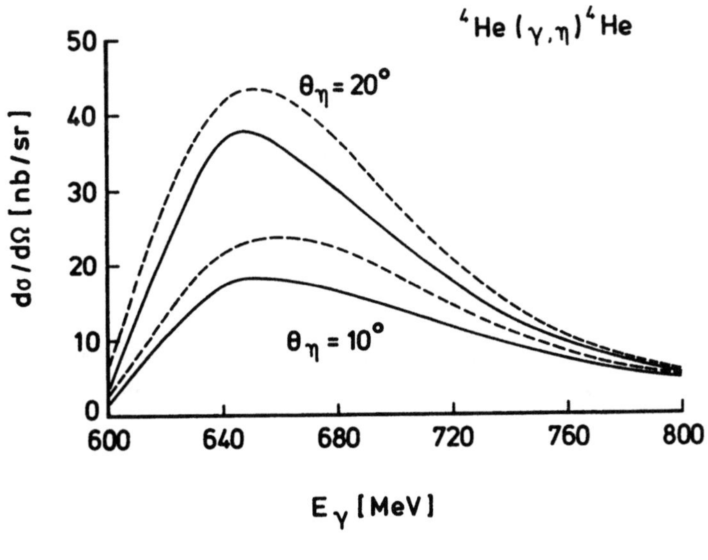

Fig. 8: Cross sections for $^4He(\gamma,\eta)^4He$ as a function of the photon energy, comparing DWIA (——) with PWIA (----) calculations.

We present in Fig. 9 the influence of the different multipoles on the differential cross section of $^4He(\gamma,\eta)^4He$. Due to nuclear Fermi motion the non spin-flip operator L defined in Eq. (7) consist not only of the magnetic multipoles M_{1-} and M_{2-} as in the elementary process, but includes a small E_{0+} contribution as well. This can be seen by inserting Eq. (9a), (10) and (8) into (7a) yielding

$$L = \frac{2\pi}{k^* M\sqrt{2W}} \left\{ (E_{0+} + 3M_{2-} + E_{2-}) \frac{W-M}{\sqrt{E'^* + M}} - (M_{1-} + 6\cos\Theta^* M_{2-}) \frac{W+M}{\sqrt{E'^* - M}} \right\} \vec{q}\cdot(\vec{\varepsilon}\times\vec{k}) \qquad (19)$$

where the expressions in the first and second brackets correspond to F_1 and F_2 of Eq. (10), respectively. Clearly, the factor multiplying F_1 is much smaller than the factor multiplying F_2, thus would the electric and magnetic multipoles be of comparable magnitude, the effect on the coherent production would not be noticeable. However, the E_{0+} dominated by the $S_{11}(1535)$ is at least an order of magnitude larger than M_{2-} and M_{1-} and, therefore, the effect is clearly visible in Fig. 9. The overall cross section is dominated by the M_{2-} multipole while the contribution of the M_{1-} which came from the $P_{11}(1440)$ state is small. This finding, however, cannot be regarded conclusive because experimental uncertainties for the isoscalar photocouplings as determined from the (γ,π) multipoles are very large[2,10]. Precise measurements of the $\gamma N \to \eta N$ process at low energies which permit a partial-wave analysis of the (γ,η) data would be highly desirable.

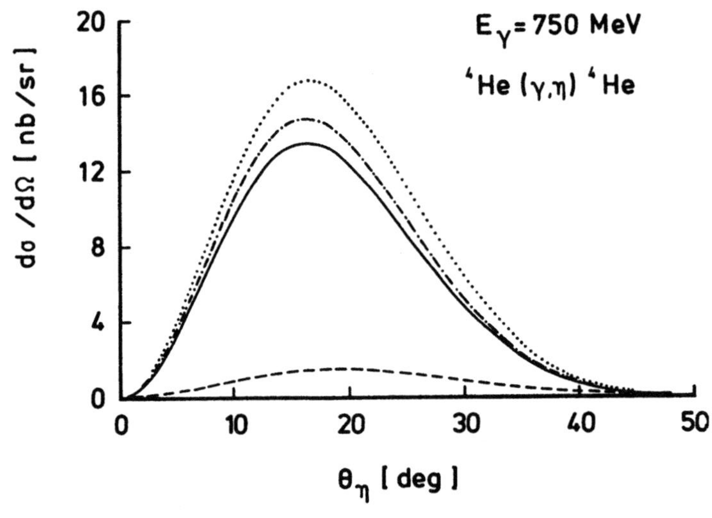

Fig. 9: Angular distribution for $^4He(\gamma,\eta)^4He$ showing the full calculation (——), only the M_{1-} multipole (---), only the M_{2-} (······), and without the E_{0+} (-·-·-).

Finally, we compare in Fig. 10 calculations using the "factorization approximation" of Eq. (16) with the full computation of Eq. (15). One can see that this approximation[11] is justified while a local frozen nucleon calculation yields very different results and should not be used in computing nuclear (γ,η) processes.

In contrast to coherent η photoproduction where the nuclear structure aspects are simple, inelastic transitions proceed from the ground state to some specified excited state and, therefore, require a detailed input of the transition density which can be specified in the matrix elements $\psi_{J;T}(a'a)$ as defined in Eq. (3). In general, there are two distinct methods to obtain those numbers. One can either employ the well-known wave functions of Cohen and Kurath[12], or resort to phenomenological methods in which the nuclear matrix elements are constrained by experimental information[13-17]. For the transitions presented in Figs. 11 and 12, the nuclear structure is given in Tables 2 and 3.

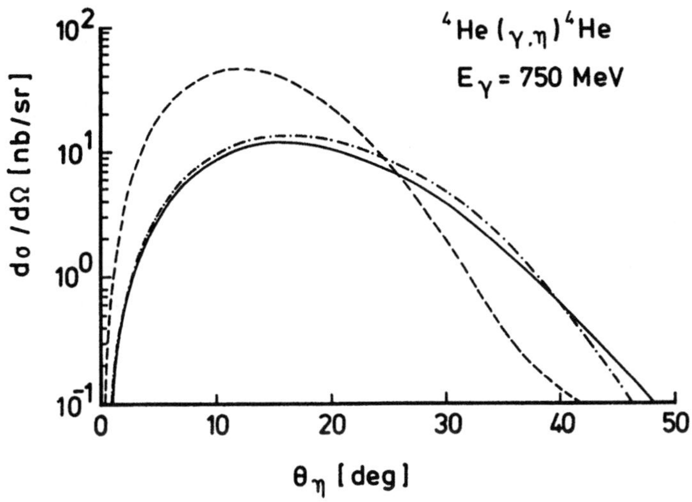

Fig. 10: Angular distributions comparing the full nonlocal calculation (——), with the "factorization approximation" (-·-·-), and the local frozen nucleon computation (----).

Table 2

Nuclear structure matrix elements $\psi_{J(LS);T}$ for the E2 and M3 transitions

reaction	J(LS) 2(20)	2(11)	2(21)	3(21)	b(fm)	ref.
$^{12}C(\gamma,\eta)^{12}C^*(2^+1;16.11)$	0.231	-0.374	-0.316	-	1.6	17
$^{12}C(\gamma,\eta)^{12}C^*(2^+0;4.44)$	-1.094	1.724	-0.282	-	1.7	18
$^{10}B(\gamma,\eta)^{10}B^*(0^+1;1.74)$	-	-	-	-0.585	1.66	18
$^{6}Li(\gamma,\eta)^{6}Li^*(3^+0;2.18)$	0.695	-0.045	0.594	0.618	2.0	18

In Fig. 11 we show three isovector transitions using the targets ^{12}C, ^{10}B and ^{14}N. Due to the large momentum transfer (Q ≈ 300 - 500 MeV/c), transitions involving large L and J are preferentially excited while M1 transitions which are dominant in charged photopion reactions achieve only small counting rates.

Table 3

Nuclear structure matrix elements $\psi_{J(LS);T}$ for the M1 transition

reaction	J(LS) 1(10)	1(01)	1(11)	1(21)	b(fm)	ref.
$^{14}N(\gamma,\eta)^{14}N^*(0^+1;2.31)$	0.339	-0.033	0.046	0.434	1.73	16 (Set H1)

In the realm of electron scattering the isovector E2 transition in the A=12 system to the 16.11 MeV level represents an exception, since longitudinal and transverse parts contribute with comparable strength.

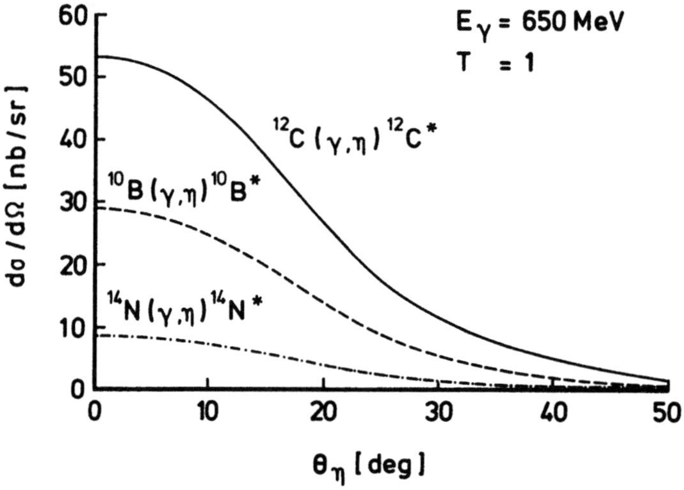

Fig. 11: Various isovector (γ,η) transitions obtained with plane wave etas.

A recent study analysed high-precision electron scattering data on the E2 and C2 form factors[17] of this level in a conventional harmonic oscillator basis and extracted nuclear structure coefficients for the 1p-shell given in Table 2. The large transverse form factor is especially significant since its nuclear matrix element multiplies the J=2, L=2, S=1 matrix element which dominates (γ,η) E2 transitions, while the coefficient extracted from the longitudinal form factor enters via the J=2, L=2, S=0 matrix element and is of minor importance. Using plane waves for the η-meson yields more than 50 nb/sr for zero degree etas.

The reaction $^{10}B(\gamma,\eta)^{10}B^*(0^+1;1.74$ MeV) involves a pure M3 transition ($J^\pi = 3^+$, $T=0\rightarrow 0^+,1$), again favored by the large momentum transfer, and leads to the isospin triplet ($^{10}B_{g.s.}$, $^{10}B(1.74)$, $^{10}C_{g.s.}$). Only one matrix element in j-j or LS coupling (see Table 2) can contribute when the basis is restricted to 1p-shell orbitals, and thus there is no flexibility in the nuclear structure, indicating that discrepancies between theory and experiment would have to come from uncertainties in the production process. The relatively small cross section of less than 30 nb/sr in the PWIA computation is mainly due to the statistical factor $(2J_i+1)^{-1}$ in Eq. (1) which lowers the strongly excited J=3, L=2, S=1 cross section by a factor of seven.

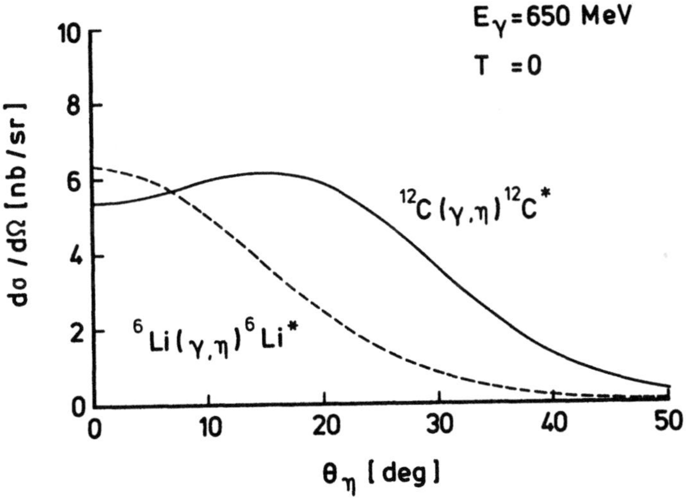

Fig. 12: Various isoscalar (γ,η) transitions obtained in the PWIA.

Finally, Fig. 11 shows an angular distribution for the reaction $^{14}N(\gamma,\eta)^{14}N^*(0^+1;2.31$ MeV). This isospin triplet ($^{14}C_{g.s.}$, $^{14}N(2.31)$, $^{14}O_{g.s.}$) has aroused considerable interest due to the retarded ^{14}C β-decay. The wave function in Table 3 agrees with the reaction $^{14}N(\gamma,\pi^+)^{14}C_{g.s.}$ but underestimates the ^{14}C life-time. We find that the cross section is dominated by the large nuclear matrix element $\psi_{1(21);1}$ and that the influence of the GT-matrix element is negligible.

Even for transitions involving large angular momentum transfer like E2 and M3 transitions, cross sections for isoscalar inelastic reactions remain below 10 nb/sr. Fig. 12 shows angular distributions for the processes $^{12}C(\gamma,\eta)^{12}C^*(2^+0;4.44$ MeV) and $^{6}Li(\gamma,\eta)^{6}Li^*(3^+0;2.18$ MeV). All the inelastic transitions, isoscalar as well as isovector, are dominated by the elementary (γ,η) E_{0+} multipole which comes from the $S_{11}(1535)$ resonance.

In conclusion, this work presented differential cross section estimates for coherent and incoherent η photoproduction and suggested targets and transitions that may deserve experimental attention. Regarding the elastic production we found that the reaction $^4He(\gamma,\eta)^4He$

achieved the highest counting rates while rapidly dropping nuclear form factors lead to smaller cross sections for heavier nuclei such as ^{12}C and ^{40}Ca. Furthermore, η distortion lowers the cross section by at most 15 % in case of a ^{4}He target whereas this effect clearly grows with the size of the nucleus. Thus, ^{4}He appears to be the most reasonable target for first generation experiments on coherent (γ,η) production. In our calculations the $D_{13}(1520)$ played the dominant role while the influence of the $P_{11}(1440)$ remained limited. As discussed before this finding is subject to improved isoscalar (γ,π) photocouplings. Due to nuclear Fermi motion the non spin-flip operator contains an E_{0+} multipole contribution which is noticeable up to 10 % in the cross sections.

The incoherent reactions can be divided into two classes corresponding to the magnitude of the differential cross section. First we found that the large momentum transfer leads to a preferential excitation of states involving large angular momentum transfer J, such as E2 and M3 transitions (or even higher once one allows contributions from outside the 1p-shell). For isovector transitions these cross sections studied in the case of ^{12}C and ^{10}B can reach up to 50 nb/sr for plane wave etas. On the other hand, all incoherent isoscalar as well as isovector M1 transitions stay below 10 nb/sr. In all inelastic transitions it was the $S_{11}(1535)$ resonance that clearly dominated the cross sections while the significance of the $P_{11}(1440)$ was small again. Therefore, elastic as well as inelastic low-energy η-photoproduction on nuclei are dominated each by one single resonance, which can be studied in the nuclear environment without interference from the Δ-resonance.

REFERENCES

1) F. Tabakin, S.A. Dytman and A.S. Rosenthal, Proc. of the Workshop "Excited Baryons - 1988" (World Scientific Singapore, 1989), p. 168
2) C. Bennhold and H. Tanabe, Proc. of the 6th Amsterdam Miniconf. at NIKHEF, Nov. 16-17, 1989, to appear in Phys. Lett. B
3) D. Halderson and A.S. Rosenthal, Nucl. Phys. A501, 856 (1989)
4) B. Delcourt et al., Phys. Lett. B29, 75 (1967)
5) C. Bacci et al., Phys. Rev. Lett. 20, 571 (1968); Nuovo Cim. 45, 983 (1966)
6) H. Ströher et al., MAMI B proposal - Photoproduction of π^0 and η from ^{12}C and ^{40}Ca, 1989 (unpublished)
7) R.S. Bhalerao and L.C. Liu, Phys. Rev. Lett. 54, 865 (1985)
8) L. Tiator and L.E. Wright, Phys. Rev. C30, 989 (1984)
9) C. Bennhold and L.E. Wright, Phys. Rev. C39, 927 (1989)
10) C. Bennhold and H. Tanabe, Mainz University preprint (1990)
11) A.A. Chumbalov, R.A. Eramzhyan and S.S. Kamalov, Z. Phys. A328, 195 (1987)
12) S. Cohen and D. Kurath, Nucl. Phys. 73, 1 (1965)
13) T.W. Donnelly and J.D. Walecka, Phys. Lett. B44, 330 (1973)
14) J. Dubach and W.C. Haxton, Phys. Rev. Lett. 41, 1453 (1978)
15) M.K. Singham, Nucl. Phys. A460, 597 (1986); L. Tiator, Phys. Lett. B125, 367 (1983)
16) R.L. Huffman et al., Phys. Rev. C35, 1 (1987)
17) L. Tiator, C. Bennhold and R. Neuhausen, Nucl. Phys. A501, 751 (1989)
18) T.-S.H. Lee and D. Kurath, Phys. Rev. C21, 293 (1980)

IV.

PHOTONUCLEAR REACTIONS

Photonuclear reactions at intermediate energies

R.C. Carrasco [1], E. Oset [1], L.L. Salcedo [2]

[1]Departamento de Fisica Teorica, Universidad de Valencia, E-46100, Burjassot(Valencia) Spain
[2]Departamento de Fisica Moderna, Facultad de Ciencias, Universidad de Granada, E-18001 Granada, Spain

Abstract: Starting from basic couplings of the photons to mesons, nucleons and isobars a microscopic manybody theory is developped which allows one to evaluate different photonuclear reactions at intermediate energies. The theory is applied to obtain the total photonuclear cross section and the separation between γ absorption and (γ, π) reaction channels.

The total photon nucleus cross section from 0 to 600 MeV has several peculiarities. In the first place at low energies one has a large contribution from the excitation of the dipole giant resonance. After that, in heavy nuclei, a flat region appears from 50 MeV to 120 MeV and immediately, a rising of the cross section in the region of the $\Delta(3/2, 3/2)$ resonance is observed, which follows closely the $\gamma N \rightarrow \pi N$ cross section. One of the peculiarities of the reaction is that σ/A is almost constant[1)] for different nuclei, although the ratio is not equal to the elementary cross section $\gamma N \rightarrow$ all (essentially Nπ), as can be seen in fig. 1

We can observe that the resonance peak is not appreciably moved with respect to its free position and that the strength at the peak of σ_γ/A is substancially reduced compared to the free value. The other characteristic around resonance is an appreciable broadening of the peak with respect to the free Δ width. Below pion threshold the cross section stays constant at an approximate value of $\sigma/A = (65-70)$ μb until it hits the dipole resonance at low energies not shown in the picture. These are some of the essential features that we will try to understand in the context of the work that we expose below. In a recent study [4)] a microscopic theory of photon nucleus interaction has been developped which allows one to evaluate total photonuclear cross sections and also split this magnitude in the different reaction channels. We will briefly describe this theory here and show results for the total cross section, the absorption cross section (with no pions in the final state) and the double differential cross section for some of the (γ, π) reactions. In addition it will be clear that a systematic experimental study of the different channels (γ, π) and γ absorption in different nuclei and at different energies can shed valuable light on pion nucleus dynamics, particularly on pion absorption. The starting point is a model for the elementary $\gamma N \rightarrow \pi N$ reaction. The model, depicted in fig. 2, is a standard one used in the literature[5,6)] which contains the direct and crossed nucleon pole term (2a, 2b), the pion pole term (2c) and the direct and crossed Δ pole term (2d, 2e).

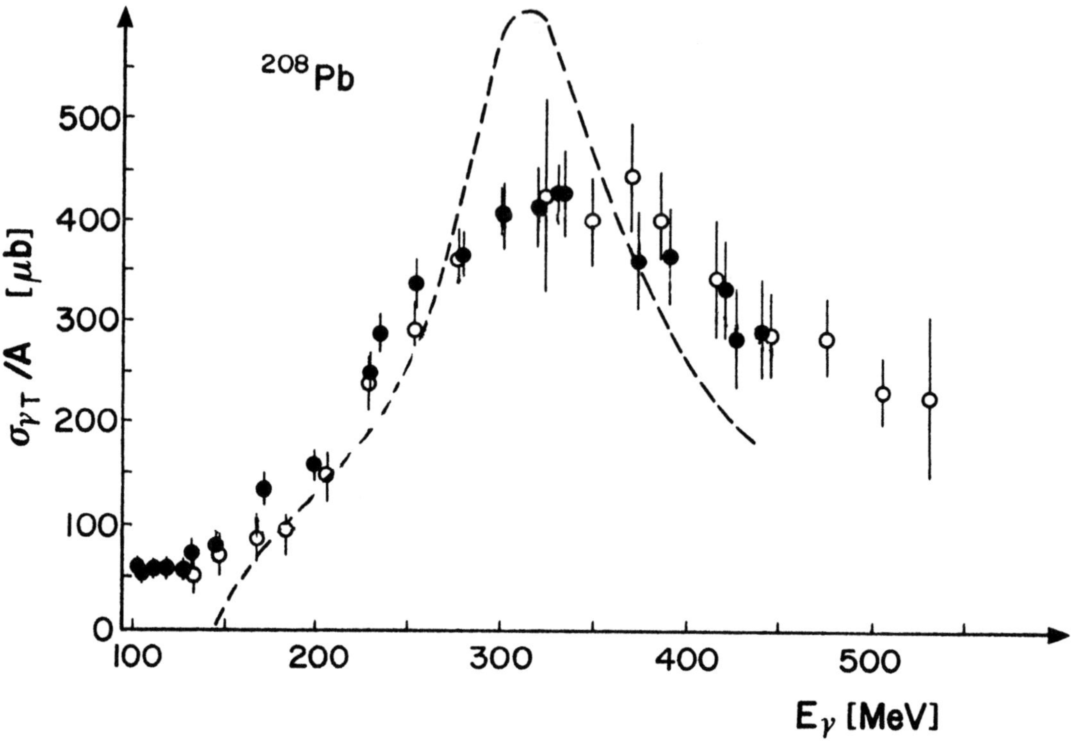

Fig.1. σ_γ/A in ^{208}Pb. The dashed line is $(Z\sigma_{\gamma p} + N\sigma_{\gamma n})/A$. The experimental points are from ref. 2 (full circles) and ref. 3 (open circles).

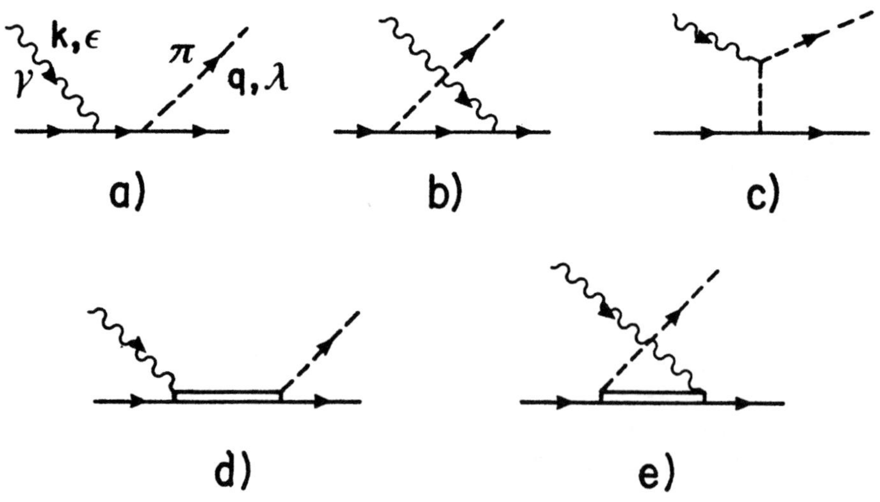

Fig.2. Diagrams contributing to pion photoproduction.

As often done, it is useful to separate the contribution from the nucleon negative energy intermediate states in figs. 2a, 2b, which gives rise to the Kroll-Ruderman term depicted in fig. 3.

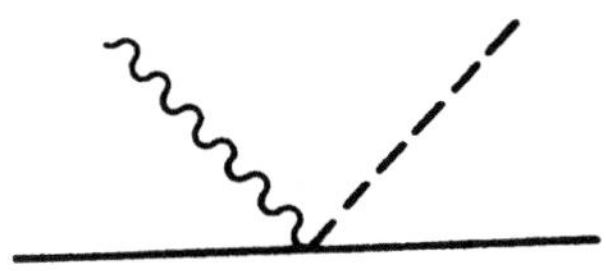

Fig.3. Kroll-Ruderman diagram.

A reasonable description of the differential cross sections in the different (γ, π) channels is obtained[4]. Next step consist in evaluating the photonuclear cross sections. For this purpose we evaluate the photon selfenergy in an infinite nuclear medium, $\Pi(\boldsymbol{k})$. A photon traveling through this medium will be removed from the elastic flux at the rate (recall that $2\omega V_{opt} \equiv \Pi$).

$$\Gamma = -\frac{1}{N}\frac{dN}{dt} = -\frac{1}{N}\frac{dN}{dl} = -2\,\mathrm{Im}\,V_{opt} = -\frac{1}{k}\,\mathrm{Im}\,\Pi(\boldsymbol{k}) \tag{1}$$

where l is the distance travelled by the photon (in natural units). In terms of cross section an element of volume d^3r would contribute to σ in

$$d\sigma = -\frac{1}{k}\,\mathrm{Im}\,\Pi(\boldsymbol{k})d^3r \tag{2}$$

Now since $\Pi(\boldsymbol{k})$ depends on the nuclear density, ρ, we make the assumption that an element d^3r in the nucleus contributes to σ in an amount given by eq. (2) with $\Pi(\boldsymbol{k})$ substituted by $\Pi\,(\boldsymbol{k}, \rho(\boldsymbol{r}))$, where $\rho(\boldsymbol{r})$ is the nucleus local density at the point $\boldsymbol{r}$. Thus

$$\sigma_\gamma = -\int \frac{1}{k}\,\mathrm{Im}\,\Pi(\boldsymbol{k}, \rho(\boldsymbol{r}))d^3r \tag{3}$$

The local density approximation is then used to obtain eq. (3). One may wonder how accurate this approximation is here. We argue that this is a case where the approximation is most accurate. Let us assume for simplicty that $\Pi(\boldsymbol{k})$ is given in infinite matter by the low density theorem[7]

$$\Pi(\boldsymbol{k}) = \tilde{t}(\boldsymbol{k}, \boldsymbol{k})\rho \tag{4}$$

where $\tilde{t}(\boldsymbol{k}, \boldsymbol{k})$ is the photon nucleon forward amplitude averaged over spin and isospin and over the Fermi motion. The local density prescription would provide a photon selfenergy $\Pi(\boldsymbol{k}, \boldsymbol{r}) = \tilde{t}(\boldsymbol{k}, \boldsymbol{k}\,, \rho(\boldsymbol{r}))$. In the same approximation the γ selfenergy in momentum space for a finite nucleus is given by

$$\tilde{\Pi}^{FN}(\boldsymbol{k}, \boldsymbol{k}') = \tilde{t}(\boldsymbol{k}, \boldsymbol{k}')\tilde{\rho}(\boldsymbol{k} - \boldsymbol{k}') \tag{5}$$

where $\boldsymbol{k}, \boldsymbol{k}'$ are the initial and final photon momenta and $\tilde{\rho}(\boldsymbol{q})$ is the Fourier transform of the nuclear density

$$\tilde{\rho}(\boldsymbol{q}) = \int d^3r \; e^{-i\boldsymbol{q}\boldsymbol{r}} \rho(\boldsymbol{r}) \tag{6}$$

By assuming $\tilde{t}$ dependent on the momentum transfer, $(\boldsymbol{k} - \boldsymbol{k}')$, we obtain a local photon selfenergy by performing the Fourier transform of eq. (5)

$$\Pi^{FN}(\boldsymbol{k}, \boldsymbol{r}) = \frac{1}{(2\pi)^3} \int d^3q \; e^{i\boldsymbol{q}\boldsymbol{r}} \, \tilde{t}(\boldsymbol{q}) \tilde{\rho}(\boldsymbol{q}) = \int d^3r' \rho(\boldsymbol{r}') t(\boldsymbol{r} - \boldsymbol{r}') \tag{7}$$

with $t(\boldsymbol{r} - \boldsymbol{r}')$ the Fourier transform of $\tilde{t}(\boldsymbol{q})$. We see that in this case, instead of the local density prescription, $\tilde{t}(\boldsymbol{k}, \boldsymbol{k}) \, \rho(\boldsymbol{r})$, we get a convolution of the nuclear density with the range of the elementary interaction. If we now integrate $\Pi^{FN}(\boldsymbol{k}, \boldsymbol{r})$ in order to obtain the cross section we have

$$\begin{aligned} \sigma_\gamma^{FN} &= -\frac{1}{k} \int d^3r \, \mathrm{Im} \, \Pi^{FN}(\boldsymbol{k}, \boldsymbol{r}) = - \int d^3q \delta^3(\boldsymbol{q}) \frac{1}{k} \, \mathrm{Im} \, \tilde{t}(\boldsymbol{q}) \tilde{\rho}(\boldsymbol{q}) = \\ &= -\frac{1}{k} \, \mathrm{Im} \, \tilde{t}(0) \tilde{\rho}(0) = -\frac{1}{k} \int d^3r \, \mathrm{Im} \, \tilde{t}(\boldsymbol{k}, \boldsymbol{k}) \rho(\boldsymbol{r}) \end{aligned} \tag{8}$$

This result is exactly the same we would obtain from the local density approximation potential of eq. (4). The reason for it, of course, is the volume contribution to the cross section. In reactions which are surface dominated, like pion nuclear reactions, taking into account the range of the interaction leads to differences with respect to the plain local density approximation[8].

The details of the model are exposed in ref.[4]. Here we make a qualitative exposition stressing the most relevant points. In order to construct the photon selfenergy in the nuclear medium, we take the model of figs. 2, 3 for the $\gamma N \rightarrow \pi N$ reaction and fold it as shown diagramatically in fig. 4.

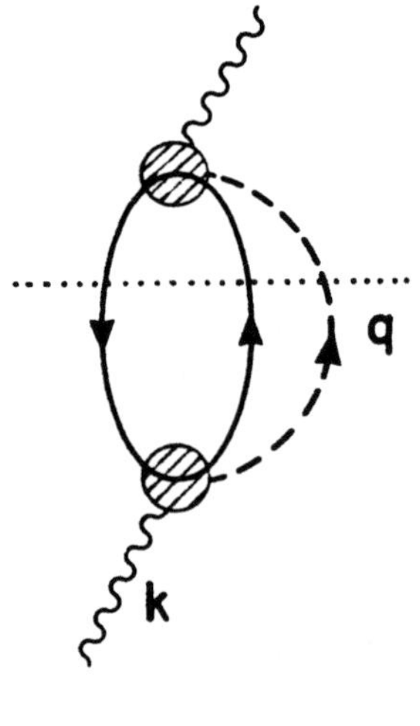

Fig.4. Photon selfenergy diagram.The dashed circle indicates any of the six terms in figs. 2, 3 for the $\gamma N \rightarrow \pi N$ reaction.

However, in order to obtain the photonuclear cross section we need only the imaginary part of Π. Using standard many body methods[9–11] we obtain, from the diagram of fig. 4.

$$-i\Pi(k) = \int \frac{d^4k}{(2\pi)^4}\frac{i}{4}U(k-q)iD_0(q)\sum_{i,j}\sum_{s_i,s_f}|T_{\gamma N_i\to\pi N_j}|^2 \tag{9}$$

where $U(k-q)$ is the Lindhard function for ph excitations and $T_{\gamma N_i\to\pi N_j}$ the T matrix for the $\gamma N_i \to \pi N_j$ reaction (the charge of the pion is fixed by N_i, N_j, the initial and final nucleons). $D_0(q)$ is the free pion propagator. The imaginary part of eq. (9) is easily obtained by means of Cutkowsky rules

$$\begin{aligned} \Pi &\to 2i\,\mathrm{Im}\,\Pi \\ U(q) &\to 2i\Theta(q^0)\,\mathrm{Im}\,U(q) \\ D_0(q) &\to 2i\Theta(q^0)\,\mathrm{Im}\,D_0(q) \end{aligned} \tag{10}$$

which leads to

$$\mathrm{Im}\,\Pi(k) = \int \frac{d^3k}{(2\pi)^3}\frac{1}{4}\Theta(k^0-q^0)\,\mathrm{Im}\,U(k-q)\frac{1}{2\omega(q)}\sum_{s_i,s_f}\sum_{i,j}|T_{\gamma N_i\to\pi N_j}|^2 \tag{11}$$

Is there any novelty in eq. (11) with respect to the impulse approximation, $\sigma_\gamma = A\,\sigma_{\gamma N}$?. In order to see this let us make use of the low density limit[12] for Im$U(k-q)$

$$\Theta(q^0)\,\mathrm{Im}\,U(q) = -\pi\rho\delta(q^0 - \frac{q^2}{2m}) \tag{12}$$

This approximation, since $\rho \to 0$, neglects the Fermi motion and the effects of Pauli blocking implicitly incorporated in $U(q)$.

Within this approximation, separating the proton and neutron contributions, and using eqs. (3), (11) we obtain[4]

$$\sigma_\gamma = Z\sigma_{\gamma p\to\pi N} + N\sigma_{\gamma n\to\pi N} \tag{13}$$

which is the impulse approximation. Hence it is clear that the treatment using eq. (11) with the exact Lindhard function takes into account the modifications to the impulse approximation arising from the Fermi motion and the Pauli blocking.

Equation (11) is obviously only the first step. Indeed, the imaginary part of a Feynman diagram comes from the contribution to the integral (eq. (9)) when the intermediate particles are placed on shell. As seen in eq. (10) this means that the ph excitation and the pion are placed on shell. Hence we are still only considering the (γ,π) part of the reaction but not the direct photon absorption. How does photon absorption come about?. We can illustrate it by starting from the diagram in fig. 4 and allowing the pion to excite a ph, as depicted in fig. 5.

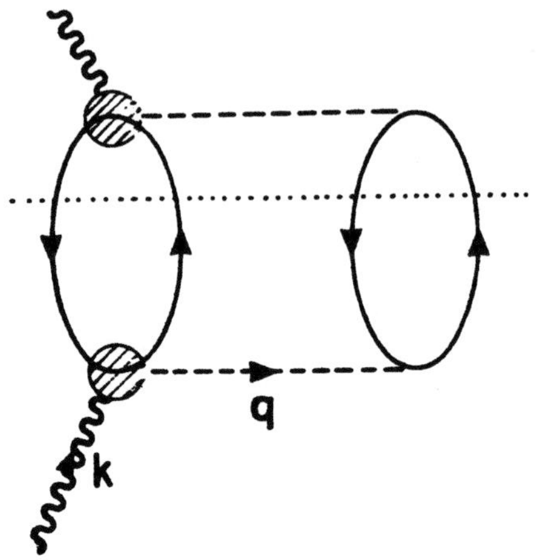

Fig.5. Diagrams contributing to photon absorption when the two ph excitations cut by the dotted line are placed on shell.

The contribution of fig. 5 to $\Pi(k)$ is readily accomplished. One obtains

$$-i\Pi(k) = \int \frac{d^4q}{(2\pi)^4}\frac{i}{4}U(k-q)(iD_0(q))^2(-i)\frac{f^2}{\mu^2}q^2U(q)\sum_{i,j}\sum_{s_i,s_f}|T_{\gamma N_i\to\pi N_j}|^2 \qquad (14)$$

but now the T matrix contains some form factors because the pions are off shell. Moreover, one of the 36 diagram implicit in fig. 5, the one corresponding to the pion pole in both dashed blubs, has to be divided by two because the diagram is symmetric with respect to a vertical axis. The use of Cutkowsky rules, where the two ph excitations are placed on shell leads us to

$$\mathrm{Im}\,\Pi(k) = -\int \frac{d^4q}{(2\pi)^4}\Theta(k^0-q^0)\,\mathrm{Im}\,U(k-q)D_0^2(q)\frac{f^2}{2\mu^2}q^2\Theta(q^0)\,\mathrm{Im}\,U(q)\sum_{i,j}\sum_{s_i,s_f}|T_{\gamma N_i\to\pi N_j}|^2 \qquad (15)$$

which by means of eq. (3) provides a contribution to the γ absorption cross section. The model of ref. 4 takes into account corrections to this formula stemming from the effect of short range correlations. It also includes ρ exchange whenever the diagram implies a p-wave coupling in each of the vertices and automatically generates, via the short range correlations corrections, the important contributions at short distances corresponding to the g' Landau-Migdal effective interaction. In addition corrections stemming from long range correlations of the RPA type are considered, as those shown in fig. 6 for the Kroll Ruderman - Kroll Ruderman term implicit in fig. 5.

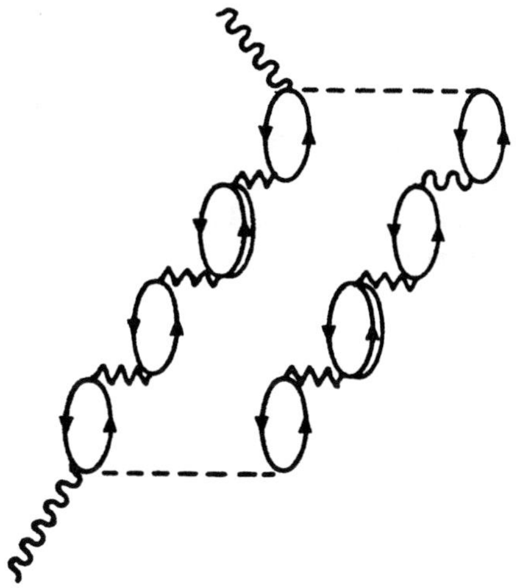

Fig.6. Series of diagrams considered in the long range correlations renormalization.

Particular emphasis is placed in the resonant piece when in both blubs of figs. 4 and 5 we place the Δ term of fig. 2d. These pieces are grouped together and iterated through Dyson's equation which gives rise to a resonant amplitude with a selfenergy correction. For this selfenergy we use the results of the calculations of ref. 13 in infinite matter, which lead to a correct description of the different inclusive pion nuclear reactions[14)] and elastic scattering[15)], are taken. The Δ selfenergy of ref. 13 also contains a contribution linked to three body absorption which was found to be[16)] in the resonance region as important as the two body absorption piece, implicit in fig. 5 for the Δ term. This is in agreement with a recent analysis of pion absorption at resonance[17]. As discussed in ref. 13, the imaginary part of the Δ selfenergy is not all related to pion absorption, being a part of it related to quasielastic scattering. In the photon case, the stament means that part of the selfenergy is related to two body and three body γ absorption and another part is related to the (γ, π) channel. This is discussed in detail in ref. 4 and the proper separation is done to include the right contribution in each one of the channels. We also consider the contribution of the diagram of fig. 7 which contains $2ph\pi$ excitation and 3 body absorption when the proper Δ selfenergy is included.This term is only appreciable at resonance and beyond.

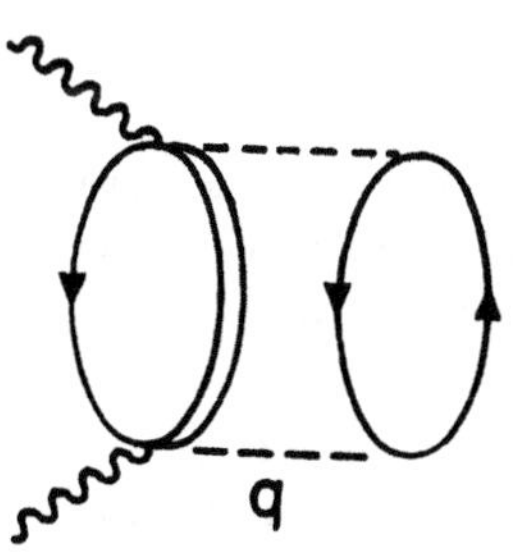

Fig.7. Photon selfenergy diagram involving the $\gamma N \rightarrow \Delta\pi$ Kroll-Ruderman term.

At energies below pion threshold the γ cross section comes mostly from the set of diagrams in fig. 5 involving the Kroll-Ruderman and pion pole terms.

With this sketchy introduction to the work of ref. 4 let us concentrate here on the discussion of the results, the implications that they have in the analysis of the experimental data and the relevant information that can be extracted from these data. In fig. 8 we show the results that we obtain for ^{12}C. The agreement with the experimental results is quite good as one can observe.

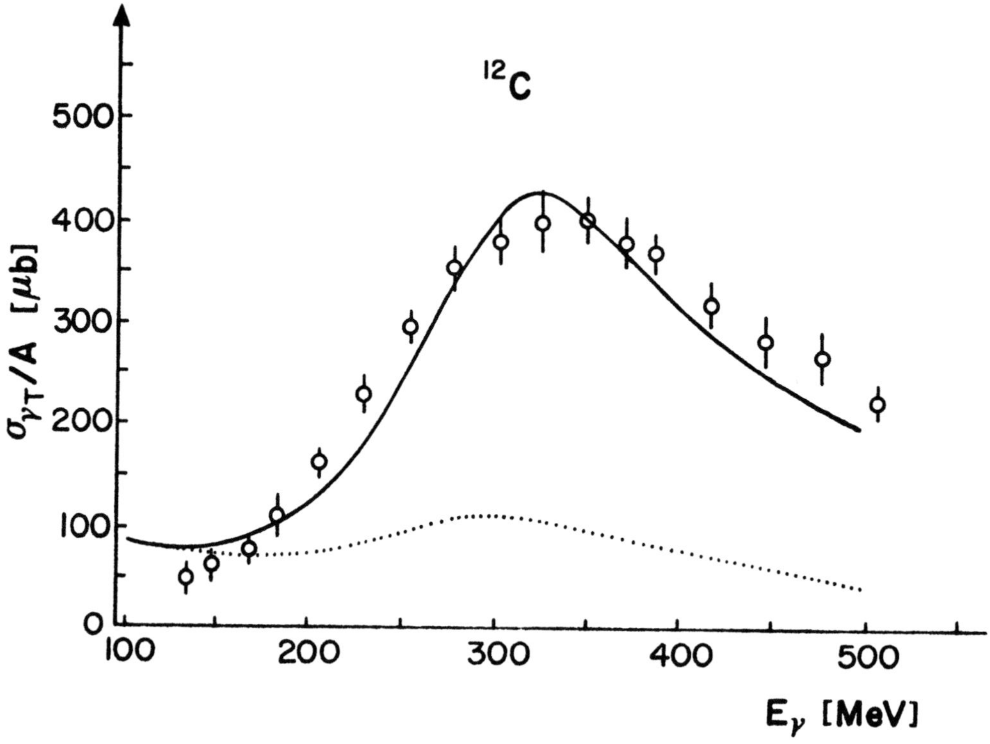

Fig.8. σ_γ/A for ^{12}C. Experimental data from ref. 3. Full curve total cross section. Dotted curve: direct photon absorption.

We also observe the decomposition between direct photon absorption and (γ,π). We call it direct photon absorption in order to differentiate it from another source of γ absorption which we call indirect and which corresponds to having originally (γ,π) events with the real pion absorbed on its way out of the nucleus. As can be seen in the figure, below pion threshold there is only photon absorption, but around resonance the most important channel is (γ,π), although some of the pions produced will be absorbed, as we will see later. In fig. 9-10 we show our results for ^{16}O and ^{208}Pb respectively. Once again the comparison with the experimental data is rather good.

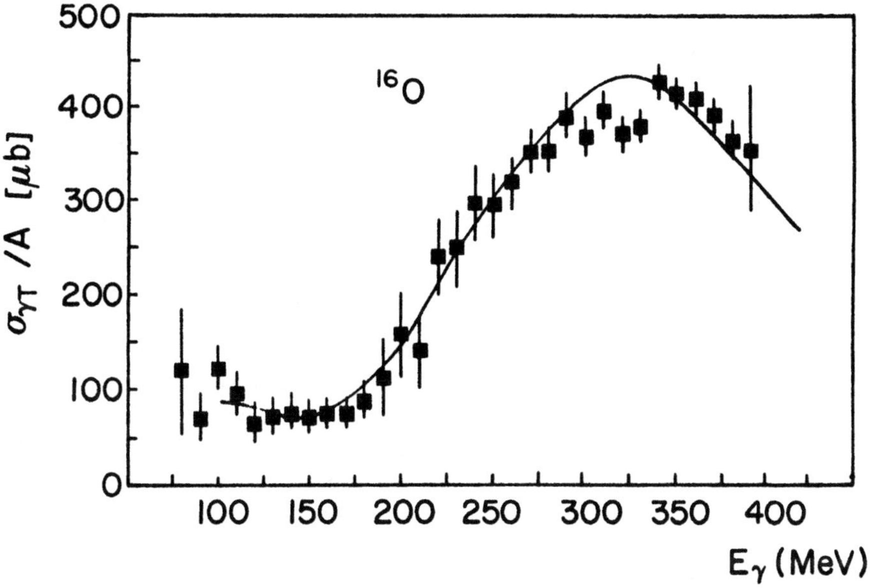

Fig.9. σ_γ/A for ^{16}O. Solid curve, theoretical total cross section. Experimental points from ref. 18.

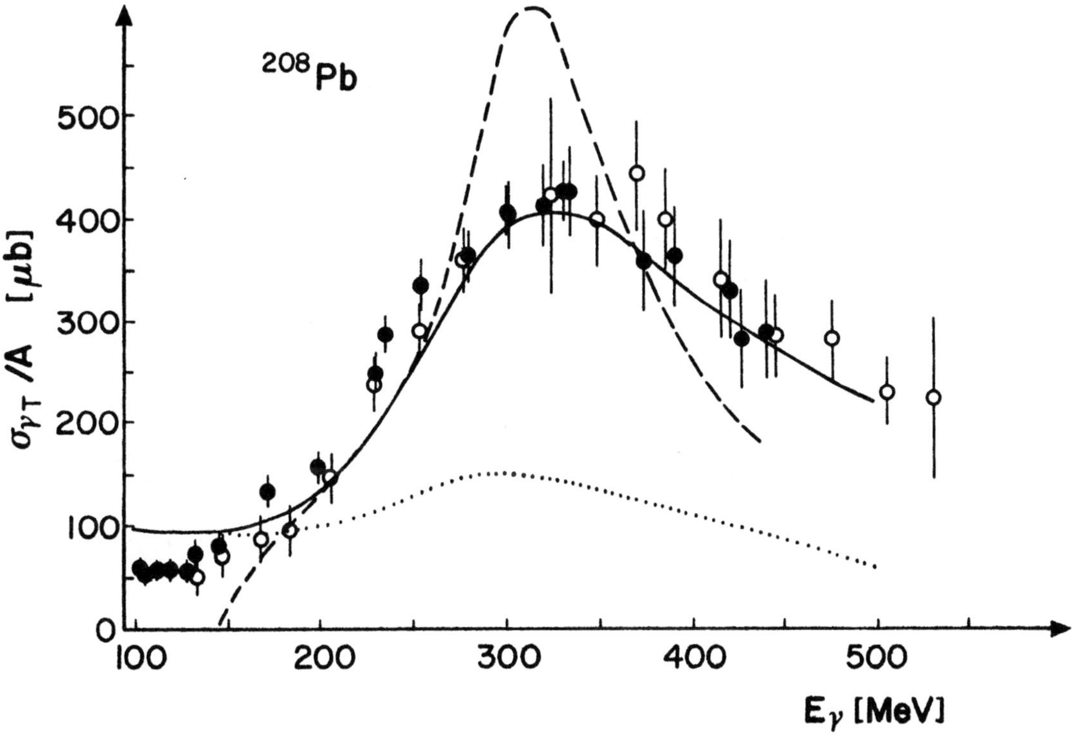

Fig.10. σ_γ/A for ^{208}Pb. Solid curve, theoretical total cross section. Dotted curve direct photon absorption. Dashed curve impulse approximation. Experimental points: full circles from ref. 2 ; open circles from ref. 3.

The reduction of σ around resonance and the corresponding broadening can now be explained. As mentioned before, the resonant part of $\Pi(k)$ becomes

$$\mathrm{Im}\, \Pi_\Delta(k) \simeq \mathrm{Im} \frac{1}{s^{\frac{1}{2}} - M_\Delta + i\frac{\tilde{\Gamma}}{2} - i\,\mathrm{Im}\,\Sigma_\Delta - \mathrm{Re}\Sigma_\Delta} \rho \tag{16}$$

where $\tilde{\Gamma}$ is the free Δ width corrected from Pauli blocking and Im Σ_Δ , ReΣ_Δ are the imaginary and real parts of the Δ selfenergy. Im Σ_Δ incorporates now the photon absorption channels and some (γ, π) corrections. While $\tilde{\Gamma}$ is smaller than Γ_{free}, the sum of $\tilde{\Gamma}/2$ and Im Σ_Δ is larger than Γ_{free} and, close to the resonant pole of eq. (16), $\mathrm{Im}\, \Pi_\Delta$ gets reduced with respect to the value with $\Sigma_\Delta = 0$ and no Pauli blocking (impulse approximation). At the same time, the Δ distribution gets broadened because the effective width $\tilde{\Gamma} + Im\,\Sigma_\Delta$ is larger than Γ_{free}. A remarkable feature of our results is that σ_γ/A is approximately constant for different nuclei, as is the case experimentally. However, this is not true for every different channel. In fig. 11 we show the results for σ_{abs}/A for ^{12}C and ^{208}Pb. We see that this cross section is larger in the case of ^{208}Pb.

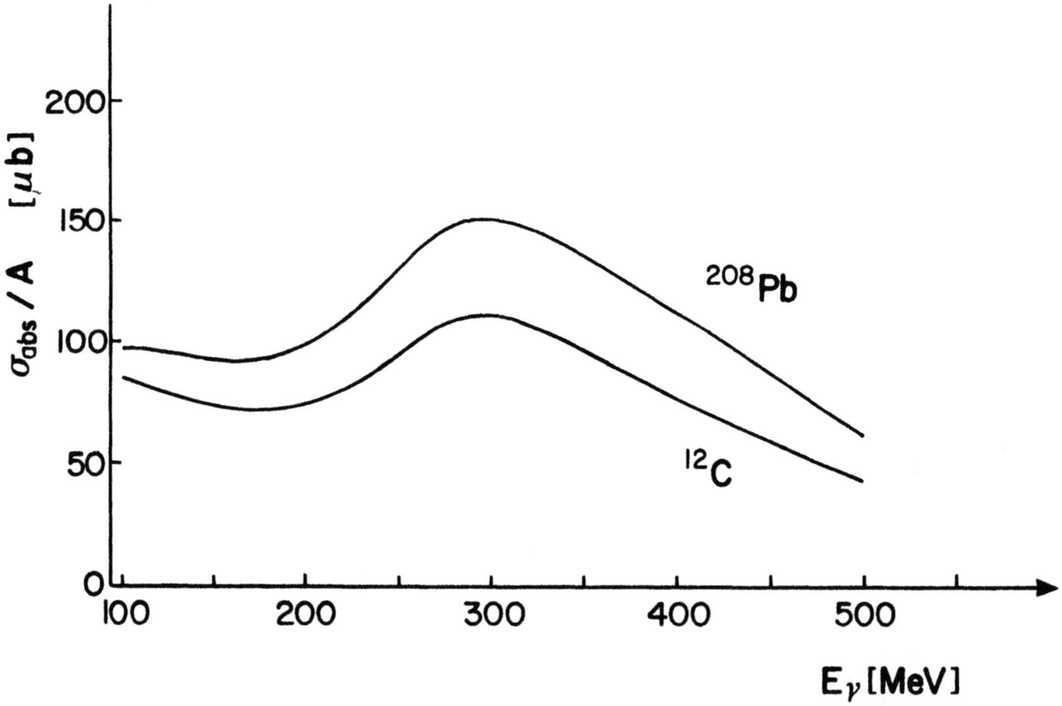

Fig.11. σ_{abs}/A for ^{12}C and ^{208}Pb.

It would be quite interesting to separate experimentally the amount of direct photon absorption from the total cross section in order to compare with the theory. Present experiments do not accomplish that. However, it is possible to separate all the different (γ, π) channels[20] and the rest of the cross section would give us the γ absorption cross section. This would contain the direct as well as the indirect photon absorption. Can we still compare our results with experiment? Awaiting further analysis what we can do is to compare our results with the measured (γ, π) reactions in the different channels. This requires some extra work, which has been already done[21], in order to follow the pions through the nucleus once they have been produced. For this we use the same theoretical ingredients which were used in ref. 14 to obtain all the inclusive pion nuclear reactions. There the theoretical input was cast in terms of reaction probabilities per unit length for each one of the different pionic channels, quasielastic, charge exchange and absorption. Then a Montecarlo computer simulation procedure was run to obtain at the end double differential cross sections for the different (π^i, π^j) reactions and the cross section for pion absorption. The agreement of the theoretical results with experiment for the different reactions and different nuclei was rather good in a broad range of energies around resonance.

In ref. 21 we use the same procedure and follow the pions through the nucleus allowing them to have quasielastic collisions, charge exchange collisions or simply be absorbed. The result is that we can obtain double differential cross sections $d^2\sigma/d\Omega\, dE$ for the pions, π^0, π^+, π^-, and determine the amount of pions that have been absorbed (indirect photon absorption).

In fig. 12 we show the photon absorption cross section for ^{12}C. The lower curve indicates the direct γ absorption while the upper curve has added to it the indirect absorption also. In fig. 13 we do the same for ^{208}Pb.

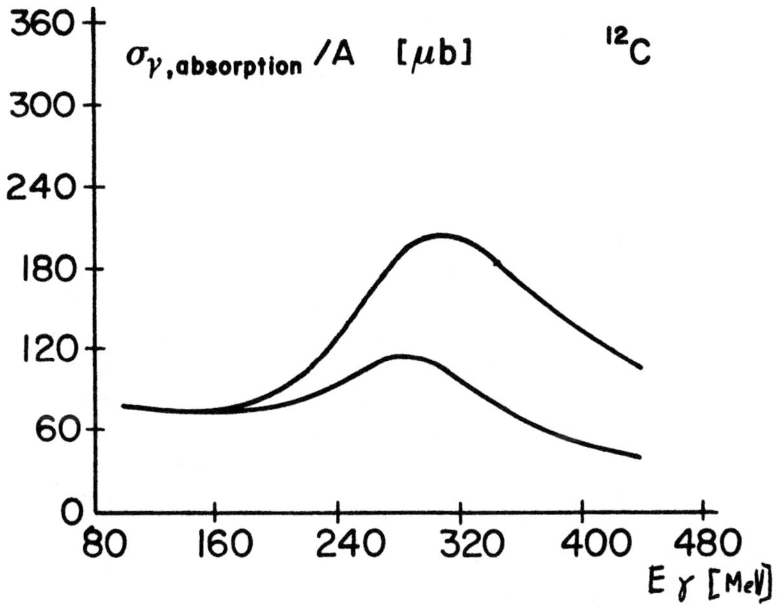

Fig.12. Direct and total photon absorption for ^{12}C

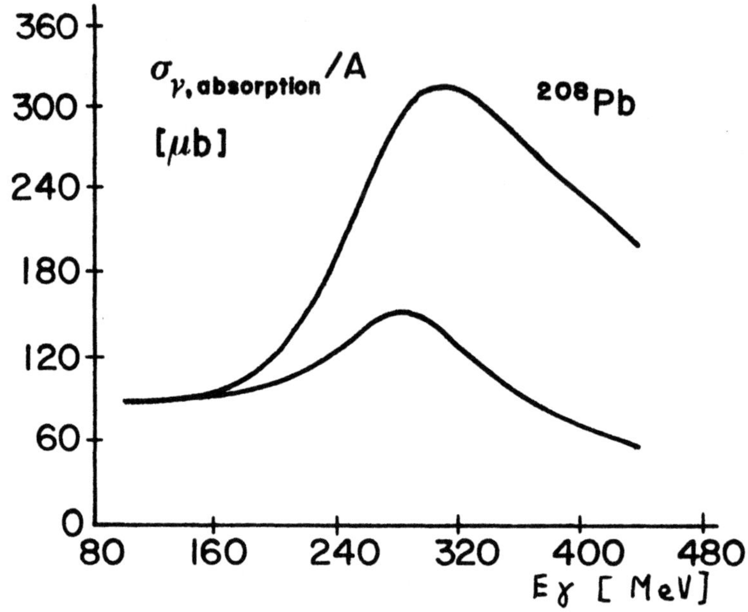

Fig.13. Direct and total photon absorption for ^{208}Pb

We can observe that σ_{abs}/A, including the indirect γ absorption changes quite a lot from one nucleus to the other. Conversely one can say that the remaining (γ,π) cross section changes appreciably from one nucleus to the other. This is obviously because in heavy nuclei the amount of indirect photon absorption is very important. Indeed, the pions are created all through the nucleus and they have a long way till they leave it, consequently most of these pions are absorbed on their way out. It is interesting to realize that the (γ,π) cross section is more sensitive to the pion absorption probability than the cross section for pion absorption. Indeed, around resonance the pion absorption probability is quite large, so much that the absorption cross section is nearly geometrical, $\sigma = \pi R^2$. In other words that the nucleus acts as a black disk. If one increases the absorption probability the pions are absorbed sooner but absorbed anyway and the cross section barely changes. In simple words we can say that a black disk can not be blacker. However, in the case of (γ,π) the more we increase the pion absorption probability the fewer pions come out, and the dependence of the cross section on this probability is strong. This means that we are about to learn more about pion absorption in nuclei by studying the (γ,π) reaction than by studying the pion absorption process itself. In any case it will offer very valuable information, complementary to the one from the pion nuclear reactions. With a theory that describes adequately the total photonuclear cross section as we have seen here, and a theory that describes adequately the different inclusive pion nuclear reactions[14] we can expect to obtain good agreement with the inclusive (γ,π) experimental cross sections. In fig. 14 we compare our results with those of ref. 22 for $\gamma\text{Pb} \rightarrow \pi^{\pm}$. We can see that the agreement with the sum of the (γ,π^+) and (γ,π^-) cross sections is good. Our results also show that the cross sections for (γ,π^+) and (γ,π^-) differ appreciably from each other. It would be interesting to have this experimental information.

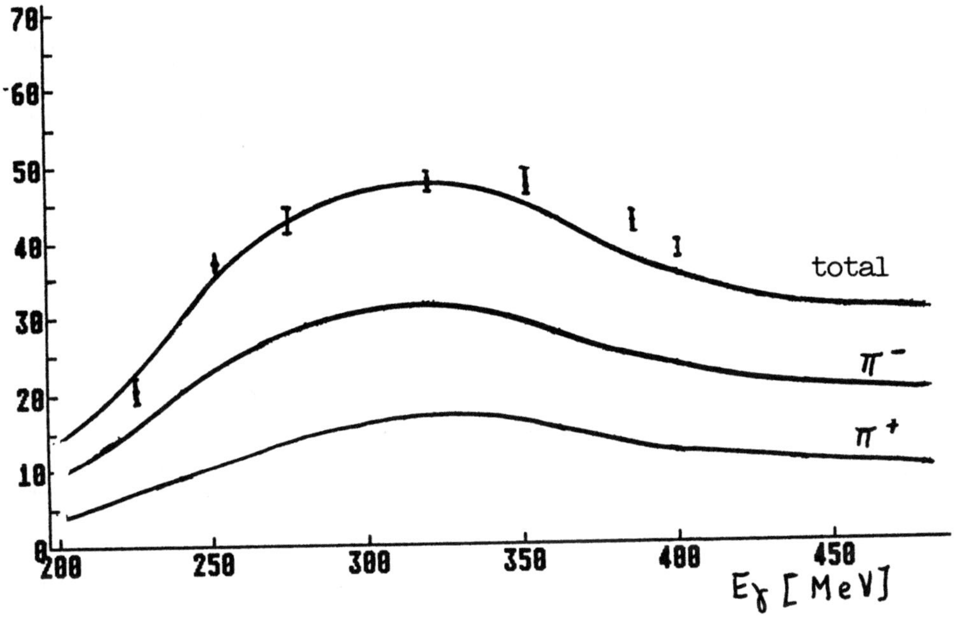

Fig.14. Charged pion photoproduction on ^{208} Pb (μb per nucleon).

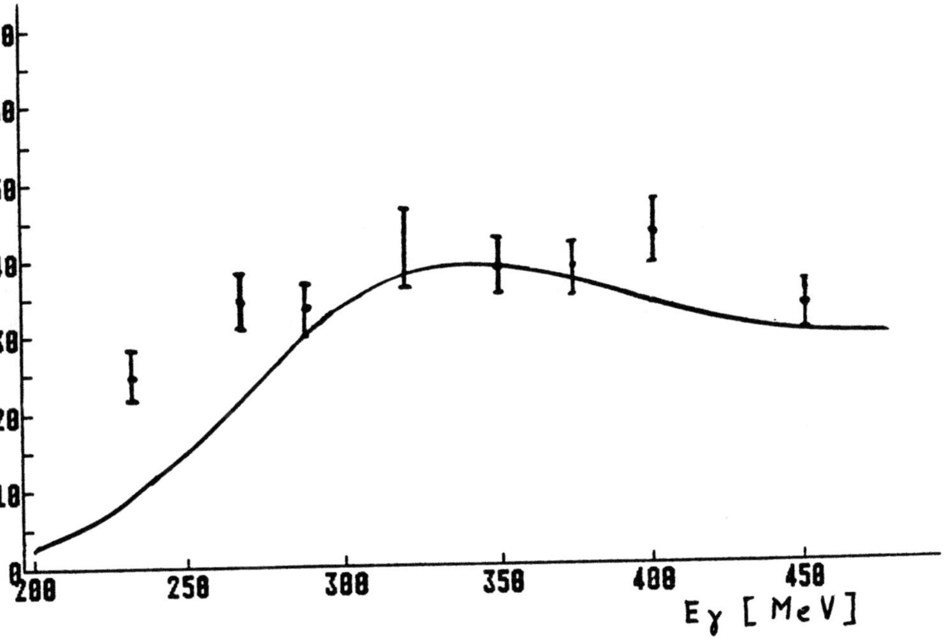

Fig.15.π^0 photoproduction on ^{208}Pb (μb per nucleon).

In fig. 15 we compare our results for (γ, π^0) inclusive with those of ref. 23. We observe a good agreement at higher energies but our results are low compared to experiment around $200 - 280 MeV$. This was to be expected since the experiment includes coherent (γ, π^0) photoproduction which is excluded from our calculations, and it peaks around these energies. In fig. 16 we can see the same features for (γ, π^0) in ^{118}Sn.

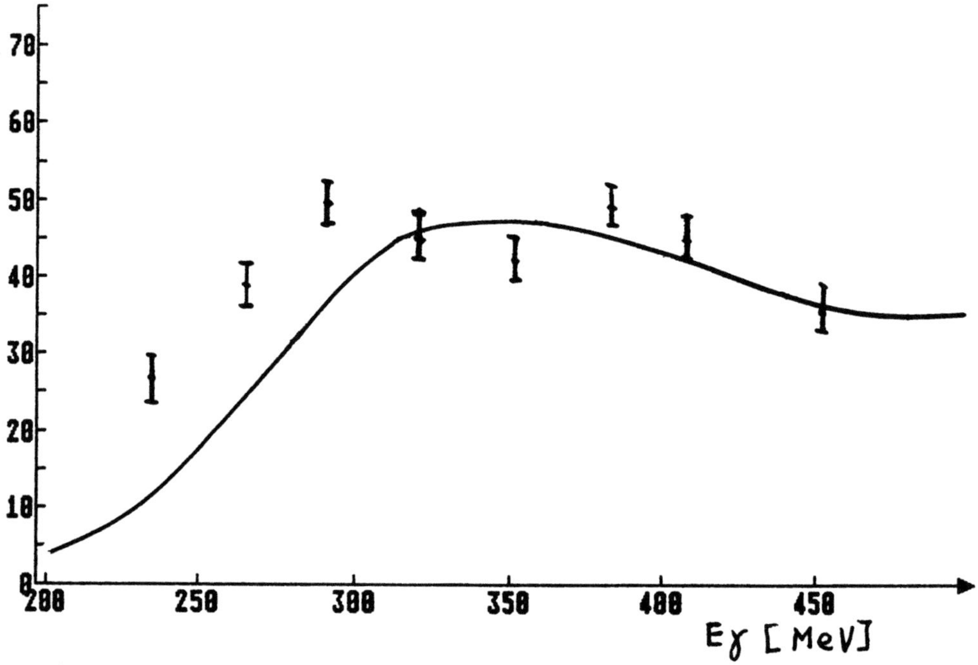

Fig.16.π^0 photoproduction on ^{118}Sn (μb per nucleon).

Finally, we show some selected differential cross sections. In fig. 17 we plot $d\sigma/d\Omega$ at $\theta = 52^0$ in ^{208}Pb for π^+ and π^-. The results compare favourably with the experimental ones of ref. 20. The double differential cross sections are also in quite good agreement with experiment for different energies, angles and nuclei. We show two selected cross sections in fig. 18 for (γ, π^+) and (γ, π^-) in ^{208}Pb at $E_\gamma = 280 MeV$ and $\theta = 52^0$

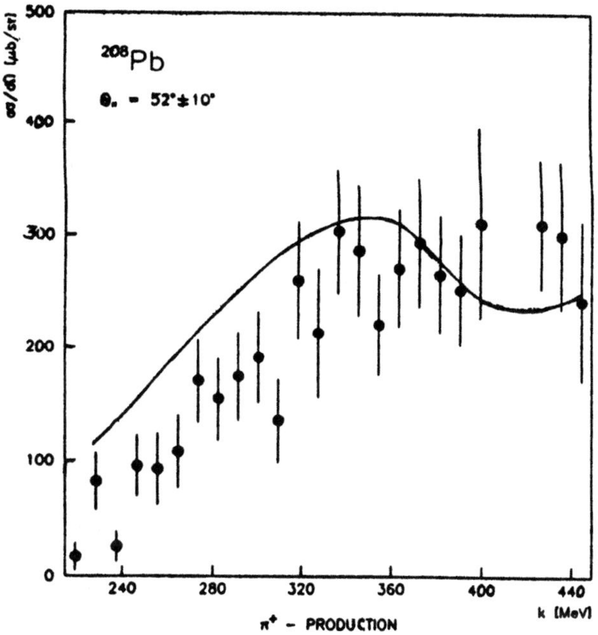

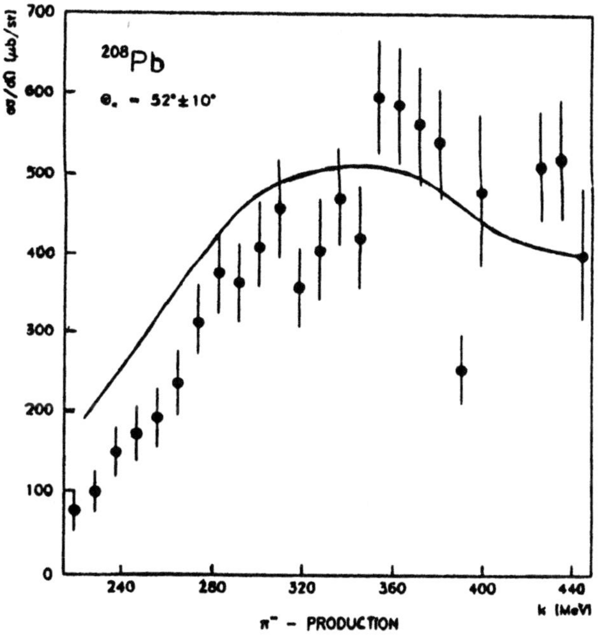

Fig.17. $\frac{d\sigma}{d\Omega}$ for charged pions at $\Theta = 52^0$ for ^{208}Pb

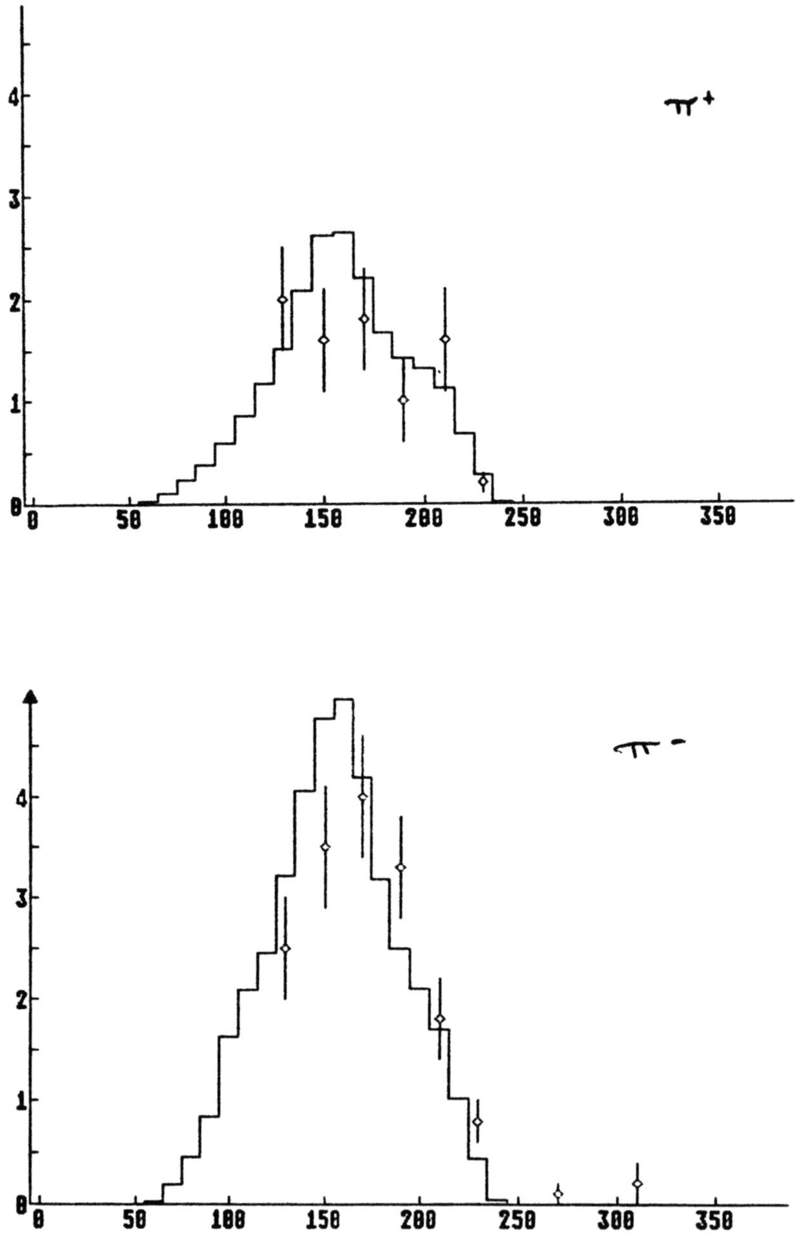

Fig.18. $\frac{d^2\sigma}{d\Omega dp}(\Theta = 52^0, E_\gamma = 280MeV)$ (μ b /(srMeV)) for $^{2}08$Pb

Some comments with respect to work published in the literature is appropriate at this point. Our work incorporates automatically the ideas of the Δh model of refs. 24, 25, however it introduces the local density approximation, duly justified at the beginning, which greatly simplifies the calculations and allows one to study heavy nuclei which were prohibitive before. In addition it includes the non resonant absorption mechanisms and at the same time provides a microscopic description for the Δ selfenergy, which is treated phenomelogically in ref.[25)]. Our work has also benefited from the first steps given, along a similar line, for the problem of electron scattering in ref. [26)]. The present work has continued those steps introducing the (γ, π) channels, considering the resonance character of the reaction and the interference between the (γ, π) and absorption channels, as well as the important effects of nuclear correlations and nuclear polarization, together with the consideration of some other excitation mechanisms.

SUMMARY

1) We have developped a microscopic many body theory which considers the different excitation channels of the nucleus explored by the photon probe. The theory provides the total photon nucleus cross section and allows a separation between the (γ, π) channels and direct photon absorption. The agreement of the result with experiment for the total cross section is remarkable. However there is no experimental information available for the separation of these two channel.

2) We found that the amount of indirect photon absorption is dominant in heavy nuclei and that the (γ, π) cross sections are very sensitive to the pion absorption probabilities. A systematic study of the different (γ, π) cross sections in different nuclei should offer simultaneous information on the probabilities of direct photon absorption and pion absorption.

The theoretical calculations of the (γ, π) reactions were carried out by following the pions through the nucleus using a theoretical framework that reproduces the different inclusive pion nucleus reactions and elastic scattering. The agreement of the calculated (γ, π) cross sections with experiment is also quite good except for puntual discrepancies in some double differential cross sections.

3) The success of these many body framework in pion and photonuclear reactions should encourage us to face in a similar way other problems involving the interaction of particles with nuclei. The interaction of electrons with nuclei, and the investigation of the different excitation mechanisms involved in the longitudinal and transverse response functions, is a good example of it. The work done here for real photons opens the way for such investigations.

4) The theoretical investigation carried out here also tells us about the richness of the photonuclear processes and the possibilities to explore details of the interaction and the nuclear excitation mechanisms so far not studied. Questions like how much direct or indirect γ absorption is there in the nucleus, how many nucleons are involved in direct photon absorption, which are the mean free paths for photon or pion absorption etc., can be properly asked here. Similar questions and investigations are presently being carried out in pion nuclear physics which evidences the common problems in the two fields and the convenience to carry out parallel studies in both of them.

References

1. J. Ahrens, Nucl. Phys. A **446** (1985) 229c.
2. P. Carlos et al., Nucl. Phys. A **431** (1984) 573.
3. L. Ghedira, PhD Thesis, University of Paris Süd, 1984.
4. R.C. Carrasco and E. Oset, University of Valencia preprint, 1989.
5. I. Blomqvist and J.M. Laget, Nucl. Phys. A **280** (1970) 405.
6. T.E.O. Ericson and W. Weise, Pions and Nuclei, Clarendon Press, 1988.
7. C.B. Dover, L. Ludeking and R.H. Lemmer, Ann. Phys. **66** (1971) 248.
8. L.L. Salcedo, E. Oset, M.J. Vicente and C. Garcia - Recio, Nucl. Phys. A **484** (1988) 557.
9. A.L. Fetter and J.D. Walecka, "Quantum Theory of Many Particle Systems", MacGraw Hill, New York, 1971.
10. G.E. Brown, "Many Body Problems", North Holland, 1972.
11. E. Oset in "Quarks, Meson and Isobars in Nuclei", ed. R. Guardiola and A. Polls, World Scientific, 1983, p. 1.
12. C. Garcia - Recio, E. Oset and L.L. Salcedo, Phys. Rev. C **37** (1988) 194.
13. E. Oset and L.L. Salcedo, Nucl. Phys. A **468** (1987) 631.
14. L.L. Salcedo, E. Oset, M.J. Vicente - Vacas and C. Garcia - Recio, Nucl. Phys. A **484** (1988) 557.
15. C. Garcia - Recio, E. Oset, L.L. Salcedo, D. Strottman and M.J. Lopez, MIT preprint 1989.
16. E. Oset, Y. Futami and H. Toki, Nucl. Phys. A **446** (1985) 584
17. S.D. Hyman et al., Phys. Rev. C **41** (1990) R409.
18. J. Ahrens et al., Nucl. Phys. A **490** (1988) 655
19. P. Carlos et al., Nucl. Phys. A **431** (1984) 573
20. R. Stenz, PhD Thesis, University of Bonn, 1986
21. R.C. Carrasco, E. Oset and L.L. Salcedo, to be published.
22. J. Arends et al. Phys. Lett. **98** B (1981) 423
23. J. Arends et al., Nucl. Phys. A **454** (1986) 579
24. E. Oset and W. Weise, Phys. Lett. B **94** (1980) 19; E. Oset and W. Weise, Nucl. Phys. A **368** (1981) 375.
25. J. Koch, E.J. Moniz and N. Ohtsuka, Ann. Phys. **154** (1984) 99.
26. W.H. Alberico, M. Ericson and A. Molinari, Ann. Phys. **154** (1984) 356.

Photonuclear Reactions at Intermediate Energies by Self-consistent RPA-SK3 Continuum Theory: the Mean-Field Quasideuteron Effect.

M. Marangoni and A.M. Saruis

ENEA , Centro Ricerche "E. Clementel", Viale Ercolani 8, 40138 Bologna (Italy)

Abstract: The reaction mechanism in (γ,p) and (γ,n) reactions at intermediate energies has been studied in the frame of the self-consistent RPA-SK3 continuum theory, fulfilling current conservation. Mean-field exchange currents, gauge-invariant with the RPA Hamiltonian, are expressed through an effective nucleon mass $m^{SK}(r) \leq m$ carried by the isovector convection current. The mass $m^{SK}(r)$ arises from proton-neutron correlations (quasideuterons) present in the mean-field Hamiltonian. The $^{16}O(\gamma,p)$ and $^{16}O(\gamma,n)$ angular distributions at energies $E_\gamma \leq 300$ MeV have been calculated in order to find an experimental signature of the mass $m^{SK}(r)$.

1. Introduction

Photonuclear reactions at intermediate energies (40 MeV $\leq E_\gamma \leq$ 300 MeV), can at present be discussed in the large context of the theoretical lines developed in the last decade for the interpretation of the data on (e,e') and (e,e'x) reactions, obtained with high duty factor electron accelerators [1]. The dynamical aspects of the nuclear interaction with real photons should emerge from the study of nuclear responses to virtual photons near the photon point: $q_\mu^2 = \omega^2 - q^2 = 0$.

We refer here to those theoretical investigations where electromagnetic (e.m.) currents are calculated consistently with nuclear states to the extent that they can be obtained by performing in the nuclear Hamiltonian the minimal substitution:

$$p \longrightarrow (p - eA). \tag{1}$$

In this case the continuity equation:

$$q_\mu J^\mu = 0 \tag{2}$$

is automatically satisfied.

Nuclear self-consistent mean-field theories which have been developed in both the non-relativistic [2,3] and relativistic sectors [4,5], can be formulated so that conditions (1) and (2) are fulfilled.

Predictions on (γ,x) and (e,e′x) reactions require the knowledge of the nuclear states in the energy continuum with the proper S-matrix boundary conditions. This requirement imposes severe restrictions on the numerical solution of the nuclear many-body problem.

In the non-relativistic sector, the full treatment of the 1p-1h energy continuum has been so far achieved in the self-consistent HF-RPA theory with a Skyrme Hamiltonian density. The theory as first developed in Bologna, with a Buck-Hill coupled-channel method, refers to a SK3 interaction [6]. It has been applied to the analysis of photonuclear reactions in ^{12}C, ^{16}O and ^{40}Ca nuclei in the energy range covering the excitation of the Giant Dipole Resonance up to the high continuum, $E_\gamma \leq 100$ MeV [7].

In the same model, an extended investigation of electronuclear reactions has been performed. Charge and transverse nuclear responses have been calculated in ^{12}C(e,e′) and ^{40}Ca(e,e′) in the quasi-elastic region. At low momentum transfer the nuclear responses have been obtained in ^{16}O(e,e′) electron scattering. In parallel to inclusive (e,e′) reactions, coincidence ^{12}C (e,e′x) and ^{16}O(e,e′x) reactions have been investigated (cf.ref.[8] and references therein).

More recently, the same HF-RPA theory has been formulated in Gent within a Green function approach for a SK2 interaction [9]. The HF-SK2 and RPA-SK2 approximations have been assumed for the analysis of photonuclear reactions at energies 40 MeV $\leq E_\gamma \leq$ 400 MeV. In addition, a systematic calculation of missing momentum distributions in ^{16}O (e,e′p) and ^{12}C(e,e′p) has been performed. The ^{58}Ni (e,e′p_i) cross-sections at moderate momentum transfer have also been obtained (cf.ref.[10] and references therein).

In the relativistic sector, quantum hadron dynamics HF-RPA theories have by now been widely discussed and tested in the calculation of the (e,e′) nuclear responses and sum-rules in the quasi-elastic region [11]. Formulations of these theories for predicting scattering nuclear states do also exist. Relativistic calculations of $(\gamma$,p) photoreactions in ^{16}O and ^{40}Ca at intermediate energies have been performed by Mc Dermott et al. with a model based on a Dirac equation for nucleons and a $\sigma - \omega$ mean-field approximation [12]. Recently, Shepard et al. have outlined a relativistic RPA theory in finite nuclei with correlations induced by σ and ω mesons. Results on quasi-elastic charge responses for ^{12}C and ^{40}Ca have been obtained [13].

A well-known property of gauge invariant e.m. currents is to produce a privileged coupling to charge-exchange processes (proton-neutron correlations) present in the many-body nuclear Hamiltonian.

The above reaction mechanism is at the origin of the so named "quasideuteron effect" which was first noticed in the analyses of photonuclear reactions with the Levinger quasideuteron model. Because of the successful predictions obtained on (γ,p) and (γ,n) data within this model, the photon absorption in nuclei was conceived as occurring on uncorrelated neutron-proton pairs (quasideuteron effect) [14]. This point of view, formulated on a full meson exchange picture of the deuteron, has motivated some recent work of Ericson, Leidemann and Orlandini on (e,e') nuclear responses and Coulomb sum-rules [15].

In mean-field self-consistent HF-RPA theories, the quasideuteron mechanism arises from current conservation which leads to the occurence of exchange currents gauge invariant with the model Hamiltonian generating the nuclear states.

The theoretical analysis of (γ,p) and (γ,n) reactions at intermediate energies, with the explicit introduction of exchange currents, gauge invariant with the nuclear Hamiltonian, was first performed by Gary and Hebach [16]. The nuclear states were calculated in the direct-semidirect model proposed by G.E. Brown. They referred to the continuity equation (2), which in the low energy limit, leads to the Siegert commutator $[V,D]$for the dipole operator D and the NN interaction V. Because of the isovector nature of the previous commutator, (γ,p) and (γ,n) angular distributions were predicted forward peaked and comparable in size, in agreement with the experimental data.

At the same time, Marangoni, Ottaviani and Saruis obtained (γ,p) and (γ,n) angular distributions below 100 MeV in agreement with experiment, within a model where e.m. transition amplitudes were calculated in the long-wave length approximation (LWA) with continuum TDA nuclear states. In the LWA the continuity equation (2) is strictly fulfilled. Therefore exchange currents, gauge invariant with TDA states, were automatically included in these calculations [17].

In the last decade, the ideas proposed in refs. [16,17] have been developed and found confirmation in more complete theoretical frames such as self-consistent mean-field theories which for their own construction fulfill current conservation. In fact they refer to a nuclear Hamiltonian where basic symmetries and conservation laws are not destroyed [3].

In the case of self-consistent continuum HF -RPA theory with a Skyrme Hamiltonian [6,9] the basic symmetry is the time-reversal invariance, which preserves current conservation. Peculiarity of the previous theory is that the gauge-invariant two-body current can be reduced to an isovector one-body density dependent current. The exchange effects are included trough an effective mass $m^{SK}(r)$ operating on the isovector convection current.

For the continuity equation (2) the reaction dynamics which affects photonuclear reactions, should also be present in the charge (e,e′) responses in the quasi-elastic region. This investigation has been the object of ref.[18], where the quenching phenomena in ^{40}Ca(e,e′) have been analyzed in self-consistent HF-SK3 and RPA-SK3 models.

In the present paper we discuss the incidence of the mean-field exchange currents on (γ,p) and (γ,n) angular distributions. We present and discuss the predictions on $^{16}O(\gamma,p)$ and $^{16}O(\gamma,n)$ cross-sections at $E_\gamma \leq 300$ MeV, obtained in the frame of the self-consistent HF-SK3 and RPA-SK3 continuum theories. We make the connection to other theoretical frames.

The plan of the paper is the following. In section 2, we define, for a Skyrme Hamiltonian,the e.m. exchange currents as well as the HF and RPA equations for continuum nuclear states. In section 3, we present and discuss the HF-SK3 and RPA-SK3 results on $^{16}O(\gamma,p)$ and $^{16}O(\gamma,n)$ at $E_\gamma \leq 300$ MeV. Finally in section 4, we draw some conclusions.

2. Self-Consistent HF-RPA Continuum Theory with a Skyrme Hamiltonian. Coupled-channel Method.

2.1 Skyrme Hamiltonian and gauge-invariant electromagnetic currents.

A Skyrme Hamiltonian is given by a one-body, two-body and three-body terms, as follows:

$$H^{SK} = \sum_i p_i^2/2m + \sum_{i\leq j} v_{ij}^{SK} + \sum_{i<j<k} v_{ijk}^{SK}. \tag{3}$$

The two-body term, which was obtained by Skyrme as a short-range expansion of a finite-range NN interaction, in coordinate space is:

$$v_{(2)}^{SK}(r_1,r_2) = t_0(1 + x_0P_\sigma)\delta(r_1 - r_2) + \frac{1}{2}t_1[\delta(r_1 - r_2)k^2 + k'^2\delta(r_1 - r_2)] + \tag{4}$$
$$+t_2k'\cdot\delta(r_1 - r_2)k + iW_0(\sigma_1 + \sigma_2)\cdot k' \times \delta(r_1 - r_2)k.$$

The three-body term takes into account short-range correlations through a two-body density dependent term as follows:

$$v_{(3)}^{SK}[\rho] = \frac{t_3}{6}(1 + x_3P_\sigma)\rho^\alpha(R)\delta(r_1 - r_2). \tag{5}$$

By performing the minimal substitution in eqs.(3)-(5) we obtain the following gauge-invariant current:

$$J(r) = j_{(1)}(r) + j_{(2)}(r), \tag{6}$$

where

$$j_{(1)}(r) = \sum_{i=1}^{A}(\frac{1+\tau_{3i}}{2})\frac{\hbar}{2m}[\delta(r-r_i)\nabla_i - \nabla'_i\delta(r-r_i)] + j_{(1)}(\sigma_i), \tag{7}$$

and

$$j_{(2)}(r) = \sum_{j,k}\frac{(\tau_{3j}-\tau_{3k})}{2}\frac{1}{4}(t_1+t_2)[\delta(r-r_j)\delta(r_j-r_k)(\nabla_j - \nabla_k) - \tag{8}$$
$$-(\nabla'_j - \nabla'_k)\delta(r_j - r_k)\delta(r-r_j)].$$

Current (7) is the convection plus spin current in impulse approximation [19]. Eq.(8) is a two body current generated by the momentum dependent terms of the Skyrme interaction (cf.eq.(4)). From the isospin dependence of eq.(8) we infer that this current is carried by an interacting neutron-proton pair (quasideuteron effect).

In addition to the momentum dependent term (8), we obtain by gauge invariance, a two-body spin-orbit current,which we here disregard. In photonuclear reactions the spin current starts indeed to be relevant at energies around 200 MeV. Therefore, in our theoretical frame the spin-current is treated in impulse approximation [19].

The Skyrme Hamiltonian (3)-(5) as well as its gauge-invariant currents (6)-(8) are operators which have a meaning in the contest of a mean-field theory. Indeed, it is in this frame that the $\delta(r)$ distributions, present in eqs.(3)-(8), lead to the (numerical well-founded) density dependent ph operators (cf.ref. [29]). This last operation will be performed in the next subsections.

2.2 Self-consistent HF-RPA continuum theory with a Skyrme Hamiltonian

General properties of non-relativistic self-consistent fields in the HF and RPA approximations have been reviewed in ref. [3] . We expose here the main properties of a self-consistent HF-RPA continuum theory formulated, in coordinate space, with a Skyrme Hamiltonian (cf.eq.(3)-(5)).

The nuclear states Ψ^c_E are continuum states which obey at large distance, to the S-matrix behaviour:

$$\Psi^c_\infty \simeq_{r\to\infty} \sum_c [I_c(k_c r)\delta_{cc'} - S_{cc'} O_{c'}(k_{c'}r)]\Phi_{c'}. \tag{9}$$

The first calculations along these lines have been performed by Bertch and Tsai [21] and Liu and Brown [22]. They were restricted to the prediction of inclusive (e,e') multipole responses in the LWA and not fully self-consistent since the spin part of the Hamiltonian density was neglected destroying the orthonomality of the RPA nuclear states.

The Skyrme Hamiltonian form (3)-(5) is so far the only expression which leads to the solution of the self-consistent HF-RPA equations in cordinate space for continuum 1p-1h scattering state (cf.eq.(9)).

In the notation of ref. [23]. we calculate the mean-field energy functional:

$$E[\rho] = <\Phi|H^{SK}|\Phi> = \int H^{SK}(r)d^3r, \tag{10}$$

where Φ is a Slater determinant.

Under the assumption of time-reversal invariance, we obtain for the Hamiltonian density $H^{SK}(r)$ an algebraic function of the nucleon densities $\rho_q(r)$, the kinetic energy density $\tau_q(r)$, the momentum density $j_q(r)$ and the corresponding spin densities. In the case of a spin saturated and charge conjugated nucleus, neglecting spin-orbit and Coulomb effects, we obtain:

$$H^{SK}(r) = \frac{\hbar}{2m}\tau + \frac{3}{8}t_0\rho^2 + \frac{1}{16}(3t_1 + 5t_2)(\rho\tau - j^2) + \tag{11}$$

$$-\frac{1}{64}(9t_1 - 5t_2)\rho\nabla^2\rho + \frac{1}{16}t_3\rho^3.$$

The density matrix $\rho(r,r) = \rho(r)$ is given by:

$$\rho(r) = \sum_{v} \Phi_v(r)\Phi_v(r), \tag{12}$$

where the index v runs over the occupied orbits (hole states) in the HF ground-state.

The time dependent Hartree-Fock (TDHF) Hamiltonian density obtained variationally from eq.(11) is the following:

$$H_q^{RPA}(r) = -\nabla \frac{\hbar^2}{2m_q^*(r)} \nabla + U_q(r) + \frac{1}{i}(\nabla \cdot I_q + I_q \cdot \nabla), \tag{13}$$

where:

$$\frac{\hbar^2}{2m_q^*(r)} = \frac{\hbar^2}{2m_q(r)} + \frac{1}{4}(t_1 + t_2)\rho(r) + \frac{1}{8}(t_2 - t_1)\rho_q(r), \tag{14}$$

$$I_q(r) = -\frac{1}{4}(t_1 + t_2)j(r) - \frac{1}{8}(t_2 - t_1)j_q(r), \tag{15}$$

and $U_q(r)$ is the static HF potential.

For ease of discussion, the spin dependence has been dropped in eq.(13). However, it is fully included in the performed calculations. In eqs.(13)-(15), the quantities t_1 and t_2 are the parameters of the momentum dependence in the Skyrme force (4) and q $= \pm 1/2$ are the proton and neutron isospin coordinates.

In the static HF limit, for the time reversal invariance $j_q = 0$, eq.(13) reduces to the static HF Hamiltonian density:

$$H_q^{HF}(r) = -\nabla \frac{\hbar^2}{2m_q^*(r)} \nabla + U_q(r). \tag{16}$$

The equations of state for operators (13) and (16) are the following, respectively:

$$i\hbar \frac{\partial \rho}{\partial t} = [H^{RPA}(r), \rho], \tag{17}$$

and

$$[H^{HF}(r), \rho] = 0. \tag{18}$$

Eq.(17), in the limit of small oscillations around the static HF state, gives the RPA equations.

We write eqs.(17) and (18) as they are actually numerically solved in our model [6]. The HF equation (18) in coordinate space, for the radial single-particle wave-functions $u_i(r)$,is as follows:

$$\frac{\hbar^2}{2m_q^*}[-\frac{d^2}{dr^2}+\frac{l(l+1)}{r^2}+U_q(r)]u_i(r)- \tag{19}$$

$$-\frac{d}{dr}[\frac{\hbar^2}{2m_q^*(r)}]\frac{du_i(r)}{dr}=\varepsilon_i u_i(r),$$

where the HF effective mass $m^*(r)$ is given by eq.(14).

The RPA equations (17) in the coupled-channel representation, for the two-component particle radial wave-functions $(f_E^c(r,c), g_E^c(r,c))$, are the following [6]:

$$[D^T-(E-\varepsilon_i)]f_E^c(r,c)+ \tag{20a}$$

$$+\sum_{c'}[M_f^T(c,c',r)f_E^c(r,c')+M_g^T(c,c',r)g_E^c(r,c')]-R_E(r,c)=0,$$

$$[D^T-(E+\varepsilon_i)]g_E^c(r,c)+ \tag{20b}$$

$$+\sum_{c'}[M_g^T(c,c',r)f_E^c(r,c')+M_f^T(c,c',r)g_E^c(r,c')]-S_E(r,c)=0.$$

We refer here to eqs.(20) in T-representation in order to discuss the properties of the RPA solutions in the T=0 and T=1 isospin channels.

The RPA nuclear states Ψ_E^c are obtained from the solutions of eqs.(20), as follows:

$$\Psi_E^c=\sum_{c(mi)}[X_E^c(m,i)a_m^+a_i-Y_E^c(m,i)\,a_i^+a_m]\,0>\;=\;\sum_{c'}[f_E^c(r,c')+g_E^c(r,c')]\phi_{c'} \tag{21}$$

where the X's and Y's are the amplitudes for creation and for annihilation of ph pairs, respectively.

In eqs.(20), the differential operator D^T,in the two cases of T=0 and T=1 solutions, has the following expression, respectively:

$$D^{T=0}=\frac{\hbar^2}{2m}[-\frac{d^2}{dr^2}+\frac{l(l+1)}{r^2}+U^T(r)], \tag{22a}$$

$$D^{T=1}=\frac{\hbar^2}{2m^{SK}(r)}[-\frac{d^2}{dr^2}+\frac{l(l+1)}{r^2}+U^T(r)]-\frac{d}{dr}[\frac{\hbar^2}{2m^{SK}(r)}]\frac{d}{dr}, \tag{22b}$$

where:

$$\frac{\hbar^2}{2m^{SK}(r)} = \frac{\hbar^2}{2m^*(r)} - \frac{1}{8}(t_1 - t_2)\rho_q = \frac{\hbar^2}{2m} + \frac{1}{4}(t_1 + t_2)\rho. \qquad (23)$$

Peculiarity of eqs.(20), is the presence of the bare nucleon mass m in the T=0 D^T operator (cf.eq.(22a)) while the effective mass $m^{SK}(r)$ appears in the T=1 operator(22b).

This behaviour can be understood by noticing that the RPA Hamiltonian density differs from the HF one for the $I_q(r)$ term (cf.eqs.(13)-(16)). The I_q operator does correspond to the ph interaction in the RPA equations and garantees time reversal invariance to the Hamiltonian (13).

As already known, the residual ph interaction in the T=0 equations produces a kinetic energy density dependent term which cancels the one appering in the static HF effective mass $m^*(r)$ (cf.eq.(14)). As a result the bare mass appears in the kinetic energy term of the $D^{T=0}$ operator (cf.eq.(22a)).

On the contrary, in the case of the RPA T=1 equations the residual interaction produces a kinetic energy term which modifies the density dependence of the HF effective mass $m^*(r)$, resulting in a mass $m^{SK}(r)$ with a stronger density dependence (cf.eq.(23)).

In Fig.1 we report, for the ^{40}Ca nucleus in HF-SK3, the r-dependence of the effective mass ratios $\frac{m^{SK}(r)}{m}$ and $\frac{m^*(r)}{m}$ as given by eqs.(23) and (14), respectively.

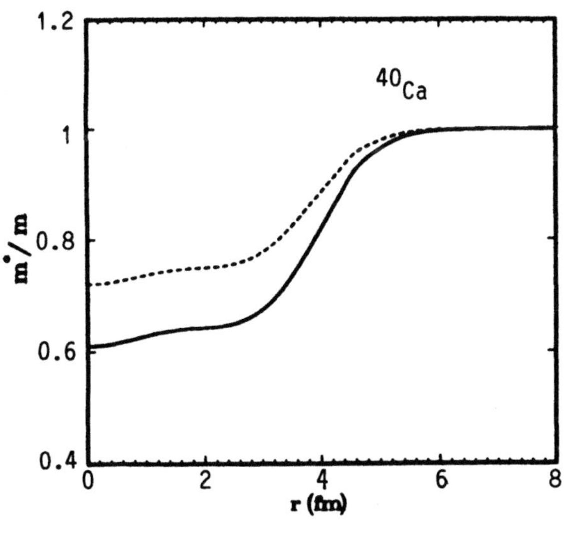

Figure 1:
The ratios $\frac{m^{SK}(r)}{m}$ (continuum curve) and $\frac{m^*(r)}{m}$ (dashed curve) for ^{40}Ca in HF-SK3.

2.3 Gauge-invariant mean-field convection current.

The gauge-invariant e.m. current J(r) (cf.eqs.(6)) enters in the e.m. transition matrix elements through the expression:

$$J^{RPA}(q) = \int e^{iq\cdot r} < \Psi^c_E | J(r) | 0 > dr, \tag{24}$$

where J(r) is given by eq.(6).

We refer first to the two body term of J(r) given by eq.(8). Its transition matrix elements for RPA nuclear states, can be written as follows:

$$< \Psi^c_E | j_{(2)}(r) | 0 > = \sum_{c(mi)} \sum_{\alpha\beta\gamma\delta} < \alpha\beta | j_{(2)} | \gamma\delta > [X^c_E(mi) < 0 | a^+_\alpha a^+_\beta a_\delta a_\gamma a^+_m a_i | 0 > + \tag{25}$$

$$- Y^c_E(mi) < 0 | a^+_\alpha a^+_\beta a_\delta a_\gamma a^+_i a_m | 0 >] = \sum_{c(mi)} [X^c_E(mi) + Y^c_E(mi)] < m | J_{ex}[\rho] | i >$$

where:

$$J_{ex}[\rho] = \sum_\nu < \nu | j_2 | \nu > = \sum_{i=1}^{A} \tau_{3i} \frac{1}{4} (t_1 + t_2) \rho(r) [\delta(r - r_i) \nabla_i - \nabla'_i \delta(r - r_i)]. \tag{26}$$

The density $\rho(r)$ is defined by eq.(12). The two body current $j_{(2)}$ reduces for RPA transitions, to an effective one-body density dependent current $J_{ex}[\rho]$ for ph transitions (cf.eq.(26)). In Fig.2 we report the vertex of the effective current $J_{ex}[\rho]$.

Figure 2:
Mean-field exchange current $J_{ex}[\rho]$.

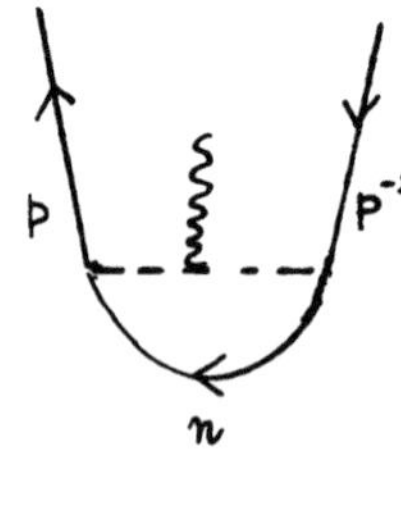

In our RPA mean-field theory, we can therefore write the gauge-invariant current operator (6) as an effective one-body ph operator given by:

$$J^{RPA}(r) = J^{RPA}_{IS} + J^{RPA}_{IV}$$

$$= \sum_i \frac{\hbar^2}{2m} [\delta(r - r_i)\nabla_i + \nabla'_i \delta(r - r_i)] + \tag{27}$$

$$+ \sum_i \tau_{3i} \frac{\hbar^2}{2m^{SK}(r)} [\delta(r - r_i)\nabla_i + \nabla'_i \delta(r - r_i)]$$

where $m^{SK}(r)$ is given by eq.(23).

We notice here that the effective mass $m^{SK}(r)$ which appears in the T = 1 RPA operator (22b), is the same effective mass brought by the isovector current in eq.(27).

We adress now the question on the m^{SK} -dependence of the RPA transition matrix $< \Psi^{\varepsilon}_{E} | J^{RPA}_{IV} | 0 >$ of the isovector current (26).

In the case of the T = 1 isospin channel, we can perform in eqs. (20) the following transformation:

$$f^T(r) = [\frac{m^{SK}(r)}{m}]^{1/2} f^L(r) \tag{28a}$$

$$g^T(r) = [\frac{m^{SK}(r)}{m}]^{1/2} g^L(r) \tag{28b}$$

By substituting eqs.(28) in eqs.(20) we obtain a set of coupled equations for the $f^L(r)$ and $g^L(r)$ solutions, where the operator $D^{T=1}$ transforms into an operator D^L given by:

$$D^L = \frac{\hbar^2}{2m} [- \frac{d^2}{dr^2} + \frac{l(l+1)}{r^2} + U^L(r)] \tag{29}$$

Since the solutions (f^L, g^L) have the same S-matrix behaviour at large distances as the T = 1 solutions (f^T, g^T) (cf.eq.(9)) the factor $[\frac{m^{SK}}{m}]^{1/2}$, in eqs.(28), is a Perey factor [24,25]. For this property, the matrix element $< \Psi^{\varepsilon}_{E} | J^{RPA}_{IV} | 0 >$ has a resulting m^{SK}-dependence given by a factor $\sqrt{\frac{m^{SK}}{m}}$ [18].

2.4 Self-consistent HF-SK3 and RPA-SK3 calculations. Numerical details.

In our calculations, the RPA equations (20) have been solved for a Skyrme 3 force (RPA-SK3). This interaction has a strong density dependence and a weak momentum dependence, relatively to other parametrization of the Skyrme force.

The RPA-SK3 nuclear states Ψ^{ε}_{E} are calculated solving eqs.(20) with a self-consistent procedure. The peculiarity of this procedure is that all the parameters of the theory are those of the effective interaction. In our calculation the SK3 parameters are kept fixed. As a result, RPA-SK3 predictions are obtained without any free parameter (cf.ref. [6]).

With boundary conditions (9) we obtain, for each energy, as many RPA solutions as there are open reaction channels c'.These reaction channels are those corresponding to a nuclear disintegration into a nucleon (proton or neutron) and a residual nucleus in its ground and excited hole states.

We call HF-SK3 the calculation performed by letting go to zero the ph matrix elements M_i in eqs.(20).

3. Photonuclear Reactions: $^{16}O(\gamma,p)$ and $^{16}O(\gamma,n)$ Angular Distributions at $E_\gamma \leq 300$ MeV.

We analyze the reaction dynamics in (γ,p) and (γ,n) angular distributions in the frame of the theory outlined in section 2.

In this context we obtain an e.m. current (cf.eqs.(6)-(8)), gauge-invariant with the Skyrme Hamiltonian (cf.eqs.(3)). The two-body current arising from the momentum dependence of the Skyrme force reduces, in the RPA transition matrix elements, to an isovector convections current carrying an effective mass $m^{SK}(r)$. This effective mass has the r-dependence given in Fig.1 in the case of ^{40}Ca and for a force SK3. Since this quantity is r-dependent, it enters dynamically in the calculation and gives contributions to the transition matrix elements which depend on the energy and on the momentum transfer.

In what follows, the incidence of the effective mass $m^{SK}(r)$ on the e.m. transition matrix elements will be tested on (γ,p) and (γ,n) angular distributions in ^{16}O.

To this aim,we have performed the calculation of the (γ,N) angular distributions [7] for the e.m. transition matrix (24) in the following cases:

i) for RPA-SK3 nuclear states and the gauge invariant current (27). $(RPA - SK3\,[m^{SK}])$

ii) for RPA-SK3 nuclear states and the convection current in impulse approximation i.e. $m^{SK}(r) = m$ (RPA-SK3 $[m]$)

iii) for HF-SK3 nuclear states and the gauge invariant current (27) (HF-SK3 $[m^{SK}]$)

The calculations have been performed with $J \leq 5$ electric and magnetic multipoles contributions to the cross-sections [7].

3.1 $^{16}O(\gamma,p)$ and $^{16}O(\gamma,n)$ angular distributions at $E_\gamma \leq 100$ MeV.

In the present subsection we analyze the $^{16}O(\gamma,p)$ and $^{16}O(\gamma,n)$ angular distributions at $E_\gamma \leq 100$ MeV.

Fig.3 shows the $^{16}O(\gamma,p_0)^{15}N$ angular distribution at $E_\gamma = 60$ MeV, with the residual nucleus left in its ground state. The continuum curve is the RPA-SK3 [m^{SK}] result. The dashed curve corresponds to the RPA-SK3 [m] prediction. Finally the dot-dashed line is the HF-SK3 [m^{SK}] calculation. The experimental data are from ref.[26].

In Fig.4 and 5 are given the $^{16}O(\gamma,n_0)$ and $^{16}O(\gamma,n_1)$ angular distributions at 60 MeV, with the residual nucleus left in its ground and first 3/2 excited state, respectively. Same convention as in Fig.3. The data are from ref.[27].

Fig.6 and 7 display the $^{16}O(\gamma, p_0)$ and $^{16}O(\gamma,n_0)$ angular distributions at 80 MeV, in the same convention of Fig.3. The data of Fig.6 are from ref.[26] while those in Fig.7 are from ref.[27].

Finally, in Fig.8 are reported the results for $^{16}O(\gamma,p_0)$ at 100 MeV with the data of ref.[26]. Same convention as Fig.3.

We notice first that, at energies below 100 MeV, the spin current gives a negligible contribution to the (γ,x) cross-sections. As a consequence, meson exchange currents arising from tensor correlations are not expected to operate here.

At these energies, the reaction mechanism outlined in section 2 for the RPA-SK3 [m^{SK}] calculation appears to be well justified.

We are specifically interested in finding the experimental signature of the mass $m^{SK}(r)$ carried by the convection current (27).

A sight at Figs.3-8, reveals that the RPA-SK3 [m^{SK}] angular distributions for (γ,p) and (γ,n) reactions are forward peaked and comparable in size, in agreement with the experiment.

The effect of the mass $m^{SK}(r)$ can be seen by comparing, in Figs.3-8, the results RPA-SK3 [m^{SK}] (continuum curve) with the RPA-SK3 [m] predictions (dashed lines). We infer that the mass $m^{SK}(r)$, gives the dominant contribution to both (γ,p) and (γ,n) cross-sections.

However, the role of the RPA correlations is essential in giving the previous effect. This point can be clarified in the light of the considerations developed at the end of subsection 2.3. There, we have shown that the RPA-SK3 [m^{SK}] matrix

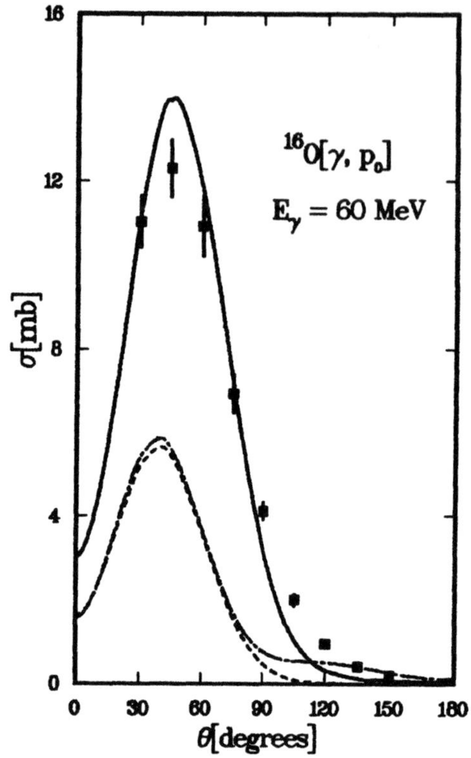

Figure 3:
$^{16}O(\gamma,p_0)$ angular distribution at $E_\gamma = 60$ MeV. The continuum curve is the RPA-SK3 $[m^{SK}]$ result. The dashed curve corresponds to the RPA-SK3 $[m]$ prediction. Finally, the dot-dashed line is the RPA-SK3 $[m^{SK}]$ calculation. Experimental data from ref. [26].

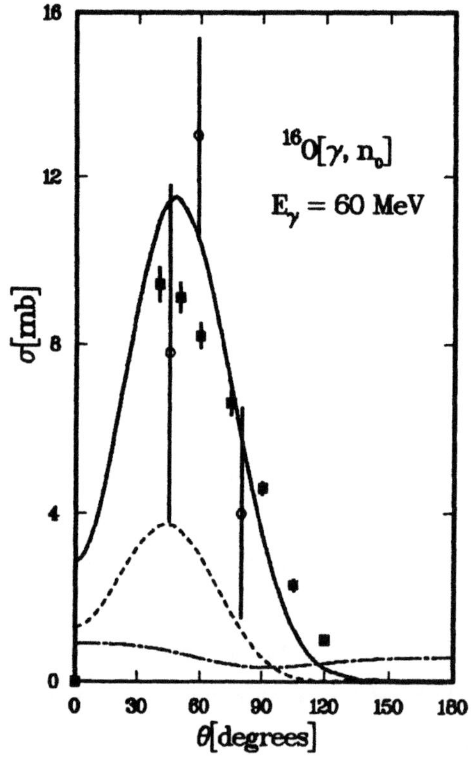

Figura 4:
$^{16}O(\gamma,n_0)$ angular distribution at $E_\gamma = 60$ MeV. Same conventions as in Fig.3. Experimental data from ref. [27].

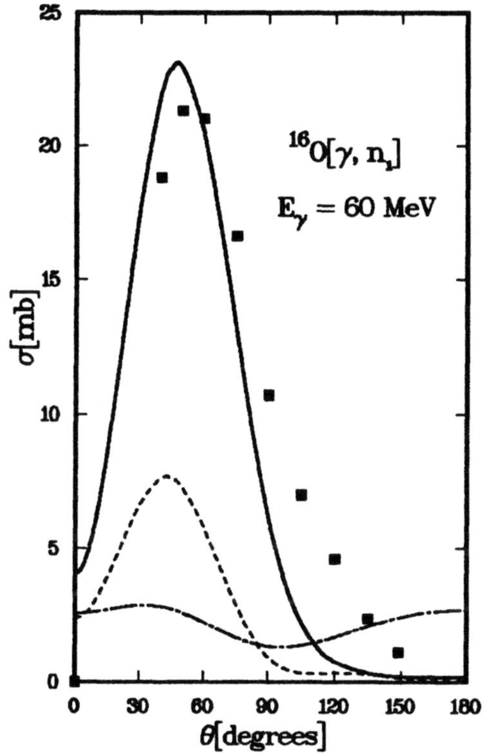

Figure 5:
$^{16}O(\gamma,n_1)$ angular distribution at $E_\gamma = 60$ MeV, with the residual nucleus left in its first 3/2 excited state. Same conventions as in Fig.3. Experimental data from ref. [27] .

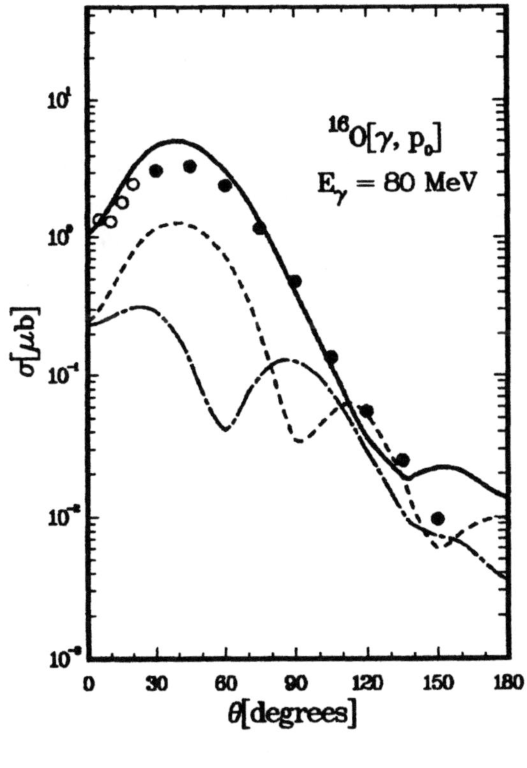

Figure 6:
$^{16}O(\gamma,p_0)$ angular distribution at $E_\gamma = 80$ MeV. Same conventions as in Fig.3. Experimental data from ref. [26].

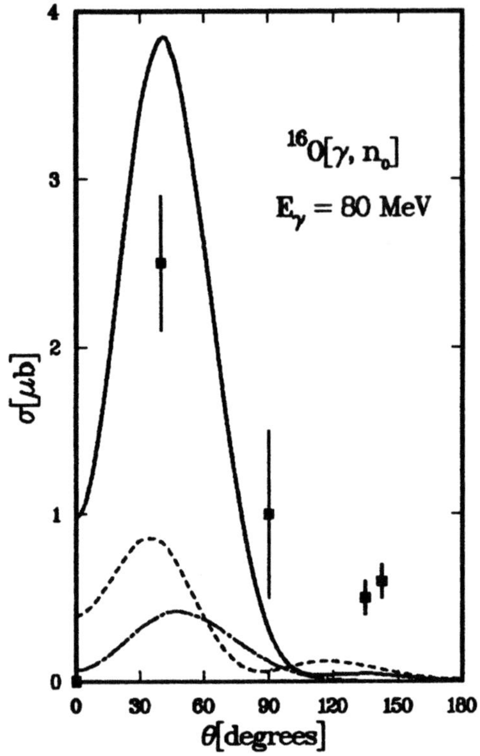

Figure 7:
$^{16}O(\gamma,n_0)$ angular distribution at E_γ = 80 MeV. Same conventions as in Fig.3. Experimental data from ref. [27].

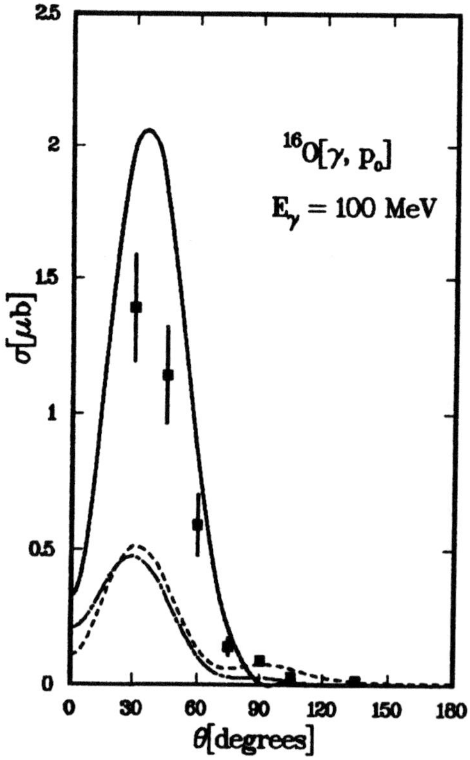

Figure 8:
$^{16}O(\gamma,p_0)$ angular distributions at E_γ = 100 MeV. Same conventions as in Fig.3. Experimental data from ref. [26].

element $<\Psi_E^c|J_{fi}^{RPA}|0>$, has a resulting m^{SK} -dependence given by a factor $\sqrt{\frac{m}{m^{SK}(r)}}$. In Figs.3-8 the enhancement of the continuum relatively to the dashed curve is produced by this factor. This enhancement results from the coerent contribution of all the channels in the RPA wave-function (cf.eq.(9)),each contributing a factor $\sqrt{\frac{m}{m^{SK}}}$.

This is not the case for the HF-SK3 [m^{SK}] calculation in Figs.3-8 (dot-dashed lines), since the HF wave-function contributes there with only the elastic channel. In particular in the case of (γ,n) angular distributions, the HF-SK3 [m^{SK}] cross-section is almost missing.

3.2 $^{16}O(\gamma,p)$ and $^{16}O(\gamma,n)$ angular distributions at 100 MeV $\leq E_\gamma \leq$ 300 MeV.

At energies above 100 MeV, spin current contribution to photoreactions becames increasingly important with the implication that the tensor components of the meson exchange potentials can no more be disregarded. In this energy region, 2p-2h meson exchange currents have to be calculated. In addition, the -excitation has also to be fully included.

So far the existing theoretical investigations care of only a particular aspect of the above reaction dynamics. As a result, any theoretical frame is able to give a systematic prediction of (γ,p) and (γ,n) data above 100 MeV.

Also our theoretical frame is insufficient at $E_\gamma > 100$ Mev. A typical result is shown in Figs.9-10, where we display the $^{16}O(\gamma,p_0)$ and $^{16}O(\gamma,n_0)$ angular distributions at 200 MeV , respectively. The continuum curve is the RPA-SK3 [m^{SK}] prediction, while the dot-dashed curve is the HF-SK3 [m^{SK}] result. The data are from ref.[28,29] in Fig.9, and from ref.[30] in Fig.10.

The quality of our predictions is comparable to that obtained by Gary and Hebach and shown in Fig.11 (continuum curve). In the same figure Van der Steenhoven [31] has also shown the relativistic prediction by Mc Dermott et al. [12] (dashed line) and the HF-SK2 calculation of Ryckebusch et al. [9] (dotted line).

A systematic analysis of data at different energies for fixed angles reveals that the disagreement is most pronounced at forward and backward angles where spin currents give the main contribution. This feature is confirmed in Fig.12, where we report the energy dependence of the $^{16}O(\gamma,p_0)$ angular distribution at 45^0. The theoretical predictions are given in the convention of Fig.3, while the data are from ref. [28]. Here the RPA-SK3 [m^{SK}] result (continuum curve) is indeed in agreement with data up to high energies, showing the observed bending

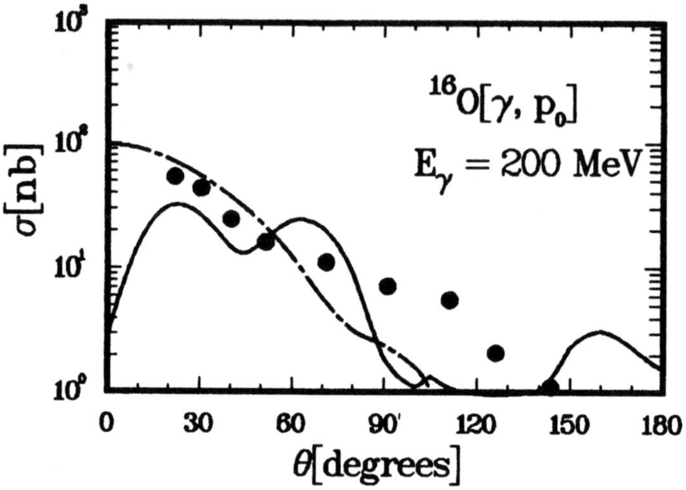

Figure 9:

$^{16}O(\gamma,p_0)$ angular distribution at $E_\gamma = 200$ MeV. Same conventions as in Fig.3. Experimental data from refs. [28,29].

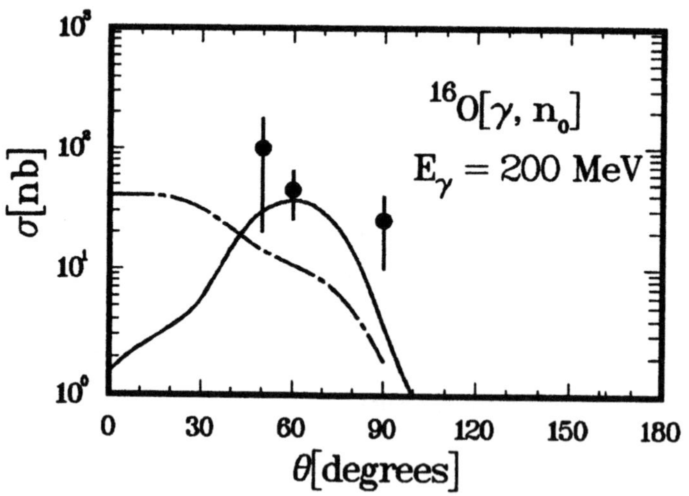

Figure 10:

$^{16}O(\gamma,n_0)$ angular distributions at $E_\gamma = 200$ MeV. Same conventions as in Fig.3. Experimental data from ref. [30].

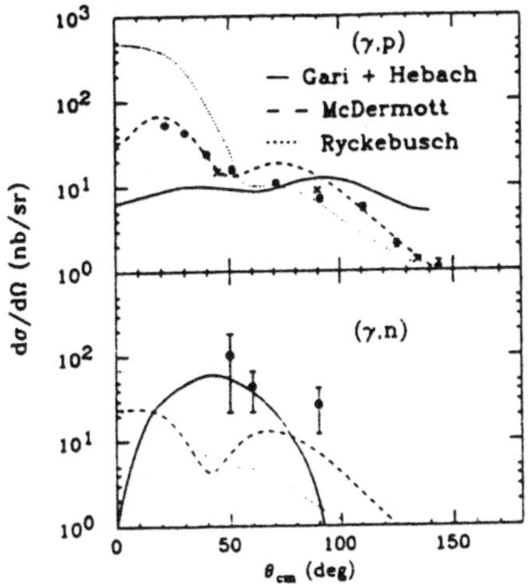

Figure 11:
$^{16}O(\gamma, p_0)$ and $^{16}O(\gamma, n_0)$ angular distributions at E_γ = 200 MeV from ref. [31].

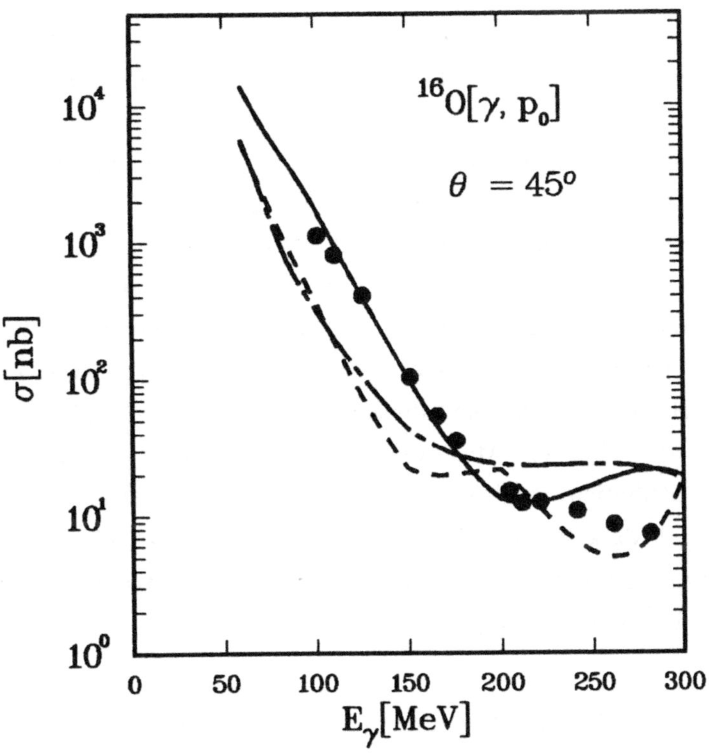

Figure 12:

Energy dependence of the $^{16}O(\gamma, p_0)$ angular distribution at $\theta = 45^0$. Same conventions as in Fig.3. Experimental data from ref. [28].

in the cross-section at 220 MeV. Similar calculations at 90^0 ana 135^0 are in disagreement with data at $E_\gamma \geq 150$ MeV.

4. Concluding Remarks.

The reaction mechanism in $^{16}O(\gamma,p)$ and $^{16}O(\gamma,n)$ photoreactions at $E_\gamma \leq 300$ MeV has been studied in the frame of the self-consistent RPA-SK3 theory.

The e.m. currents are obtained by minimal substitution on the SK3 Hamiltonian. Exchange currents, originated from the mean-field proton-neutron correlations (quasideuterons), are incorporated into the isovector convection current through an effective mass $m^{SK}(r)$ lower than the bare mass m (cf.Fig.1).

The calculations have been performed in order to find the experimental signature of the mass $m^{SK}(r)$. To this aim, the $^{16}O(\gamma,p)$ and $^{16}O(\gamma,n)$ angular distributions at energies $60 \leq E_\gamma \leq 300$ MeV, have been predicted in the three theoretical schemes: a) RPA-SK3 $[m^{SK}]$, b) RPA-SK3 $[m]$, c) HF-SK3 $[m^{SK}]$ (cf.section3).

Confronting calculations a) and b), we verify at energies $E_\gamma \leq 100$ MeV (cf.Figs.3-8) that the mass $m^{SK}(r)$ causes an enhancement of both (γ,p) and (γ,n) angular distributions (quasideuteron effect). They are predicted forward peaked and comparable in size, in agreement with exprerimental data.

Confronting calculations a) and c), we infer that the exchange currents introduced through the mass $m^{SK}(r)$ are strictly connected to the RPA correlations Indeed for HF nuclear states the enhancement is strongly reduced.

At energies $E_\gamma > 100$ MeV, the mean-field exchange currents (cf.eq.(27)) give still contribution to $^{16}O(\gamma,p)$ and $^{16}O(\gamma,n)$ angular distributions, but the agreement with data is poor. At these energies theoretical schemes where the baryon-meson dynamics in the nucleus is formulated explicitly, appear to be more appropriate for a reliable analysis of data.

References.

1. Drechsel D., Giannini M.M, Rep.Prog.Phys. 52, 1083 (1989).
2. Negele J.W., Rev.Mod.Phys. 54, 913 (1982).
3. Blaizot J.P., Ripka G., Quantum Theory of Finite Systems, The MIT Press (Cambridge Ma) 1986.
4. Danos M., Gillet V., Cauvin M., Methods in Relativistic Nuclear Phisics, North-Holland 1984.

5. Serot B.D., Walecka J.D., Adv.Nucl.Phys. eds. Negele J.W. and Vogt E. (Plenum, New York 1986).
6. Cavinato M., Marangoni M., Saruis A.M., Nucl.Phys. A373, 445 (1982).
7. Cavinato M., Marangoni M., Saruis A.M., Nucl.Phys. A422, 237 (1984).
8. Cavinato M., Marangoni M., Saruis A.M., Z.Phys. A335, 401 (1990).
9. Ryckebusch J. et al., Phys.Lett. B194, 453 (1987); Nucl.Phys. A476, 237 (1988).
10. Ryckebusch J. et al., Nucl.Phys. A503, 694 (1989).
11. Horowitz C.J., Phys.Lett. B208, 8 (1988).
12. McDermott et al., Phys.Rev.Lett. 61, 814 (1988).
13. Shepard J.R., Rost E., Mc Neil J.A., Phys.Rev. C40, 2320 (1989).
14. Goringer H., Schoch B., Luhrs G., Nucl.Phys. A384, 414 (1982).
15. Ericson M., Leidemann W., Orlandini G., Z.Phys. A331, 369 (1988).
16. Gari M., Hebach H., Phys.Rep. 72, 1 (1981).
17. Marangoni M., Ottaviani P.L., Saruis A.M., Report CNEN RT/FI/76, Bologna (1976).
18. Cavinato M., Marangoni M., Saruis A.M., Phys.Lett. B235, 15 (1990).
19. Ciofi degli Atti C., Progr.Part. and Nucl.Phys. 3, 163 (1980).
20. Cavinato M., Marangoni M. and Saruis A.M., Phys.Lett. B213, 111 (1988).
21. Bertsch G., Tsai Phys.Rep. 18C, 125 (1975).
22. Liu K.F., Brown G.E., Nucl.Phys. A265, 385 (1976).
23. Vautherin D., Brink D.M., Phys.Rev. C5, 626 (1972).
Engel Y.M. et al., Nucl.Phys. A249, 215 (1975).
24. Dover C.B., Van Giai N., Nucl.Phys. A190, 373 (1972).
25. Perey F.G., Saruis A.M., Nucl.Phys. 70, 225 (1965).
26. Findlay D.J.S., Owens R.O., Nucl.Phys. A 279, 385 (1977).
27. Schier H., Schoch B., Nucl.Phys. A229, 93 (1974).
28. Leitch et al., Phys.Rev. C31, 1633 (1985).
29. Turley R.S. et al., Phys.LETT. 157B, 19 (1985).
30. Beise E.J., PH.D. Thesis MIT (1988) and submitted to Phys.Rev.Lett. (1989).
31. Van der Steenhoven G., Proc. 4th Workshop on Perspectives in Nuclear Physics at Intermediate Energies. Eds. Boffi, Ciofi_degli Atti, Giannini. Trieste 1989. World Scientific Singapore.

Dipole Sum Rule Enhancement in Nuclei

R.C. Carrasco [1], E. Oset [1], W. Weise [2]

[1]Departamento de Fisica Teorica, Universidad de Valencia, E-46100, Burjassot(Valencia) Spain
[2]Institut für Theoretische Physik, Universität Regensburg, D-8400 Regensburg, Fed. Rep. of Germany

Abstract: In this work we apply a microscopic many-body approach to photonuclear reactions[1] in order to improve the understanding of the dipole sum rule. At the same time, the sum rule provides us with test of consistency of the underlying theory at low photon energies

To begin with, let us recall the main steps in the derivation of the dipole sum rule. The cross section for absorption of a dipole photon by a nucleus is given by [2]:

$$\sigma_{E1}(w) = 4\pi^2\alpha \sum_n \omega| < n|\, \boldsymbol{D}\epsilon\, |0> |^2 \delta(\omega - (E_n - E_o)) \tag{1}$$

Where $\alpha = \frac{1}{137}$, and

$$\boldsymbol{D} = \frac{N}{A}\sum_{p=1}^{Z} \boldsymbol{r}_p - \frac{Z}{A}\sum_{n=1}^{N} \boldsymbol{r}_n \tag{2}$$

is the dipole operator, ϵ the polarization vector, $|0>$ the ground state, and $|n>$ any excited nuclear state; E_o and E_n are their respective energies. By integrating (1) one obtains the nuclear dipole sum rule in terms of the double commutator of the Hamiltonian with the dipole operator:

$$\int_0^\infty dw\ \sigma_{E1}(w) = 2\pi^2\alpha < 0|\, [D_z, [H, D_z]\, |0> \tag{3}$$

$$(\epsilon = \hat{z})$$

Assuming that the nuclear Hamiltonian is expressed as a sum of kinetic energy and static potential:

$$H = T + V = \sum_{i=1}^{A} \frac{p_i^2}{2M} + V \tag{4}$$

we find that (3) reduces to

$$\int_0^\infty dw\ \sigma_{E1}(w) = 2\pi^2\alpha \frac{NZ}{AM}(1+\kappa) \equiv (1+\kappa)\, \Sigma \tag{5}$$

where we have introduced the sum rule unit $\Sigma \simeq 60\text{MeV.mb}\frac{NZ}{A}$. The enhancement κ over the classical Thomas-Reiche-Kuhn dipole sum is

$$\kappa = \frac{AM}{NZ} < 0|[D_z, [V, D_z]|0 > \tag{7}$$

and we have used

$$< 0|[D_z, [T, D_z]|0 >= \frac{NZ}{AM} \tag{7}$$

The dipole sum rule enhancement κ is non-zero if the potential has charge exchange terms proportional to $\tau_1\tau_2$. In a meson-exchange model such terms arise e.g. from the exchange of charged pions. They contribute to the mesonic exchange current as illustrated in fig.1. Hence, κ is a measure of such exchange currents in nuclei.

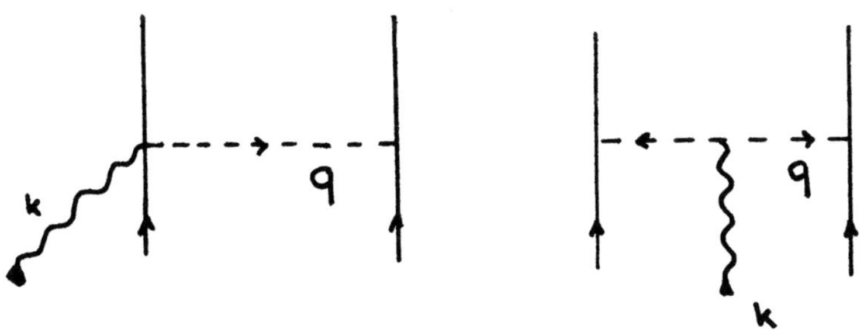

Fig. 1. Kroll-Ruderman and pion pole exchange currents

The integral in (5) extends up to infinite energy, whereas the empirical sum rule is usually taken as the integral over the measured total photon-nucleus cross section up to the pion production threshold:

$$\int_0^{m_\pi} dw\, \sigma(w) = (1 + \bar{\kappa})\Sigma \tag{10}$$

The relationship between the theoretical κ and the empirical $\bar{\kappa}$ has been a subject of longstanding discussions to wich we would like to contribute in the following.

Let us first discuss the assumptions that have been made in the derivation of the sum rule (5). As mentioned before, exchange currents modify the nuclear electromagnetic interaction, so that the total nuclear current now consists of the nucleonic current $\boldsymbol{j}_0$ and the exchange current $\boldsymbol{j}_{ex}$: $\boldsymbol{j} = \boldsymbol{j}_0 + \boldsymbol{j}_{ex}$. Furthermore, the total charge density generally has an exchange part, $\rho = \rho_0 + \rho_{ex}$. The dipole sum rule is conceptually based on Siegert's low energy theorem, wich implies $\rho_{ex} = 0$ in the limit of point-like, static nucleons. If we, for instance, consider the pion pole diagram (fig. 1b) the ratio between $\rho_{ex} = j^0_{ex}$ andj_{ex} is

$$\frac{\rho_{ex}}{j_{ex}} = \frac{q^0}{q} \simeq \frac{1}{2}\sqrt{\frac{\omega}{M}} \tag{11}$$

wich vanishes as $\omega \to 0$. However, it cannot be neglected at high photon energies ω, for wich the static potential picture fails because of retardation effects. Furthermore, once the pion production is reached, real pions contribute to the charge density. Hence, pion production must be excluded from the sum rule evaluation. This is the reason why comparisons are commonly made with the integral (10) which terminates at $\omega = m_\pi$. The aim of our work is to examine, in a microscopic model, to what extent such a comparison between the theoretical sum rule (5) and the empirical sum rule (10) is justified.

In order to calculate the nuclear cross section we use the framework of ref.[1] We start from the standard couplings between photons, pions, nucleons, and Δ particles which accurately describe the (γ, π) reactions on free nucleons. The relevant processes are shown in fig.2

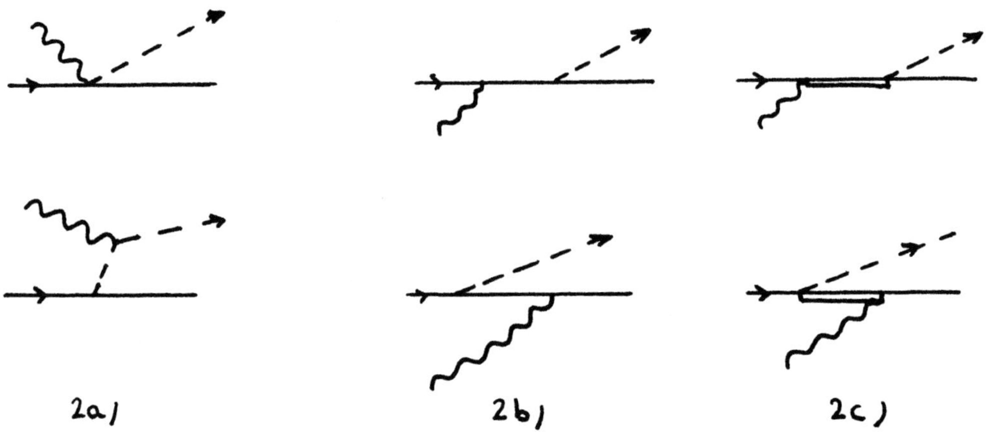

Fig. 2. 2a)Kroll-Ruderman and pion pole diagrams. 2b)Nucleon pole diagrams. 2c)Δ-diagrams.

The Kroll-Ruderman term is the non vanishing part of the photoproduction amplitude as $q \to 0$. Together with the pion pole term and the convection current part of the nucleon pole terms, it is responsible for the dipole cross-section. We calculate the nuclear photoabsorption cross-section by including these diagrams in a many body description using the expression

$$\sigma = -\frac{1}{\omega} \int d\boldsymbol{r} \, \mathrm{Im}\, \Pi(\omega, \rho(\boldsymbol{r})) \tag{12}$$

Where the local density approximation is used in the photon self-energy $\Pi(\omega, \rho)$ in the nuclear medium. The self-energy diagrams leading to photon absorption are drawn in fig. 3. The shadowed bubble stands for any of the (γ, π) diagrams of fig. 2.

The detailed calculations include the Δ self-energy[3], short-range correlations[1] and a variety of medium polarization effects which lead to considerable improvements over previous work. As a result, we arrive at a good description of photonuclear reactions at intermediate energies for medium and heavy nuclei[1].

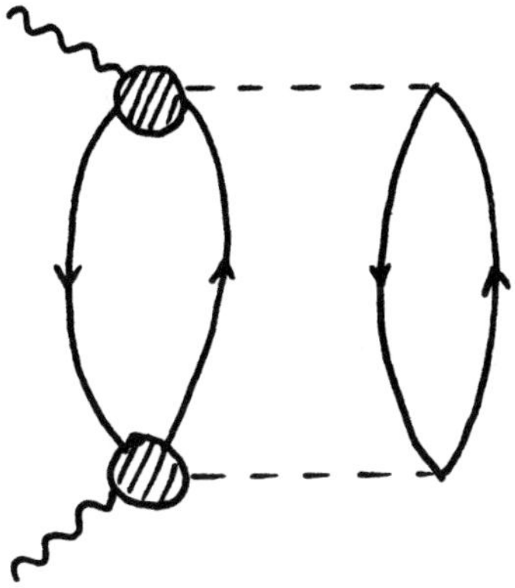

Fig. 3. Photon selfenergy diagrams leading to two-nucleon absorption.

Our model gives a realistic description of photon absorption by means of two body mechanisms mediated by isovector forces. It does not include the low-energy absorption of photons by a nuclear system resulting from isoscalar, velocity independent forces. However, these parts of the Hamiltonian just contribute one unit of the sum rule, so that our results correspond directly to the enhancement κ. In fact, the sum rule strength from isoscalar potentials is located almost completely below the pion threshold $\omega = m_\pi$. Hence, a direct comparison of our calculations with the empirical $\bar{\kappa}$ is indeed justified. In fig. 4 we compare our theoretical result for $\bar{\kappa}_{th}$, defined by the the integral

$$\int_0^{m_\pi} d\omega\ \sigma_{th}(\omega) = \bar{\kappa}_{th}\Sigma \tag{13}$$

over the calculated absorption cross section, with empirical values for $\bar{\kappa}$. The agreement with data for heavy nuclei is quite satisfactory. Discrepancies of 30-40 % persist for light and medium heavy nuclei. We should point out that the local density approximation used in the present approach should be particularly good for heavy nuclei.

We have extracted the part of $\bar{\kappa}_{th}$ wich arises from isovector elecric dipole mechanisms at the level of the basic $\gamma\pi N$ interactions shown in fig.2. This includes absorption driven by the Kroll-Ruderman and pion pole terms and the convection current of the nucleon pole terms. We denote these contributions to $\bar{\kappa}_{th}$ by $\bar{\kappa}_{dip}$ and list them in table 1. We see that about 80 % of $\bar{\kappa}_{th}$ comes from the leading electric dipole absorption mechanisms. The remaining 20 % is mainly due to M1 processes related to the Δ and nucleon pole terms.

Next, let us return to our basic question about the theoretical sum rule (3): how much of the dipole absorption in

$$\int_0^{\infty} d\omega\sigma_{E1}(\omega) = \kappa\Sigma \tag{14}$$

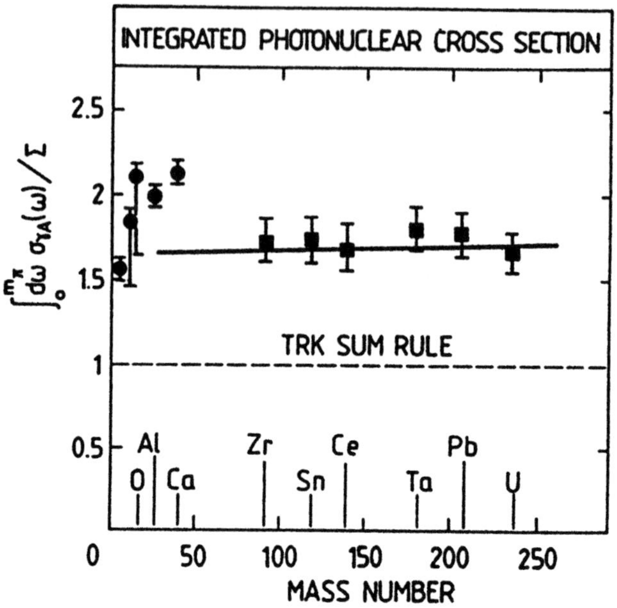

Fig. 4. Total photonuclear cross section integrated up to the pion mass. Solid line:our result. Data from ref. [4] and [5]

is located in the interval from pion threshold to infinity ? We see from table 2 that this high energy contribution is by no means negligible. We show it here in comparison with calculations based on the double commutator in (3) using static potentials. The differences were expected, mainly due to a combination of retardation effects (explicitily present in our approach) and higher order multipoles wich arise in the transformation from the γN to the γA frame.

Table 1. Theoretical prediction for $\bar{\kappa}_{dip}$ and $\bar{\kappa}_{th}$ in different nuclei.

	$\bar{\kappa}_{dip}$	$\bar{\kappa}_{th}$
C	0.51	0.66
O	0.51	0.67
AL	0.52	0.68
S	0.53	0.68
Ca	0.55	0.69
Cu	0.56	0.70
Ce	0.57	0.71
Pb	0.57	0.71

In summary, the integral of the photoabsorption cross section from pion threshold to infinity is far from negligible. This should be taken into account when comparing the theoretical κ with the empirical $\bar{\kappa}$. On the other hand, the absorption cross section integrated up to m_π includes about 20 % of absorption in $\bar{\kappa}$ wich is not of the E1 type.Finally, we were able to compare the values for κ and $\bar{\kappa}$ provided by our theory. Agreement with the experimental $\bar{\kappa}$ was overall good while agreement with theoretical calculations for κ hold at the qualitative level, as expected. This tests the consistency of our theory, wich can be used to study a variety of photonuclear processes involving complex nuclei.

Table 2. Theoretical results for κin different nuclei. Left: our prediction. Right: those from refs. [6],[7],[8]

O	1.31	1.27
Ca	1.53	1.15
Nuc. matter	2.2	1.80

References

1. R.C.Carrasco, E.Oset: Universidad de Valencia preprint (1989). Proc. Int. Workshop on Two Nucleon Emission, Elba(1989). World Scientific, to be published.
2. T.E.O.Ericson, W.Weise, "Pions and Nuclei" Clarendon Press (1988)
3. E.Oset, L.L.Salcedo: Nucl. Phys A **468** 631 (1987)
4. J.Ahrens et al. Nucl. Phys. A **251** 479 (1975);
   J.Ahrens, Nucl. Phys. A **446** 229c (1985)
5. A. Lepretre et al. Nucl. Phys. A **367** 237 (1981)
6. W.T.Weng et al.: Phys. Lett. **46B** 329 (1973)
7. A.Arima et al.: Nucl. Phys. A **205** 27 (1973)
8. A.Fabroccini, S.Fantoni: Nucl. Phys. A **435** 448 (1985)

Lecture Notes in Mathematics

Vol. 1236: Stochastic Partial Differential Equations and Applications. Proceedings, 1985. Edited by G. Da Prato and L. Tubaro. V, 257 pages. 1987.

Vol. 1237: Rational Approximation and its Applications in Mathematics and Physics. Proceedings, 1985. Edited by J. Gilewicz, M. Pindor and W. Siemaszko. XII, 350 pages. 1987.

Vol. 1250: Stochastic Processes – Mathematics and Physics II. Proceedings 1985. Edited by S. Albeverio, Ph. Blanchard and L. Streit. VI, 359 pages. 1987.

Vol. 1251: Differential Geometric Methods in Mathematical Physics. Proceedings, 1985. Edited by P.L. García and A. Pérez-Rendón. VII, 300 pages. 1987.

Vol. 1255: Differential Geometry and Differential Equations. Proceedings, 1985. Edited by C. Gu, M. Berger and R.L. Bryant. XII, 243 pages. 1987.

Vol. 1256: Pseudo-Differential Operators. Proceedings, 1986. Edited by H.O. Cordes, B. Gramsch and H. Widom. X, 479 pages. 1987.

Vol. 1258: J. Weidmann, Spectral Theory of Ordinary Differential Operators. VI, 303 pages. 1987.

Vol. 1260: N.H. Pavel, Nonlinear Evolution Operators and Semigroups. VI, 285 pages. 1987.

Vol. 1263: V.L. Hansen (Ed.), Differential Geometry. Proceedings, 1985. XI, 288 pages. 1987.

Vol. 1265: W. Van Assche, Asymptotics for Orthogonal Polynomials. VI, 201 pages. 1987.

Vol. 1267: J. Lindenstrauss, V.D. Milman (Eds.), Geometrical Aspects of Functional Analysis. Seminar. VII, 212 pages. 1987.

Vol. 1269: M. Shiota, Nash Manifolds. VI, 223 pages. 1987.

Vol. 1270: C. Carasso, P.-A. Raviart, D. Serre (Eds.), Nonlinear Hyperbolic Problems. Proceedings, 1986. XV, 341 pages. 1987.

Vol. 1272: M.S. Livšic, L.L. Waksman, Commuting Nonselfadjoint Operators in Hilbert Space. III, 115 pages. 1987.

Vol. 1273: G.-M. Greuel, G. Trautmann (Eds.), Singularities, Representation of Algebras, and Vector Bundles. Proceedings, 1985. XIV, 383 pages. 1987.

Vol. 1275: C.A. Berenstein (Ed.), Complex Analysis I. Proceedings, 1985–86. XV, 331 pages. 1987.

Vol. 1276: C.A. Berenstein (Ed.), Complex Analysis II. Proceedings, 1985–86. IX, 320 pages. 1987.

Vol. 1277: C.A. Berenstein (Ed.), Complex Analysis III. Proceedings, 1985–86. X, 350 pages. 1987.

Vol. 1283: S. Mardešić, J. Segal (Eds.), Geometric Topology and Shape Theory. Proceedings, 1986. V, 261 pages. 1987.

Vol. 1285: I.W. Knowles, Y. Saitō (Eds.), Differential Equations and Mathematical Physics. Proceedings, 1986. XVI, 499 pages. 1987.

Vol. 1287: E.B. Saff (Ed.), Approximation Theory, Tampa. Proceedings, 1985–1986. V, 228 pages. 1987.

Vol. 1288: Yu. L. Rodin, Generalized Analytic Functions on Riemann Surfaces. V, 128 pages, 1987.

Vol. 1294: M. Queffélec, Substitution Dynamical Systems – Spectral Analysis. XIII, 240 pages. 1987.

Vol. 1299: S. Watanabe, Yu.V. Prokhorov (Eds.), Probability Theory and Mathematical Statistics. Proceedings, 1986. VIII, 589 pages. 1988.

Vol. 1300: G.B. Seligman, Constructions of Lie Algebras and their Modules. VI, 190 pages. 1988.

Vol. 1302: M. Cwikel, J. Peetre, Y. Sagher, H. Wallin (Eds.), Function Spaces and Applications. Proceedings, 1986. VI, 445 pages. 1988.

Vol. 1303: L. Accardi, W. von Waldenfels (Eds.), Quantum Probability and Applications III. Proceedings, 1987. VI, 373 pages. 1988.

Lecture Notes in Physics

Vol. 342: M. Balabane, P. Lochak, C. Sulem (Eds.), Integrable Systems and Applications. Proceedings, 1988. VII, 342 pages. 1989.

Vol. 343: M. Month, S. Turner (Eds.), Frontiers of Particle Beams; Observation, Diagnosis and Correction. Proceedings, 1988. IX, 509 pages. 1989.

Vol. 344: M. Rascle, D. Serre, M. Slemrod (Eds.), PDEs and Continuum Models of Phase Transitions. Proceedings, 1988. VII, 229 pages. 1989.

Vol. 345: H. Holden, A. Jensen (Eds.), Schrödinger Operators. Proceedings, 1988. V, 458 pages. 1989.

Vol. 346: D. Lüst, S. Theisen, Lectures on String Theory. VII, 346 pages. 1989.

Vol. 347: Ph. Blanchard, J.-P. Dias, J. Stubbe (Eds.), New Methods and Results in Non-linear Field Equations. Proceedings. 1987. VII, 133 pages. 1989.

Vol. 348: A. Bohm, M. Gadella, Dirac Kets, Gamow Vectors and Gel'fand Triplets. VII, 119 pages. 1989.

Vol. 349: Y.A. Kubyshin, J.M. Mourão, G. Rudolph, I.P. Volobujev, Dimensional Reduction of Gauge Theories, Spontaneous Compactification and Model Building. X, 80 pages. 1989.

Vol. 350: G. Tenorio-Tagle, M. Moles, J. Melnick (Eds.), Structure and Dynamics of the Interstellar Medium. Proceedings, 1989. XXI, 537 pages. 1989.

Vol. 351: A. Dervieux, B. Larrouturou (Eds.), Numerical Combustion. Proceedings, 1989. VII, 481 pages. 1989.

Vol. 352: K.B. Wolf (Ed.), Lie Methods in Optics II. Proceedings, 1988. XII, 197 pages. 1989.

Vol. 353: M. Barthes, J. Léon (Eds.), Nonlinear Coherent Structures. Proceedings, 1989. X, 280 pages. 1990.

Vol. 354: J. De Coninck, F. Dunlop (Eds.), Wetting Phenomena. Proceedings, 1988. VII, 112 pages. 1990.

Vol. 355: R. Lima, L. Streit, R. Vilela Mendes (Eds.), Dynamics and Stochastic Processes. Proceedings, 1988. V, 298 pages. 1990.

Vol. 356: R. Wehrse (Ed.), Accuracy of Element Abundances from Stellar Atmospheres. Proceedings, 1988. V, 109 pages. 1990.

Vol. 357: J.L. Lumley (Ed.), Whither Turbulence? Turbulence at the Crossroads. Proceedings, 1989. V, 535 pages. 1990.

Vol. 358: J. Kovalevsky, Astrométrie moderne. VI, 294 pages. 1990.

Vol. 359: K. Kirchgässner (Ed.), Problems Involving Change of Type. Proceedings, 1988. XII, 207 pages. 1990.

Vol. 360: Y. Mellier, B. Fort, G. Soucail (Eds.), Gravitational Lensing. Proceedings 1989. XV, 315 pages. 1990.

Vol. 361: P. Gaigg, W. Kummer, M. Schweda (Eds.), Physical and Nonstandard Ganges. Proceedings, 1989. IX, 310 pages. 1990.

Vol. 362: N.E. Kassim, K.W. Weiler (Eds.), Low Frequency Astrophysics from Space. Proceedings. XII, 280 pages. 1990.

Vol. 363: V. Ruždjak, E. Tandberg-Hanssen (Eds.), Dynamics of Quiescent Prominences. Proceedings, 1989. XI, 304 pages. 1990.

Vol. 364: T.T.S. Kuo, E. Osnes, Folded-Diagram Theory of the Effective Interaction in Nuclei, Atoms and Molecules. V, 175 pages. 1990.

Vol. 365: M. Schumacher, G. Tamas (Eds.), Perspectives of Photon Interactions with Hadrons and Nuclei. Proceedings. IX, 251 pages. 1990.

GPSR Compliance

The European Union's (EU) General Product Safety Regulation (GPSR) is a set of rules that requires consumer products to be safe and our obligations to ensure this.

If you have any concerns about our products, you can contact us on ProductSafety@springernature.com

In case Publisher is established outside the EU, the EU authorized representative is:

Springer Nature Customer Service Center GmbH
Europaplatz 3
69115 Heidelberg, Germany

Batch number: 10620881

Printed by Printforce, the Netherlands